ELECTRON MICROSCOPY IN BIOLOGY

ADVISORY BOARD

ELECTRON MICROSCOPY IN BIOLOGY

Volume 1

EDITED BY

Jack D. Griffith

The University of North Carolina
Chapel Hill

A WILEY-INTERSCIENCE PUBLICATION

JOHN WILEY AND SONS • New York • Chichester • Brisbane • Toronto

This publication is designed to provide accurate and authoritative information in regard to the subject matter covered. It is sold with the understanding that the publisher is not engaged in rendering legal, accounting, or other professional service. If legal advice or other expert assistance is required, the services of a competent professional person should be sought. *From a Declaration of Principles jointly adopted by a Committee of the American Bar Association and a Committee of Publishers.*

ISSN 0275–5262

ISBN 0-471-05525-5

Printed in the United States of America

10 9 8 7 6 5 4 3 2 1

AUTHORS

Timothy S. Baker

The Rosenstiel Basic Medical
Sciences Research Center
Brandeis University
Waltham, Massachusetts

Thomas R. Broker

Cold Spring Harbor Laboratory
Cold Spring Harbor, New York

Louise T. Chow

Cold Spring Harbor Laboratory
Cold Spring Harbor, New York

Mircea Fotino

Department of Molecular, Cellular, and
Developmental Biology
University of Colorado
Boulder, Colorado

Barbara A. Hamkalo
Department of Molecular Biology and Biochemistry
Department of Developmental and Cell Biology
University of California at Irvine
Irvine, California

Kenneth R. Miller

Division of Biology and Medicine
Brown University
Providence, Rhode Island

Claire L. Moore

Cancer Research Center
and Department of Bacteriology and Immunology
University of North Carolina
Chapel Hill, North Carolina

J. B. Rattner

Department of Molecular Biology and Biochemistry
University of California at Irvine
Irvine, California

SERIES PREFACE

Over the past 40 years, electron microscopy in biology has changed from a frontier of cell biology to a routine tool of fields as diverse as protein chemistry and chemical carcinogenesis. In the early days of electron microscopy most studies employed thin sectioning, and a microscopist could keep abreast of the major preparative procedures and advances in the field. Now, with so many diverse applications and subspecialties, few electron microscopists have more than a reading knowledge of the preparative techniques outside their own area of study. The past few years have also seen the cost of state-of-the-art electron microscopes rise beyond the usual resources of an individual laboratory. This has forced the growth of central EM facilities serving from a few research groups to one or more departments. Those with primary interests in biological electron microscopy are being called on to supervise these facilities, and to counsel colleagues who face research problems very different from their own. Following the literature in a single field is difficult enough; scanning journals for articles of possible interest to colleagues is impossible. Furthermore, the literature of biological electron microscopy appears primarily in journals dedicated to the area of application, making it even more likely that one would miss articles outside one's normal reading.

A yearly series of invited articles can help solve this dilemma. A well-executed series can bring together articles reviewing techniques and advances in all areas of biological electron microscopy. Properly written, each article should be useful to the expert as a brief refresher and resource of reference material, and to the novice as an introduction. A collection of these volumes spanning several years should provide a valuable resource for introducing colleagues to new applications of electron microscopy and to the expert as an on-hand reference library. The series also would provide a needed format for electron microscopists to communicate among themselves in ways that the journals cannot accommodate. Most journal articles focus on the end rather than the means, and often important details are omitted, simply because of the style dictated by the format of a journal article. A discussion of techniques can be very important in a technique-oriented field such as biological electron microscopy, and a critical review of techniques, illustrated with successful applications, makes a valuable contribution to our colleagues.

JACK D. GRIFFITH

Chapel Hill, North Carolina
March, 1981

PREFACE

This volume contains excellent articles that combine discussion of techniques and applications with critical appraisal of the interpretation of electron micrographs. The chapter by Miller on freeze-etching of chloroplast membranes provides state-of-the-art electron micrographs of chloroplasts and a valuable discussion of our growing understanding of their structure based on such data. The chapter on chromatin ultrastructure by Hamkalo and Rattner provides an overview by acknowledged experts in this field and clues into fruitful directions for future research. Moore, in Chapter 3, discusses the visualization of RNA by electron microscopy and provides examples from the work of the best laboratories in this field. The visualization and mapping of genes by the R-loop technique described by Broker and Chow is very timely. These authors have contributed greatly to the development of this very powerful technique now finding use in many laboratories. The chapter by Fotino discusses potential for biological applications of high-voltage electron microscopy in a way that is highly informative to novice and expert alike. Finally, the review by Baker of the literature of image processing serves as an excellent introduction to the field for the novice, including a complete and well-cross-referenced literature survey that will provide valuable direction for the beginner and a handbook for the expert in the field.

As editor, I thank my panel of advisors for a great deal of encouragement and counseling and for aiding with the selection of authors for this and future volumes.

JACK D. GRIFFITH

Chapel Hill, North Carolina
March, 1981

CONTENTS

1 **Freeze-Etching Studies of the Photosynthetic Membrane** 1
Kenneth R. Miller

2 **Visualization of Chromosomes and Chromatin** 31
J. B. Rattner and Barbara A. Hamkalo

3 **The Electron Microscopy of Ribonucleic Acids** 67
Claire L. Moore

4 **Experimental Studies on Resolution in High-Voltage Tranmission Electron Microscopy** 89
Mircea Fotino

5 **Mapping RNA:DNA Heteroduplexes by Electron Microscopy** 139
Louise T. Chow and Thomas R. Broker

6 **Image Processing of Biological Specimens: A Bibliography** 189
Timothy S. Baker

Index 291

ELECTRON MICROSCOPY IN BIOLOGY

FREEZE–ETCHING STUDIES OF THE PHOTOSYNTHETIC MEMBRANE

1

Kenneth R. Miller
Division of Biology and Medicine
Brown University
Providence, Rhode Island

CONTENTS

1 Introduction

2 General organization of the thylakoid membrane

3 Freeze-fracture studies of the photosynthetic membrane

4 Preparation of membranes for freeze-fracturing

5 Fracture faces of the thylakoid membrane

6 Deep etching of the photosynthetic membrane

7 Structures on fractured and etched surfaces: their interrelationship

References

1 INTRODUCTION

Of all the biochemical and functional adaptations of living things, none is more fundamental than photosynthesis, the ability to transduce light into chemical energy. Because of this basic importance, the chloroplast, the organelle of photosynthesis, has been intensively studied since the development of biological electron microscopy. The very first electron microscope studies (17, 19, 41) indicated the presence of an extensive membrane system in the chloroplast. These flattened membranes were later termed "thylakoids" by Menke (22). Because these structures carry out the light reaction of photosynthesis, they are also often called photosynthetic membranes. These membranes were also among the first biological structures to be studied extensively with the freeze-etching technique methodology of Moor and Mühlethaler (28).

In this review I summarize some of the recent literature relating to freeze-etching of the photosynthetic membrane, and present a comprehensive view of the organization of the thylakoid membrane as revealed by this technique. Space limitations and the nature of this volume do not permit a detailed review of the broader question of structure and function in the photosynthetic membrane. For this more general topic, the reader is referred to any number of recent reviews that treat this question from several perspectives (1, 5, 6, 35, 40).

2 GENERAL ORGANIZATION OF THE THYLAKOID MEMBRANE

Thin sections of the chloroplast illustrate at once the extensiveness of the thylakoid lamellar system (Figure 1). Thylakoid membranes are piled into large stacks called grana, and individual grana are connected by a network of

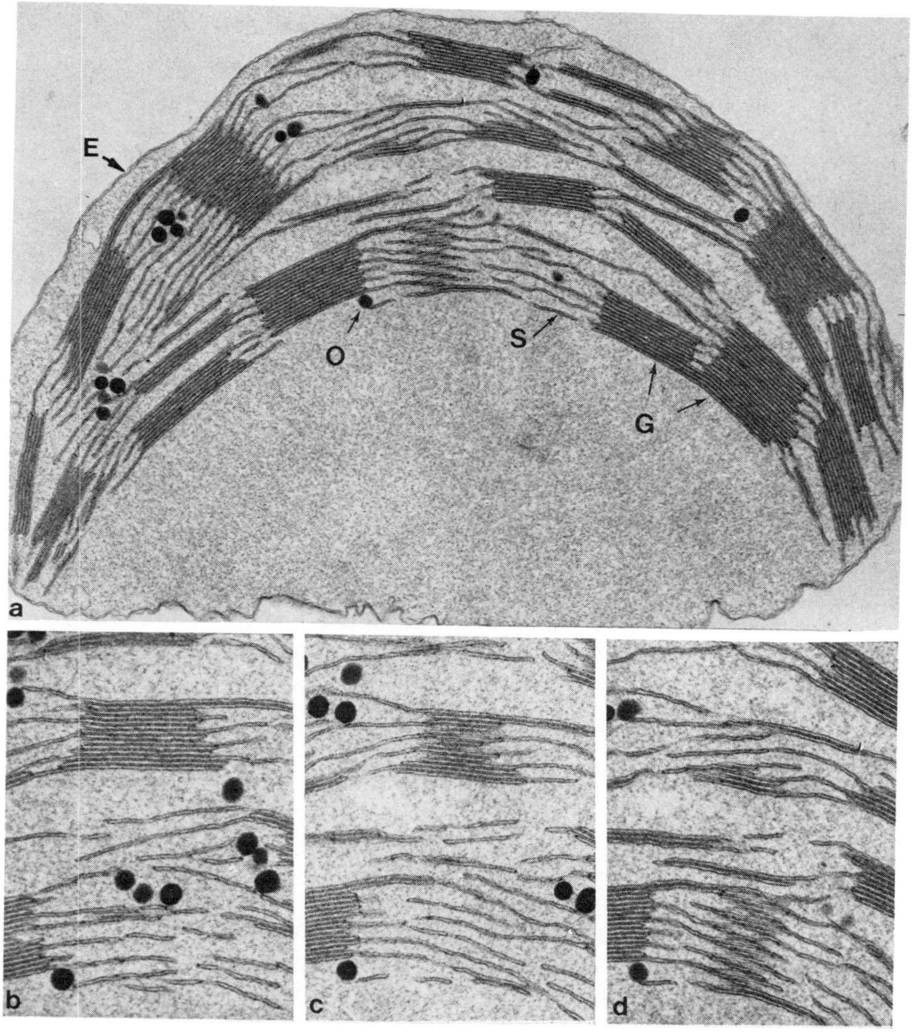

Figure 1 The structure of an intact chloroplast as visualized in thin section. (*a*) The entire organelle in an intact leaf. The chloroplast is enclosed by a pair of envelope membranes (*E*) and shows a number of dense osmiophilic granules (*O*), which probably represent lipid droplets stored in the organelle. The internal membranes of the chloroplast, which are termed thylakoids, contain the elements of the light reaction of photosynthesis and are stacked in several regions to form larger structures known as grana (*G*). (*b*), (*c*), and (*d*) A series of sequential sections taken through the central region of the chloroplast shown in (*a*) illustrates the relationship of the unstacked stromal (*S*) thylakoid membranes to the stacked membranes that comprise the grana. By following membranes across the three photographs, one can visualize the manner in which unstacked stromal membranes seem to spiral around the grana. The membranes in a single chloroplast, therefore, seem to represent a complex, interconnected network organized into two phases: stacked membranes and unstacked membranes. Magnification: (*a*): 40,000×. (*b–d*): 45,400×.

unpaired membranes that links the entire membrane system of the organelle. Studies of the membrane system by serial sectioning techniques have suggested that the apparently separate membranes of any individual granum are in fact linked by the unpaired membranes that appear at the edge of the granum (18, 32). The net result is a complicated three-dimensional network of paired and unpaired membranes that may in fact result from the intricate folding and refolding of a single thylakoid membrane sac.

It should be noted that distinctive stacking of thylakoid membranes into grana can disappear under certain experimental conditions. Izawa and Good (20) were the first to show that when chloroplasts are isolated in buffers lacking divalent cations (and containing only small amounts of monovalent cations), the thylakoid membrane unfolds and membrane stacking is no longer observed. When divalent cations are returned to the medium, the membranes begin to reassociate, and the characteristic grana morphology is once again observed. One of my reasons for suggesting that the complex membrane system of the chloroplast may be merely the elaboration of a single thylakoid is the observation that the unstacked membranes do not dissociate into small vesicles and disperse: rather, they remain as a single whorl of membrane observable under the light microscope, as might be expected if the thylakoid system were in fact composed of a single membrane sac.

3 FREEZE–FRACTURE STUDIES OF THE PHOTOSYNTHETIC MEMBRANE

Chloroplast thylakoid membranes were among the first structures to be systematically studied with the freeze-etching device developed by Moor and Mühlethaler (28), and the first report of the appearance of these membranes in freeze-fracture was published in 1965 (29). The complexity of structure in the photosynthetic membrane was immediately apparent to these workers, and they suggested that the various classes of particles observable in the replicas were manifestations of a series of enzyme-pigment complexes on the two surfaces of the thylakoid membrane.

The suggestion that particles visible in freeze-fracture replicas of the thylakoid were complexes on the surface of the membrane derived from the belief that freeze-fracturing resulted in a separation of the membrane from the surrounding ice in such a way that the membrane surface was exposed. Branton (10), in contrast, suggested that the fracturing process might actually split biological membranes, and maintained that the particles thus visualized were derived from structures embedded in the membrane, rather than exposed at its surface. A study of fracture faces in the photosynthetic membrane was reported and discussed with this interpretation in mind (13). Later reports showed that biological membranes do indeed seem to undergo splitting during the freeze-fracturing process (see reference 11 for a summary), and Branton's suggestion of membrane splitting during freeze-fracturing is now generally accepted.

However, the four fracture faces found in freeze-fractured chloroplast thylakoid membranes still presented a severe problem of interpretation, and Branton and Park (13) attempted to resolve it by suggesting that perhaps fracturing could occur at several levels within the thylakoid membrane. It then became necessary to explain why chloroplast membranes should fracture differently (i.e., at several levels) from other biological membranes, and this problem was solved in 1971 by Goodenough and Staehelin (16) in their study of fracture faces in stacked and unstacked thylakoid membranes from *Chlamydomonas*.

It now appears that the photosynthetic membrane fractures in roughly the same way as other biological membranes, and that the additional fracture faces are the result of differences in membrane internal structure between stacked and unstacked regions of the thylakoid membrane system. In the remainder of this chapter I outline our understanding of the way in which the photosynthetic membrane behaves during the fracturing process, commenting briefly about the functional organization of the membrane in the light reaction of photosynthesis.

4 PREPARATION OF MEMBRANES FOR FREEZE–FRACTURING

The membranes of the chloroplast can be examined *in situ,* whether in leaf tissue or intact algae, although this does present some disadvantages. Leaf tissue is particularly difficult to handle and fracture successfully, and algae may possess thick cell walls or other structures that tend to prevent the fracture plane from exposing the photosynthetic membrane in enough cases to provide a large sample for the microscopist. Therefore, it is quite common to isolate chloroplasts by standard techniques (43) prior to freezing. The isolated chloroplasts are infiltrated with glycerol to a concentration of 20–25% v/v for 1 hour, and frozen in liquid Freon or some suitable medium. Once frozen, the membrane samples can be stored indefinitely in liquid nitrogen until used. Replicas are prepared by standard techniques on one of several different freeze-etching devices. Freeze-fracture replicas are those that have been prepared by shadowing the frozen surface immediately following the cleavage event, or following the last pass of a fracturing knife. Such replicas show details of membrane fracture faces only.

5 FRACTURE FACES OF THE THYLAKOID MEMBRANE

Freeze-fracture replicas of stacked thylakoid membranes show four fracture faces, as illustrated in Figures 2 and 3. Two of these fracture faces are derived from membrane splitting in stacked regions of the thylakoid system, and two

Figure 2 Freeze-fractured isolated thylakoid membranes. The fracturing process reveals internal details of membrane organization, which can then be copied by the replica technique and observed in the microscope. Several different fracture faces are evident in this low-power micrograph. Magnification: 29,600×.

from the splitting of unstacked membranes. The labeling scheme now used for such fracture faces is illustrated in Figure 4. Detailed analysis of each of the four fracture faces can be carried out by careful measurement of particle sizes and distribution densities (see, e.g., references 16 and 39), and such information can be used to supplement the qualitative observation that the four fracture faces differ in terms of both the average diameters of particles present and the density with which the particles are distributed. One important piece of information gleaned from such analyses is that a wide range of particle sizes is present on any one fracture face, and it would therefore be misleading to assume a priori that the particles found on any single fracture face are biochemically or energetically homogeneous.

The effect of membrane stacking on the image of the membrane in freeze-fracture is illustrated in Figure 5. Isolated membranes that have been unstacked by washing in a buffer of low ionic strength show only two fracture faces. When such unstacked membranes are incubated with small (1–2 mm) amounts of Mg^{2+}, restacking of the membranes occurs, and four frac-

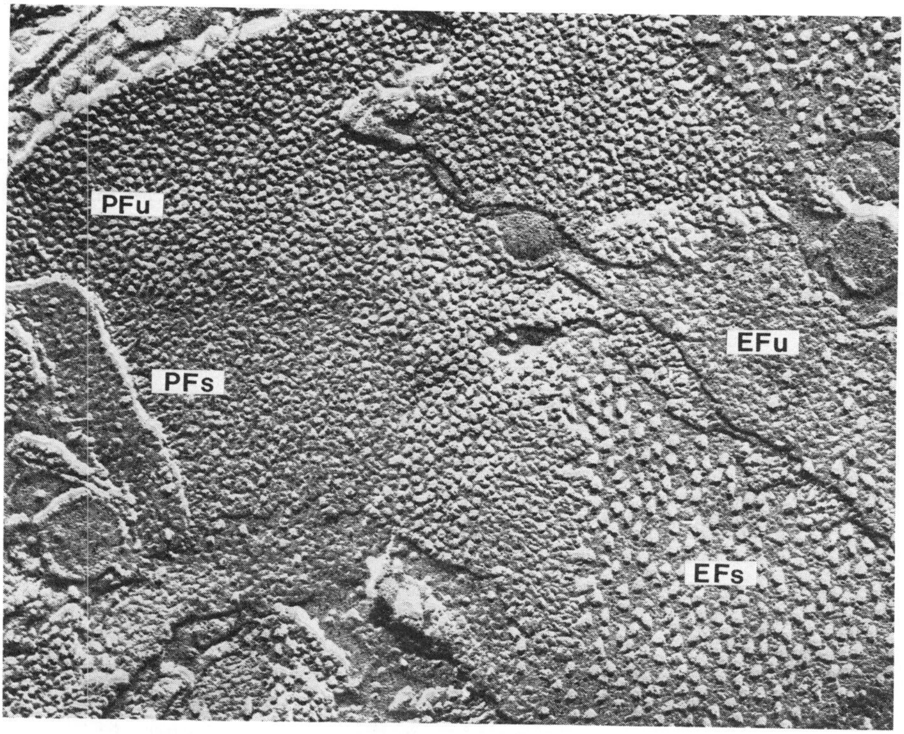

Figure 3 The four fracture faces formed by freeze-fracturing of stacked chloroplast membranes. For the details of the labeling system, see Figure 4. The particles visualized in replicas such as this represent structures revealed by membrane splitting during the fracturing process. Magnification: 103,500×.

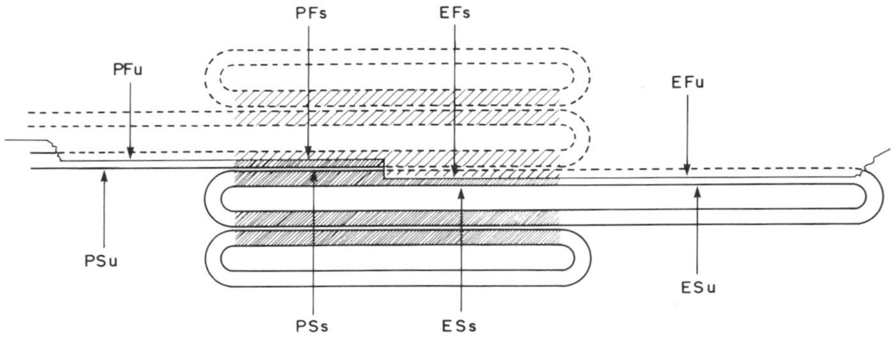

Figure 4 Four fracture faces are formed during freeze-fracturing of isolated thylakoid membranes. The membranes of a granum are shown as they might behave during freeze-fracturing. The membranes split away by the fracturing process are shown with a dotted line, while the remaining membranes are shown with solid lines. The region of membrane stacking (granum) in the center of the diagram is shaded. The fracture faces are labeled as follows: PFu, protoplasmic face, unstacked; PFs, protoplasmic face, stacked; EFs, ectoplasmic face, stacked; EFu, ectoplasmic face, unstacked. In addition, four types of membrane surface can be distinguished with this system of nomenclature, even though these surfaces cannot be observed in freeze-fractured preparations (etching, the sublimination of ice from the fractured surface, is required for that purpose): PSu, protplasmic surface, unstacked; PSs, protoplasmic surface, stacked; ESs, ectoplasmic surface, stacked; ESu, ectoplasmic surface, unstacked.

ture faces are again observable. It is of course of interest to know just what sort of molecular rearrangements this change in fracture face appearance involves. This question was first addressed by Ojakian and Satir (30), who carried out a careful study of particle behavior during unstacking and restacking in the thylakoid membrane. These authors concluded that lateral migrations of certain particle size classes were responsible for the observable changes. This finding has since been confirmed (39). Two general effects of the stacking process can be noted from these freeze-fracture studies: (1) there is a concentration of the E fracture face particle into stacked regions (see Figure 6); and (2) there is a segregation of particles on the P fracture face, with the larger particles being found in unstacked regions of the membrane and smaller particles concentrated in stacked regions (see Figure 7).

6 DEEP ETCHING OF THE PHOTOSYNTHETIC MEMBRANE

One of the principal advantages of the freeze-etching technique for the study of biological membranes is its versatility: it is possible to examine not only membrane fracture faces (in freeze-fractured samples) but also the true outer and inner surfaces of a membrane, by means of a technique known as deep etching. A sample of isolated chloroplast photosynthetic membranes is frozen in either distilled water or a very dilute (less than 10 mM) buffer without

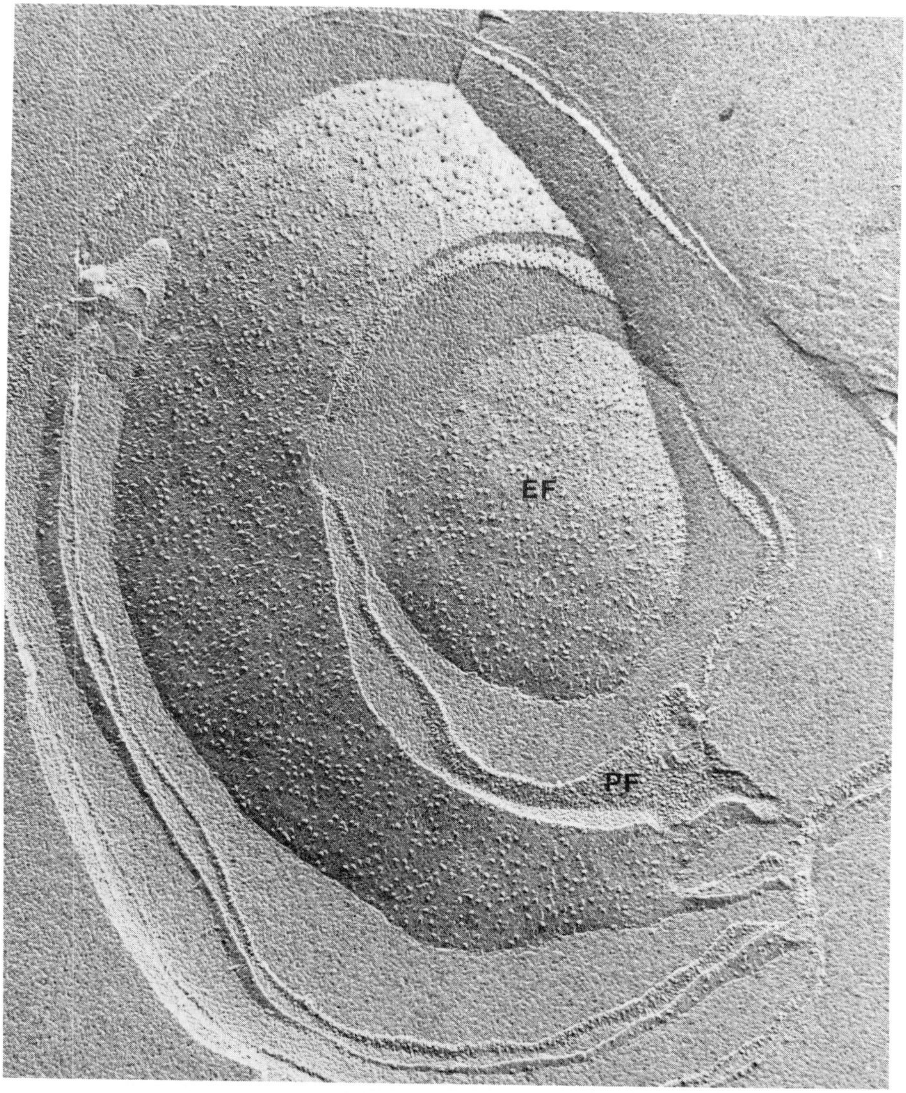

Figure 5 Suspension of isolated membranes in a buffer with low cation concentration (in this case, 50 m*M* Tricine) results in the "unstacking" of the associations that hold membranes together in grana. Stacked regions can no longer be observed as the membranes unfold, and only two fracture faces can be observed (labeled PF and EF only, because the membrane is no longer divided into stacked and unstacked regions). Particles on both the fracture faces are now randomly distributed. Micrograph courtesy of L. Andrew Staehelin, University of Colorado, Boulder. Magnification: 47,000×.

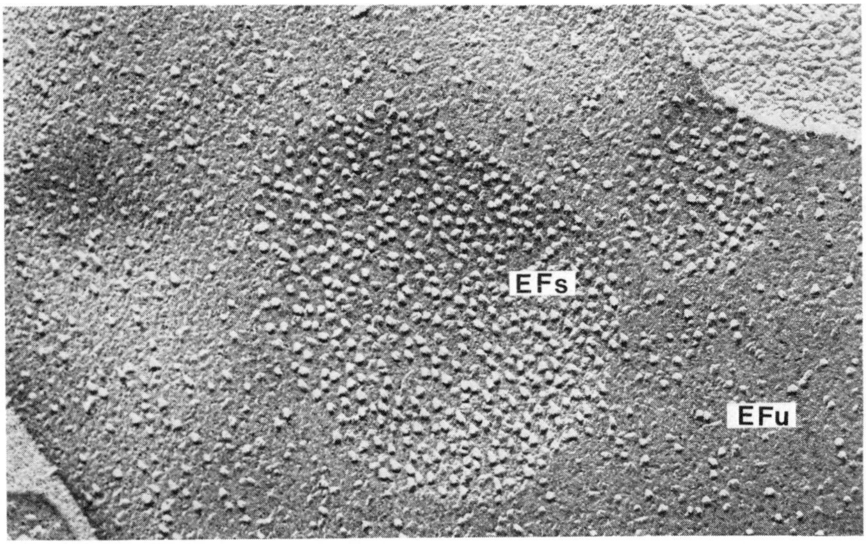

Figure 6 Restacking of an artificially unstacked membrane preparation can easily be achieved by the addition of small amounts of divalent cations (2 *mM* Mg^{2+} in this case). Stacked regions are formed, and can easily be recognized even when only one membrane fracture face is observed. Here the stacked (EFs) and unstacked (EFu) regions of the E fracture face are shown. Note the concentration of particles in the stacked areas. Magnification: 80,000×.

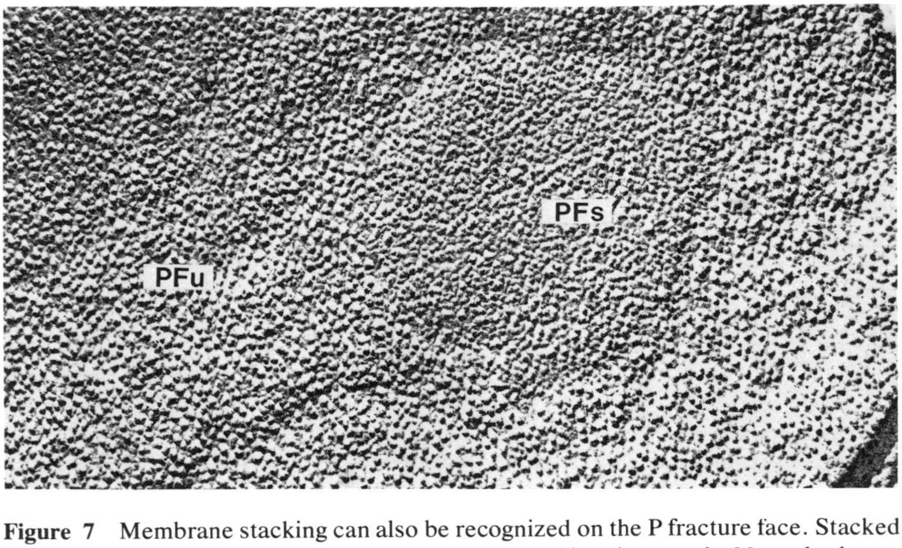

Figure 7 Membrane stacking can also be recognized on the P fracture face. Stacked (PFs) and unstacked (PFu) regions are marked in this micrograph. Note the larger apparent diameter of particles on the unstacked regions of the fracture face. Particles on the PFu average 110 Å in diameter; those on the PFs average 80 Å in diameter. Magnification: 80,000×.

the addition of a cryoprotectant. This sample is placed in a freeze-etching device, and following fracturing, the frozen surface is allowed to sublime (or "etch") in the vacuum chamber for 1–5 minutes. During this time, enough ice sublimes from the frozen surface to expose large membrane surfaces that previously were hidden beneath the ice. The manner in which this technique exposes thylakoid membrane surfaces is illustrated in Figure 8.

By combining freeze-fracturing of a membrane with deep etching, it is possible to obtain four views of that membrane: the two fracture faces, and two etched surfaces. In the case of the photosynthetic membrane, even more information can be gathered: two etched and two fractured images in stacked regions, and two etched and two fractured views derived from unstacked membrane regions. By combining these views, it is possible to make a detailed structural model of the photosynthetic membrane at the limit of resolution of the shadowing techniques used in freeze-etching.

The outer surface of the photosynthetic membrane, as revealed by deep etching, is shown in Figure 9. Two types of particles are visible at the surface of the thylakoid membrane: large particles averaging 120 Å in diameter, and smaller particles, about 80 Å in diameter. The large particles have been shown in a number of studies to be associated with the coupling factor, an enzyme responsible for ATP synthesis in the functional membrane (9, 15, 27, 31). The smaller particles have not been definitively identified.

When a fracture through frozen membrane samples breaks open the sac of a thylakoid, it is possible to observe the inner surface of the thylakoid membrane by deep etching. Figure 10 illustrates the nature of this surface in a stacked membrane preparation. The membrane is studded by a number of unusual particles that seem to be composed of four or more subunits (Figure 11). These particles are not uniformly distributed, but seem to be confined to certain regions of the inner surface. Unstacking experiments (Figure 12) indicate that these regions of particle concentration correspond to stacked membrane regions, and the areas surrounding them correspond to unstacked regions. Note that one of the effects of membrane stacking is the apparent migration of these particles into stacked regions.

Although I have shown micrographs of stacked and unstacked regions on the inner surface of the thylakoid membrane, I have not distinguished two regions in this way on the outer surface. This is because membrane stacking involves the close apposition of membrane outer surfaces; therefore the membrane outer surface in stacked regions does not seem to be accessible to the deep-etching technique. Nonstacked regions are well exposed, however, and Figure 9 illustrates the appearance of a nonstacked (PFu) membrane at the outer surface. Attempts have been made, however, to examine the PSs surface by unstacking membranes at temperatures near 0°C where the mobility of membrane components is limited, in the hope that surfaces that previously were involved in membrane stacking will be exposed. In Figure 13, which seems to illustrate just such a surface, both large and small outer

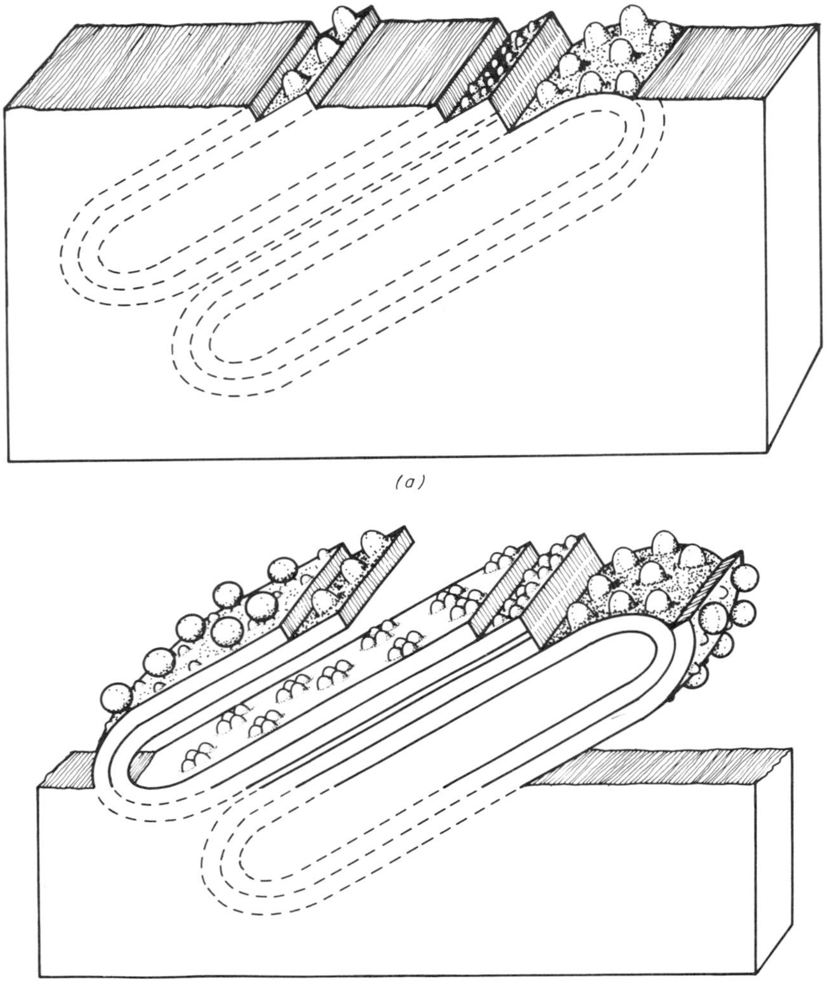

(b)

Figure 8 Diagrammatic representation of the processes of fracturing and etching. (*a*) Formation of fracture faces by the splitting of frozen thylakoid membranes. Note that the actual surfaces of each of the two membranes in this diagram are still obscured beneath the frozen material. Deep etching, which allows the sublimination of ice from the fractured surface, lowers the level of ice around the membrane and exposes the surfaces that previously were hidden, as seen in (*b*). In the case of the photosynthetic membrane, a number of structures on the surface of the membrane are made visible.

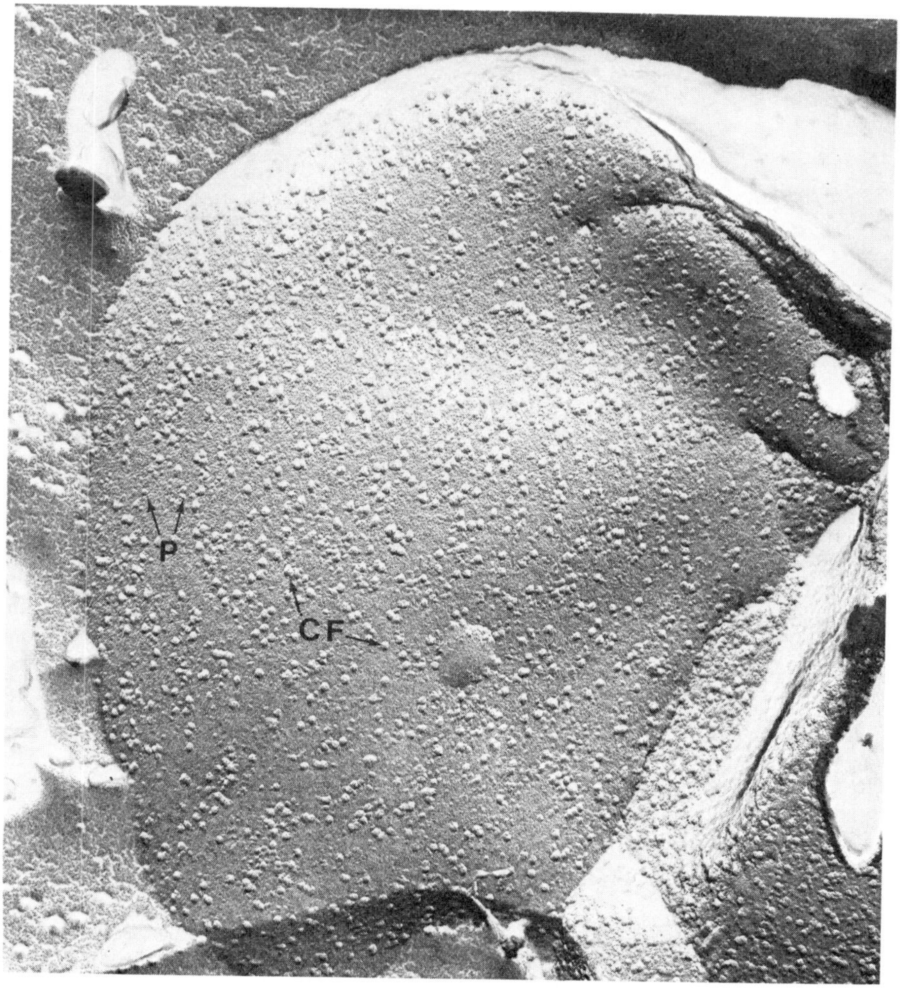

Figure 9 The outer surface of the chloroplast thylakoid membrane, revealed by the deep-etching technique. The membrane surface (as opposed to fracture face) visible in this micrograph was exposed by etching away frozen buffer in a fashion similar to that outlined in Figure 8. Large particles (120 Å average diameter) and smaller particles (P, ~ 80 Å) are visible on the membrane. A lipid droplet, similar to those in Figure 1, seems also to have collapsed on the surface of the membrane. The large particles (CF) have been biochemically identified in several studies as coupling factor molecules, enzymes responsible in part for the photophosphorylation of ATP. Magnification: 90,000×.

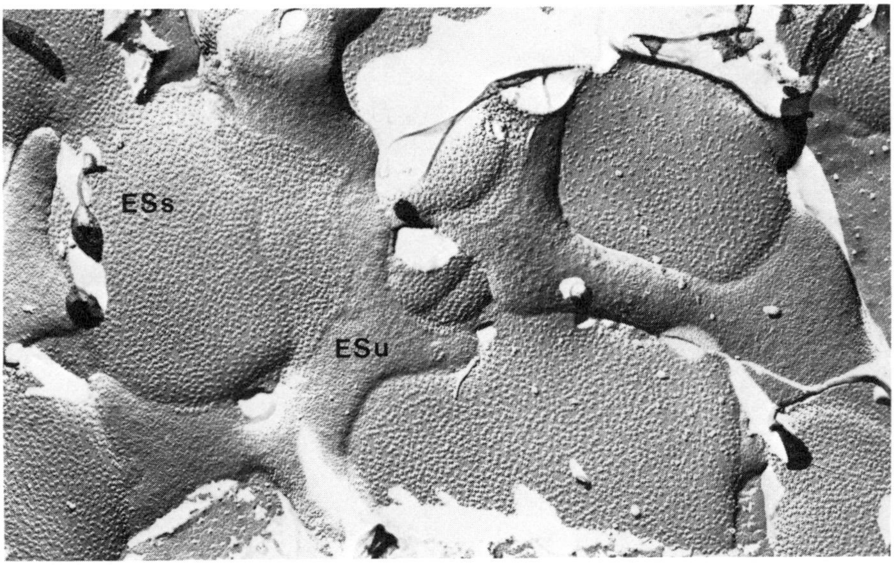

Figure 10 The inner surface of the thylakoid membrane can also be exposed by the deep-etching technique, as shown in this micrograph. Both stacked (ESs) and unstacked (ESu) regions of the inner surface are clearly visible. The stacked regions contain high concentrations of unusual particles. Magnification: 34,000×.

surface particles seem to be absent from these regions of possible membrane contact.

7 STRUCTURES ON FRACTURED AND ETCHED SURFACES: THEIR INTERRELATIONSHIP

Having considered the appearance of the thylakoid membrane in freeze-fractured and deep-etched preparations, we now attempt to integrate the information available on each of these views of the membrane into a coherent structural picture. The process of membrane stacking (and unstacking) provides one way to realize this goal. For example, if a certain structure was visible at one surface of the membrane (by deep etching) and also was found to be anchored deep within the thylakoid (made visible by freeze-fracturing), we would expect stacking and unstacking of the thylakoid membrane system to affect the two views of this structure in the same way. In other words, if a class of particles on the membrane surface migrates the same way as a class of particles seen embedded in the membrane, we can begin to guess that these particles might represent two different views of the same structure. Exactly such a case is presented by particles observed on the EFs and ESs views of the thylakoid. Membrane stacking results in the concentration of these particles in stacked regions, viewed from either the inner membrane surface (Figure 10) or the E fracture face (Figure 6). When

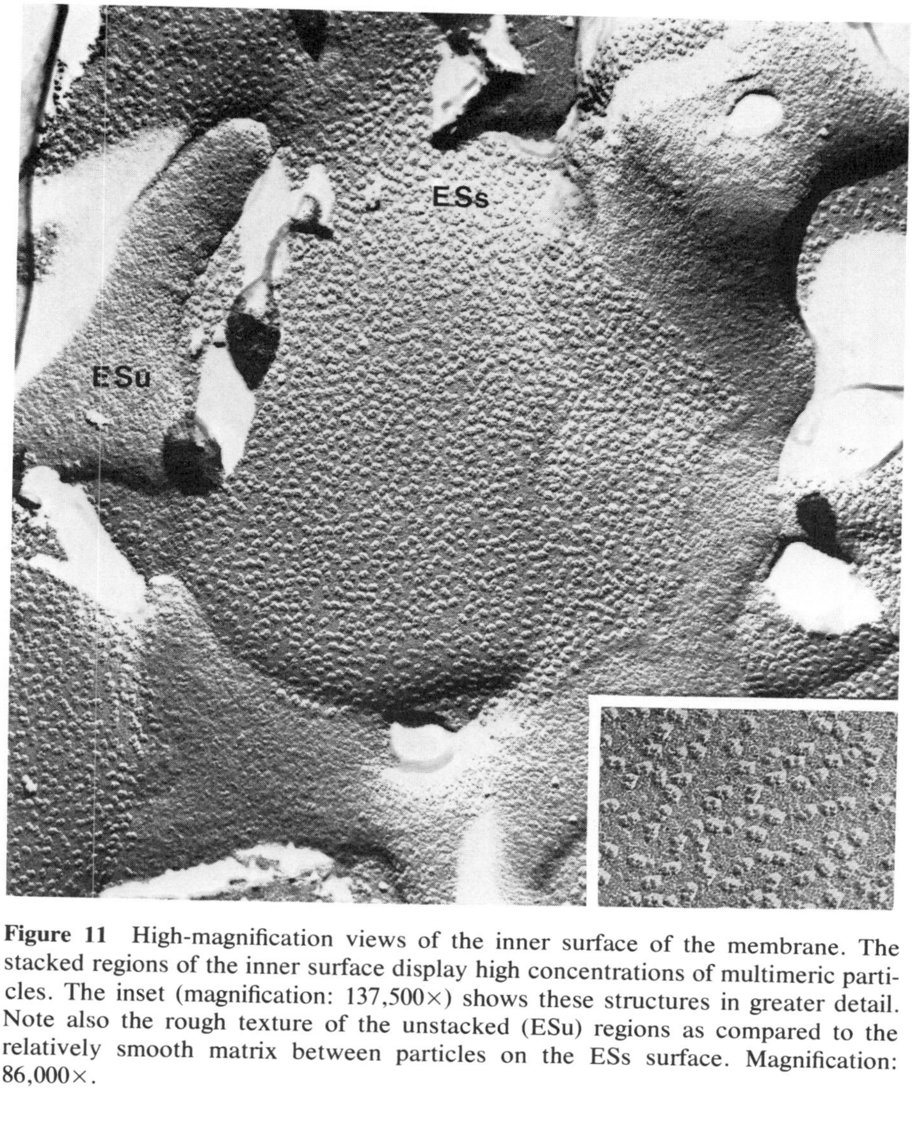

Figure 11 High-magnification views of the inner surface of the membrane. The stacked regions of the inner surface display high concentrations of multimeric particles. The inset (magnification: 137,500×) shows these structures in greater detail. Note also the rough texture of the unstacked (ESu) regions as compared to the relatively smooth matrix between particles on the ESs surface. Magnification: 86,000×.

thylakoid membranes are artificially unstacked, these particles become randomly distributed throughout the plane of the membrane (16), and when the membranes are restacked by the addition of divalent cations, they are again found concentrated in stacked regions. This observation suggests, at least superficially, that these structures may be two separate views of the same complex in the photosynthetic membrane.

More detailed observations concerning the interrelationships of particle structures within the thylakoid can be made by taking advantage of a widely observed phenomenon: the occurrence of regular lattices of particles in the

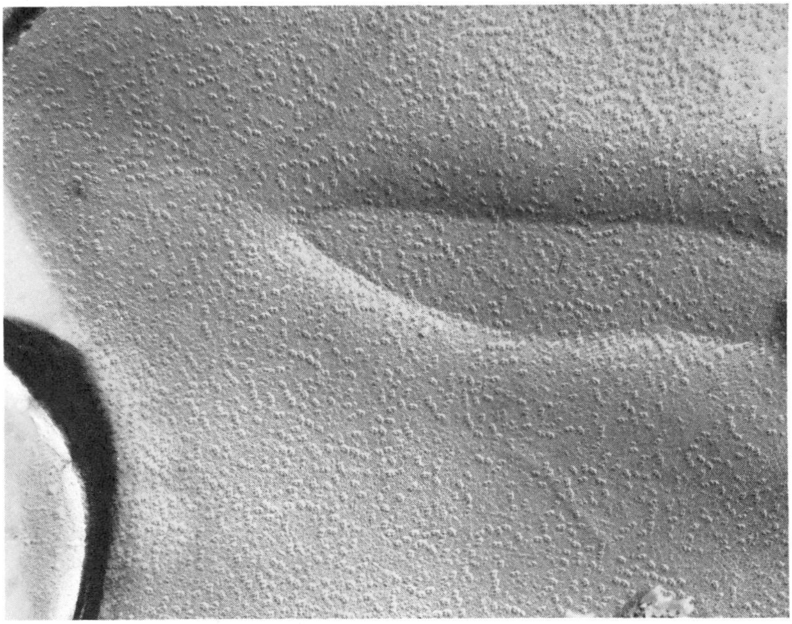

Figure 12 Artificial unstacking of the membrane produces changes in particles distribution that parallel those seen on fracture faces. The particles that were once seen concentrated in stacked regions on the inner surface of the membrane become uniformly distributed after the membrane has been artificially unstacked, as here. Micrograph, courtesy of L. Andrew Staehelin, University of Colorado, Boulder. Magnification: 65,500×.

photosynthetic membranes. Such structures were first observed by Park and Biggins (33) in their studies of air-dried and shadowed photosynthetic membranes. Park and Biggins termed the structures present in such lattices "quantasomes," and suggested that all the elements of the photosynthetic apparatus might be present in such regular structures. Since these observations were reported, it has become clear that the photosynthetic membrane in fact contains several distinct and separate structures, all of which may be required for complete photosynthetic activity. Nonetheless, these regular lattices still afford a unique opportunity for structural analysis. Next I attempt to show how the regularity of these structures allows us to make detailed studies of the associations between particles on fracture faces and etched surfaces.

In freeze-fractured preparations, particle lattices can be observed both in stacked and unstacked membranes (24). In stacked regions, lattices can be observed both on the E and P fracture faces (Figure 14). From similar micrographs, Staehelin (38) pointed out that there were two possible arrangements of large and small particles: one where large particles were aligned with large particles on adjacent membranes, and small particles with small

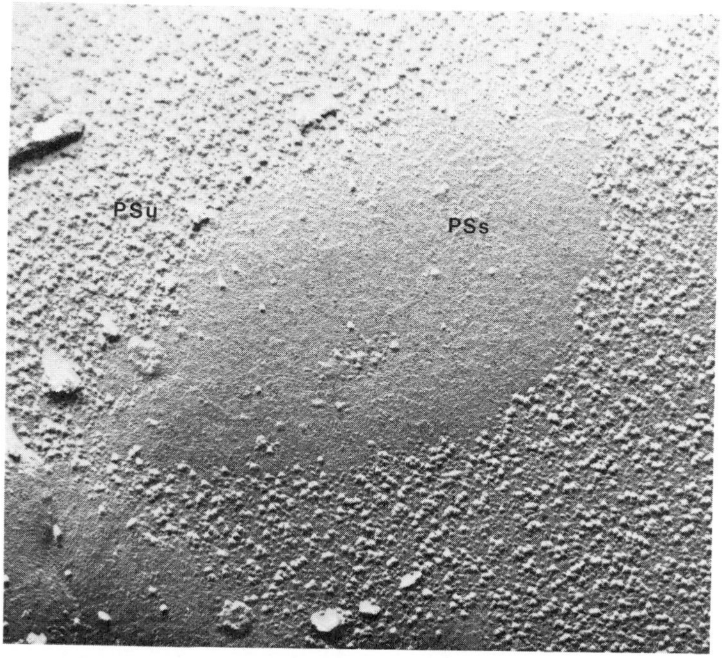

Figure 13 Because the outer surface of two adjacent membranes are pressed close together in stacked regions of the membrane, it generally is not possible to observe PSs membrane surfaces (see Figure 4). However, in this preparation, membranes have been artificially unstacked at low temperatures that inhibit lateral mobility of membrane components, and a region corresponding to the PSs surface may have been preserved even though the adjacent membrane has separated. Both the coupling factor molecules and smaller particles of the P surface seem to be absent from this region of presumed membrane contact, which is extremely smooth. (Micrograph courtesy of L. Andrew Staehelin, University of Colorado, Boulder. Magnification: 62,000×.)

particles (Figure 15*a*), and a second where large particles were aligned with small ones on the adjacent membrane (Figure 15*b*). Analysis of micrographs of adjacent E and P faces in stacked regions (Figure 16) shows that the second possibility is the case. Also, the geometry of lattices visible in the micrographs shows four PF particles associated with each EF particle.

These regular lattices of particles are visible on the etched surfaces of the membrane as well. On the inner surface, regular lattices of EF particles (Figure 17) have been observed by many authors, and these seem to be identical to the structures responsible for the "quantasome" pictures produced some years earlier (33). Very similar lattices also appear on the outer surface of the membrane (Figure 18), and the virtually identical spacings observed at each surface and in the E fracture face led to the suggestion that each lattice type is merely a different view of the same structure, which must therefore span the photosynthetic membrane (24).

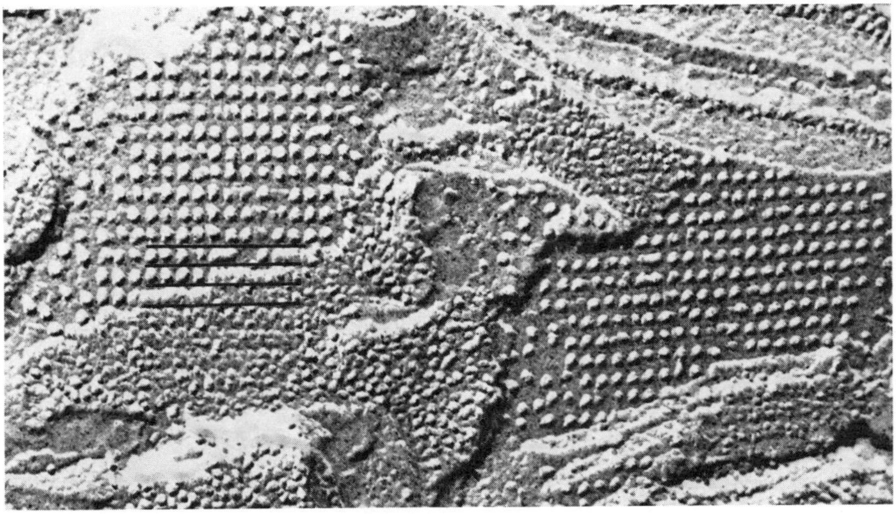

Figure 14 Regular lattices of membrane particles appear from time to time in preparations of photosynthetic membranes, and the spacings in these lattices can be used to analyze the interrelationships of particles on different fracture faces. This replica, for example, shows a lattice of particles on both E and P faces in a granum. Note that each row of large particles on the EFs face seems to be directly aligned with a double row of small particles on the PFs face. Horizontal bars drawn illustrate this alignment more clearly. Magnification: 108,000×.

If indeed these structures observed on each surface of the membrane are related to the internal lattices seen by freeze-fracture, two important predictions should be realized: in membranes where a fractured E face and etched P surface are adjacent, the lattices on each should be contiguous; and in membranes where PF and ES views are adjacent, the lattices should be contiguous, but out of register by half a lattice spacing, as predicted by the model developed by Staehelin (38) (Figure 15*b*). These key predictions are in fact realized. In Figure 19 adjacent EF and PS views of the membrane show the predicted contiguity of the lattices. And Figures 20 and 21 show EF, PF, and ES views of the membrane, where the lattice spacings precisely match those predicted by the model.

It is therefore established that at least in membrane regions where these regular lattices occur, the structures within the membrane behave in a way that is consistent with the model outlined in Figure 15*b*, and are not consistent with any number of other models that might, for example, have large (EF) particles paired at the point of membrane stacking. These micrographs indicate several other features of the membrane that should also be pointed out.

1 The small (80 Å) particles seen on the outer membrane surface (Figure 9) are excluded from regions where these regular lattices of particles occur (Figures 18 and 19).

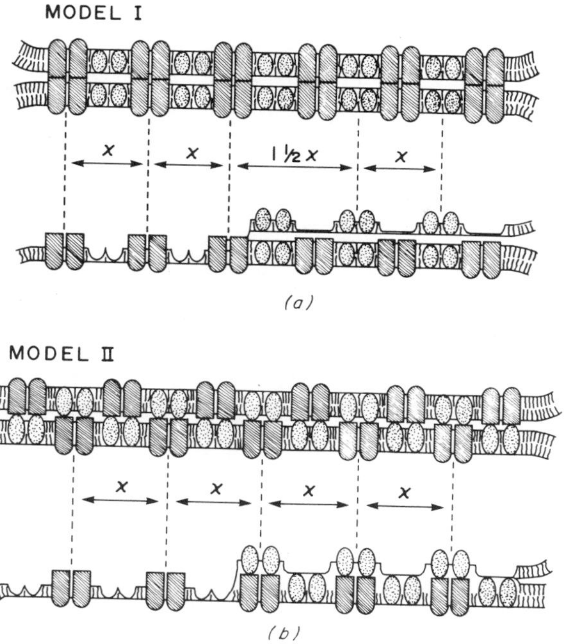

Figure 15 Because regular particle lattices seem to be aligned in micrographs such as Figure 14, one can draw two models based on how large and small particles might be aligned within the two stacked membranes. (*a*) Large particles on one membrane are paired with large particles on the adjacent membrane, and small particles are paired with small particles. (*b*) The other possible case: large particles are paired with small ones on the adjacent membranes. These two models give different predictions on the spacings that would be observed when EFs and PFs faces appear side by side in the same lattice. (Redrawn from reference 38.)

2 The small granular particles visible on the ESu surface are excluded from stacked regions of the E surface, and also from regions where regular lattices occur (Figures 10, 11, and 17).

3 The PF particles visible in the lattices are of the smaller size (about 80 Å average diameter) typical of stacked regions on the P fracture face.

4 Although the particle lattices are most often rectangular, and the particles in such lattices on the PF face are arranged in a rectangular fashion (Figures 14 and 16), the subunits of particles visible on the inner surface of the membrane are not always arranged in line with the rectangular lattice (Figure 17).

A structural model for the photosynthetic membrane that can be derived from these studies is summarized in Figure 22. A minimum of four classes of particles are associated with the thylakoid membrane: (1) the coupling factor, localized at the outer surface of the membrane, visualized as a single large particle, and removable from the membrane surface by appropriate

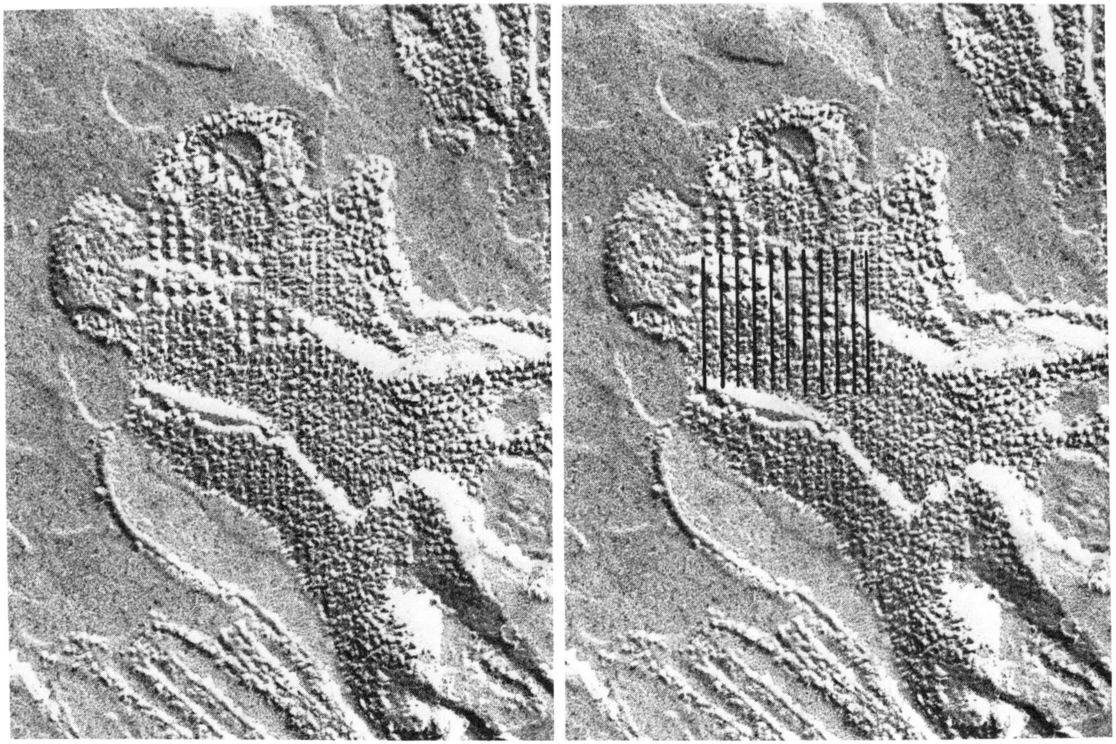

Figure 16 Extensive alignment of lattices on E and P faces is evident in this micrograph, which has been used to test the two models proposed by Staehelin (38) and outlined in Figure 15. The prediction suggested in model II seems to be realized, as shown by the markings in the right-hand micrograph. It is also of interest that the general direction of the lattice seems to be the same across several pairs of membranes in the granum. Why lattices on more than two membranes should be lined up so precisely is not understood. Magnification: 92,500×.

Figure 17 Regular particle lattices are also visible on the inner (ES) surface of the membrane. Here the particles visible on the inner surface are aligned in regular arrays with an interparticle spacing of about 225 Å in each direction. Magnification: 126,000×.

chemical treatments; (2) a large membrane-spanning structure, visible as the EF particles, most of which are localized in stacked regions, also observable on the membrane inner surface as a tetramerlike structure and at the membrane outer surface as well; (3) a small particle, associated with the large membrane-spanning structure in regular PF lattices, visible in stacked regions as the PFs particle and possibly visible at the membrane outer surface in the regular P surface lattices; (4) a small membrane-spanning particle, visible on the PFu fracture face, and possibly accounting for the small PS particles, and the small particles visible on unstacked regions of the membrane inner surface (ESu).

It is in principle possible to divide this classification still further. One might suggest, for example, that there are two types of large membrane-spanning structures, one found in stacked regions, and one in unstacked regions (5). One might also suggest that the small membrane-spanning particles in fact represent several classes of biochemically distinct structures.

How are these structures in the membrane related to the light reaction of photosynthesis? Although it is beyond the scope of this chapter to offer a detailed critique of structural and biochemical studies that have addressed this question, it is important, if only for developing an idea of how membrane structure may relate to function, to deal with this problem briefly.

Some early suggestions along this line were summarized in 1971 by Park and Sane (35). Several lines of evidence, including physical (21, 36) and chemical (2, 7, 8) techniques of separation, indicated that membrane prep-

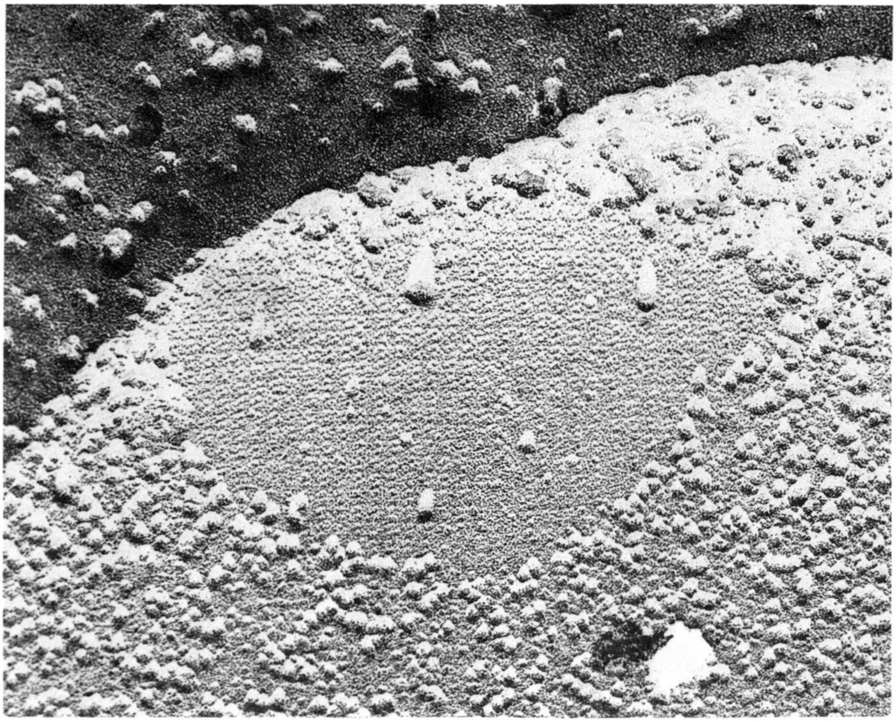

Figure 18 Particle arrays are also seen on the outer (PS) surface of the membrane. Here a regular array of particles on the PS surface of thylakoid has been exposed by the deep-etching technique. The spacings of this lattice are identical to the spacings on the inner surface of similar membranes (Figure 17). Note the exclusion of both coupling factor particles and smaller particles from the regions of the lattice. Magnification: 129,000×.

arations derived primarily from grana (stacked) thylakoids were enriched in photosystem II activity, whereas those derived from stroma (unstacked) membranes were enriched in photosystem I. From these observations it seemed logical to suggest that the large EF particles (which I have called the large membrane-spanning particle) were associated with photosystem II activity, since these particles, like photosystem II activity, are associated with stacked regions of the membrane. Similar arguments were advanced to associate the PF particles with photosystem I activity, and the particles on the PFs regions were thought to account for the fraction of photosystem I activity that could be detected in grana preparations.

In general the results of more recent studies have tended to support these ideas, although considerable detail has now been added to our picture of the membrane. The suggestion that the particles visible on the EFs fracture face

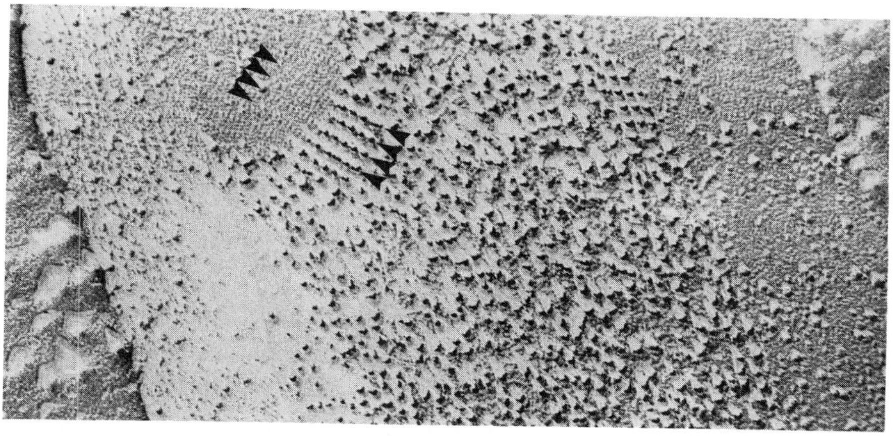

Figure 19 It has been suggested that these regular lattices are all related to a single structure that spans the photosynthetic membrane. If this were true, one would expect views of the same lattice region to appear occasionally in membranes where fractures and etched regions are adjacent. Such regions are in fact observed, and the alignment is as predicted. This micrograph, provided by Roderick B. Park of the University of California, shows the continuity of a lattice across the PS and EF views of the membrane in artificially unstacked thylakoids. From (34). Magnification: 78,000×.

might be related to photosystem II has held up rather well. Staehelin (39) showed that the concentration of these particles into stacked regions parallels the ability of certain treatments to separate photosystem I and photosystem II. Studies with developing chloroplasts (4) and mutants missing membrane components associated with photosystem II (25, 26, 37) have also tended to indicate that these structures could be associated with photosystem II activity.

One point of recent controversy has been the position of a major chlorophyll-binding protein in the membrane. This protein, sometimes

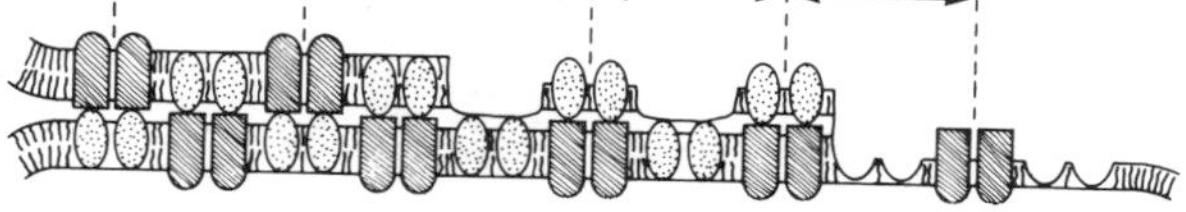

Figure 20 The model for particle organization in stacked regions outlined in Figure 15 also predicts a certain spacing of particle rows where PSs, EFs, and ESs views of the membrane are visible for the same lattice. These spacing predictions are visible here. Figure 21 is a micrograph showing these views.

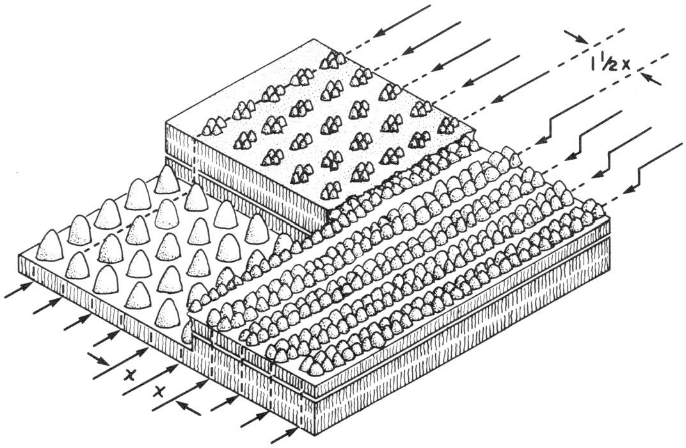

(b)

Figure 21

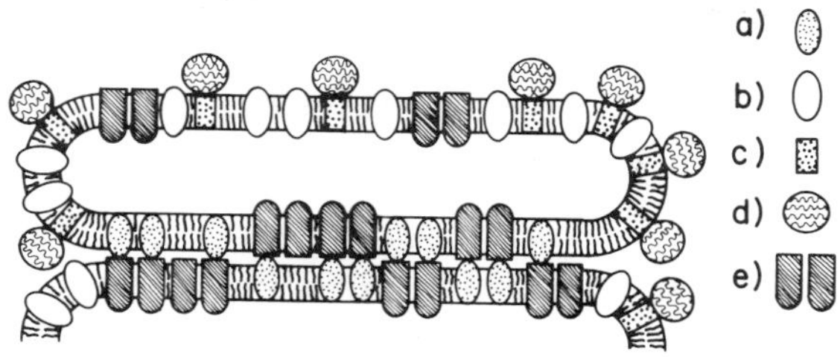

Figure 22 Highly schematic representation of the size classes of particles revealed by freeze-etching in the photosynthetic membrane, put together to illustrate the distribution of these structures in stacked and unstacked regions of the membrane. (*a*) Small particles visible on the PFs face. These small particles align with large particles in regions where lattices occur. (*b*) Large particles visible on the PFs fracture face. These particles seem to be confined to unstacked regions of the membrane, and it is possible that they are exposed at the outer surface of the membrane to form the small particles visible on the PSu surface. Some studies have suggested they may be related to photosystem I. (*c*) A particle visible on the PFu fracture face that may represent the binding site of the coupling factor to the membrane. (*d*) The coupling factor; confined to unstacked membrane surfaces. (*e*) A large, membrane-spanning structure found primarily in stacked regions, which seems to be related to photosystem II.

termed "chlorophyll-protein complex II" (CP II), binds equimolar amounts of chlorophyll a and chlorophyll b, and is composed of two polypeptides, with apparent molecular weights of 23,000 and 25,000 daltons. Miller et al. (26) first reported, in a study of a barley mutant apparently devoid of the complex, reductions in the size of particles visible on the EFs fracture face, along with reductions in the apparent diameter of particles visible on the inner (ESs) surface of the thylakoid sac. These results suggested that (1) the large membrane-spanning complex is photosystem II, since CP II is closely associated with photosystem II in the membrane, and (2) CP II is situated at the periphery of the photosystem II complex.

Armond et al. (4) subsequently reported on membrane organization in pea seedlings that had been greened under unusual lighting conditions, resulting in a complete absence of CP II in the thylakoid membrane. These workers also found a reduction in EF particle diameters, and they presented a model of membrane organization in which CP II was exposed at the inner surface of the membrane. They further suggested that the tetrameric appear-

---

Figure 21 (*a*) Fractured and etched chloroplast thylakoid membranes showing PFs, EFs, and ESs views of the membrane in a region where a single-particle lattice dominates the organization of the thylakoid. Spacings are marked on the thylakoid, and a drawing of the membrane in perspective (*b*) is included to clarify the observation. Note that the observed spacings correspond exactly to Staehelin's model given in Figure 15. Magnification: (*a*): 200,000×.

ance of the membrane-spanning complex at the inner surface (ESs) was the result of the binding of four molecules of CP II to a single photosystem II complex (4). These workers have also reported that there is not a complete absence of CP II in the barley mutant used by Miller et al. (26) for their studies; rather, at least one of the two CP II polypeptides is still present in the mutant, albeit in an unpigmented form (14).

Very recently Simpson and his associates (personal communication) have reexamined the barley mutant and confirmed many of the observations of the earlier work. However, this study has also showed that substantial changes exist on the PFs fracture face of the mutant, and these workers have argued that CP II may in fact be located in the membrane in such a way as to appear on the P fracture face after membrane splitting. This observation would help to explain the persistence of tetrameric particles on the inner (ESs) surface of the thylakoid membrane in this barley mutant (26); it would also be consistent with the role assigned to CP II in mediating thylakoid membrane stacking (5). At this juncture, the association between CP II and the large membrane-spanning complex seems clear, but the exact location of this pigment protein complex in the membrane and its behavior in freeze-fracturing remains to be settled.

There have been a few studies with photosystem I mutants or photosystem I–deficient systems that have produced some data about the location of photosystem I in the thylakoid membrane. Two studies (23; Arntzen, personal communication) have shown that the incorporation of purified photosystem I reaction complexes in artificial membranes results in the appearance of particles in freeze-fracture that have approximately the same size as those found on the PFu fracture face in the native thylakoid. And recently a study in my laboratory (24a) of a photosystem I–deficient mutant of maize has shown that the absence of photosystem I results in a measurable decrease in particle diameters on the PFu fracture face, while other structures in the membrane are unaffected.

If these studies are supported by subsequent work, we will be faced with a model of photosynthetic membrane architecture in which there is a substantial spatial separation between photosystem II centers in stacked region of the thylakoid system, and photosystem I centers in unstacked regions. The difficulties this system would present for integrated electron transport beweeen the photosystems are clear, but it would be wise to remember that in all cases what we have been studying with electron microscopy has fundamentally been an artificial system. That is, chloroplasts have been opened and their internal membranes suspended in a synthetic buffer solution, and this procedure may result in a substantial change (especially in terms of membrane stacking) from what is seen in the native thylakoid. It has been argued (1, 3, 5) that thylakoid membrane stacking *in vivo* may serve as a method of regulating energy distribution and electron transport between the photosystem, and this does not seem to be an unreasonable suggestion.

As useful as freeze-etching has been in developing information about the organization of this and other biological membranes, its applicability is clearly limited to certain situations and certain levels of detail. One of the problems in interpreting studies of thylakoid membranes missing certain components has been that in most cases the macromolecules missing have proved to be small parts of much larger structures in the membrane. The inability of freeze-etching to resolve these structures into their component parts is due to the distortion induced by the fracturing process and the limited resolution of the shadowing replication technique.

A clear demonstration of this limitation can be found in attempts to use image analysis techniques to enhance structural detail in micrographs of regular lattices in the photosynthetic membrane. Image analysis techniques, which are reviewed in detail in Chapter 6, are useful in analyzing regular structures because they allow averaging the images of individual particles in such arrays, which can result in a dramatically improved signal-to-noise ratio. Such analysis can be carried out by optical or computer methodology, and the result is an improved image of regular particle lattices at both surfaces of the membrane (Figures 23 and 24). The amount of detail obtainable by such techniques is limited by several factors, including the absolute resolution of the image itself, and the disorder of the structures within the

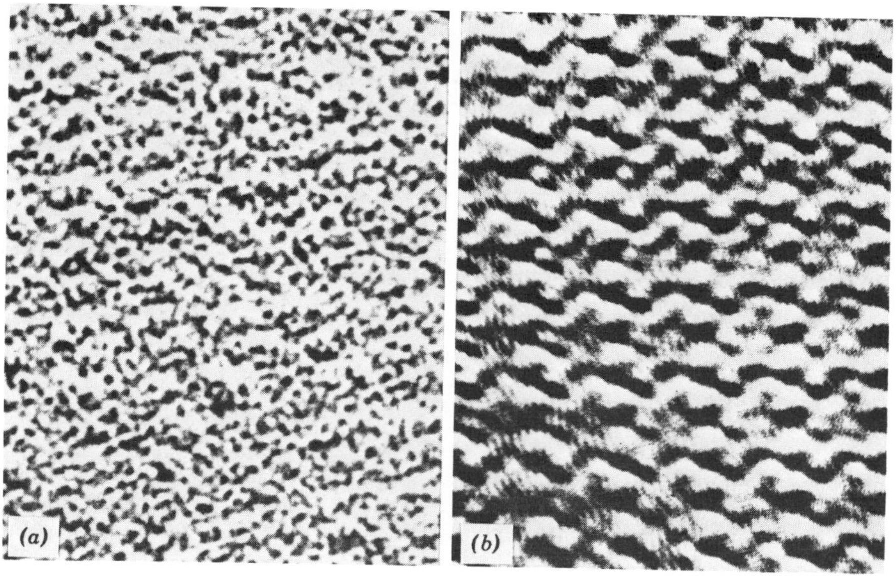

Figure 23 More detailed views of the surface of the photosynthesis membrane can be obtained by techniques of image analysis (see Chapter 6 for a detailed discussion of the methodology). Here, high-magnification views of the membrane outer surface where particle lattices are evident are compared before (*a*) and after (*b*) optical filtration. (Note: the micrograph used for these figures is the same as that for Figure 18). Magnification: 575,000×.

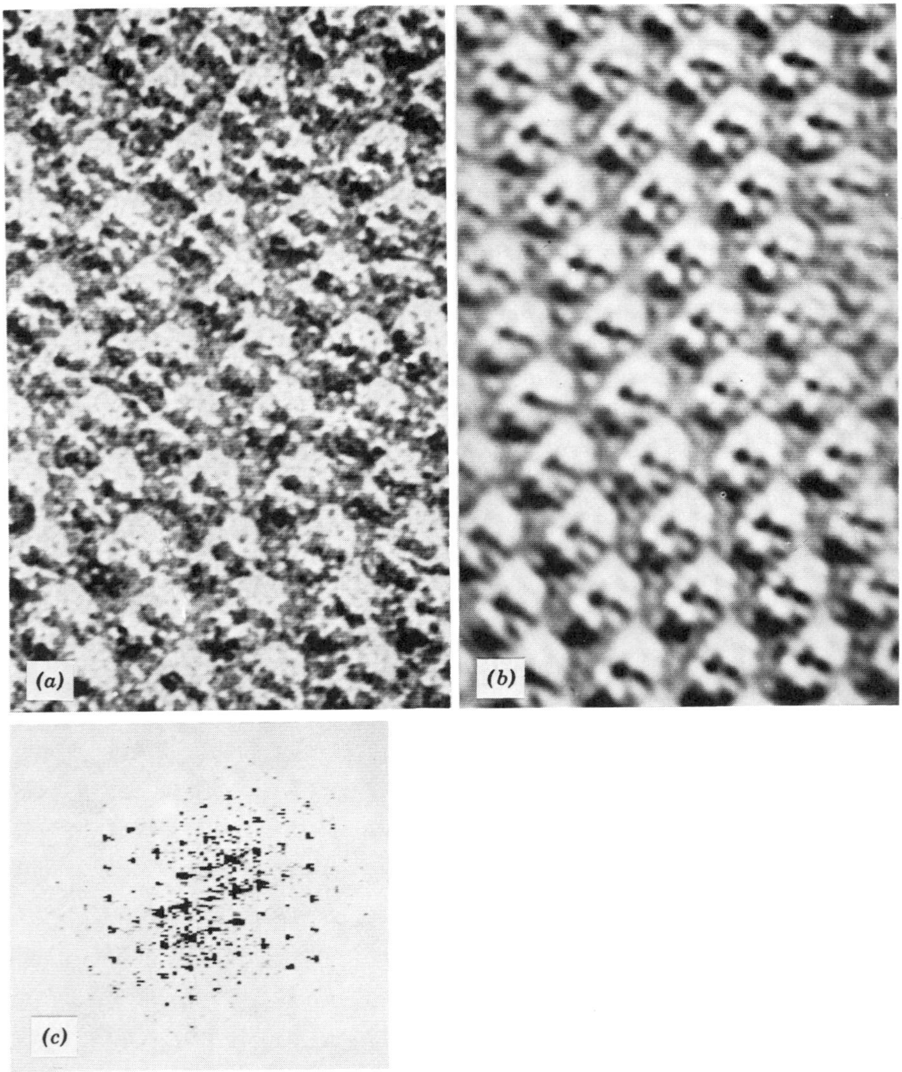

Figure 24 The filtration method displayed in Figure 22 can also be carried out by computer. (*a*) Regular lattice of particles on the ESs surface of the membrane. (*b*) Computer-generated filtered image formed by Fourier analysis techniques of the same image. (*c*) Computer-generated "diffraction pattern" prepared by Fourier analysis of Figure 23*a*. The pattern of spots corresponds to the center-to-center spacing of the particles, and the faint regular spots at the edge of the photograph indicate a resolution of approximately 40 Å in the replica. Magnification: 523,000×.

regular lattice. In the case of freeze-fracture replicas that have been analyzed here, the absolute resolution (which can be "read" directly from the most distant points in the optical or digital transform) is limited to about 35 Å. It is clear, therefore, that this level of detail is one of the principal limitations of freeze-etching in investigating chloroplasts and other membrane systems.

REFERENCES

1 Anderson, J. M. (1975) *Biochim. Biophys. Acta* **416:**191.

2 Anderson, J. M. and Boardman, N. K. (1966) *Biochim. Biophys. Acta* **112:**403.

3 Anderson, J. M., Goodchild, D. J., and Boardman, N. K. (1973) *Biochim. Biophys. Acta* **325:**573.

4 Armond, P. A., Staehelin, L. A., and Artzen, C. J. (1977) *J. Cell Biol.* **73:**400.

5 Arntzen, C. J. (1978) *Curr. Top. Bioenerg.* **8:**111.

6 Arntzen, C. J. and Briantais, J.-M. (1974) In *Bioenergetics of Photosynthesis* (Govindjee, Ed.), Academic Press, New York, p. 51.

7 Arntzen, C. J., Dilley, R. A., and Crane, F. L. (1969) *J. Cell Biol.* **43:**16.

8 Arntzen, C. J., Dilley, R. A., Peters, G. A., and Shaw, E. R. (1972) *Biochim. Biophys. Acta* **256:**85.

9 Berzborn, R. J., Kopp, F., and Muhlethaler, K. (1974) *Z. Naturforsch.* **29c:**694.

10 Branton, D. (1966) *Proc. Natl. Acad. Sci. U.S.A.* **55:**1048.

11 Branton, D. and Deamer, D. W. (1972) *Membrane Structure, Protoplasmatologia*, Vol. 2, Section E, Part 1, Springer-Verlag, Vienna.

12 Branton, D., Bullivant, S., Gilula, N. B., Karnovsky, M. J., Moor, H., Muhlethaler, K., Northcote, D. H., Packer, L., Satir, B., Satir, P., Speth, V., Staehelin, L. A., Steere, R. L., and Weinstein, R. S. (1975) *Science* **190:**54.

13 Branton, D. and Park, R. B. (1967) *J. Ultrastruct. Res.* **19:**283.

14 Burke, J. J., Steinback, K. E., and Arntzen, C. J. (1979) *Plant Physiol.* **63:**237.

15 Garber, M. P. and Steponkus, P. L. (1974) *J. Cell Biol.* **63:**24.

16 Goodenough, U. W. and Staehelin, L. A. (1971) *J. Cell Biol.* **48:**594.

17 Granick, S. and Porter, K. R. (1947) *Am. J. Bot.* **34:**545.

18 Heslop-Harrison, J. (1962) *Planta* **60:**243.

19 Hodge, A. J., McLean, J. E., and Mercer, F. V. (1955) *J. Biophys. Biochem. Cytol.* **1:**605.

20 Izawa, S. and Good, N. E. (1966) *Plant Physiol.* **41:**544.

21 Jacobi, G. (1969) *Z. Pflanzenphysiol.* **61:**203.

22 Menke, W. (1962) *Annu. Rev. Plant Physiol.* **13:**27.

23 Miller, K. R. (1978) In *Chloroplast Development* (G. Akoyunoglou et al., Eds.), Elsevier/North-Holland Biomedical Press, Amsterdam, p. 17.

24 Miller, K. R. (1976) *J. Ultrastruct. Res.* **54:**159.

24a Miller, K. R. (1980) *Biochim. Biophys. Acta.* **592:**143.

25 Miller, K. R. and Cushman, R. A. (1979) *Biochim. Biophys. Acta* **546:**481.

26 Miller, K. R., Miller, G. J., and McIntyre, K. R. (1976) *J. Cell Biol.* **71:**624.

27 Miller, K. R. and Staehelin, L. A. (1976) *J. Cell Biol.* **68:**30.

28 Moor, H. and Muhlethaler, K. (1963) *J. Cell Biol.* **17:**609.

29 Muhlethaler, K., Moor, H., and Szarkowski, J. W. (1965) *Planta* **67:**305.

30 Ojakian, G. K. and Satir, P. (1974) *Proc. Natl. Acad. Sci. U.S.A.* **71:**2052.

31 Oleszko, S. and Moudrianakis, E. N. (1974) *J. Cell Biol.* **63:**936.

32 Paolillo, D. J. (1970) *J. Cell Sci.* **6:**243.

33 Park, R. B. and Biggins, J. (1964) *Science* **144:**1009.

34 Park, R. B. and Pfeifhofer, A. O. (1969) *J. Cell Sci.* **5:**299.

35 Park, R. B. and Sane, P. V. (1971) *Annu. Rev. Plant Physiol.* **22:**395.

36 Sane, P. V., Goodchild, D. J., and Park, R. B. (1970) *Biochim. Biophys. Acta* **216:**162.

37 Simpson, D. J., Lindberg-Moller, B., and Hoyer-Hansen, G. (1978) In *Chloroplast Development* (G. Akoyunoglou et al, Eds.), Elsevier/North-Holland Biomedical Press, Amsterdam, p. 507.

38 Staehelin, L. A. (1975) *Biochim. Biophys. Acta* **408:**1.

39 Staehelin, L. A. (1976) *J. Cell Biol.* **71:**136.

40 Staehelin, L. A., Armond, P. A., and Miller, K. R. (1976) *Brookhaven Symp. Biol.* **28:**278.

41 Steinman, E. and Sjostrand, F. S. (1955) *Exp. Cell Res.* **8:**15.

42 Trebst, A. (1974) *Annu. Rev. Plant Physiol.* **25:**423.

43 Walker, D. A. (1971) In *Methods in Enzymology*, Vol. 23 (A. San Pietro, Ed.), Academic Press, New York, p. 211.

VISUALIZATION OF CHROMOSOMES AND CHROMATIN

2

J. B. Rattner
Department of Molecular Biology and Biochemistry
University of California at Irvine
Irvine, California

Barbara A. Hamkalo
Department of Molecular Biology and Biochemistry and
Department of Developmental and Cell Biology
University of California at Irvine
Irvine, California

CONTENTS

1 Introduction

2 Chromatin structure

  2.1 The nucleosome organization of chromatin and the basic fiber

  2.2 Chromatin structure and transcription

    Ribosomal precursor RNA genes

    Nonribosomal transcription

    Interpretation of ultrastructural data

  2.3 Chromatin structure and replication

3 Higher order chromatin structure

  3.1 Fiber organization

  3.2 Chromosome organization

4 Concluding remarks

Acknowledgments

References

1 INTRODUCTION

An understanding of the molecular mechanisms involved in eukaryotic gene expression and chromatin replication requires detailed knowledge of the structural organization of chromatin and the way this organization may be modified as a prelude to chromatin function or as a consequence of it. In recent years this area of chromatin research has been under intense investigation primarily as a result of the discovery of a basic unit of chromatin organization, the nucleosome (Woodcock, 1973; Hewish and Burgoyne, 1973; Olins and Olins, 1974; Kornberg, 1974; Sahasrabuddhe and Van Holde, 1974). Our current understanding of this particle containing DNA and histone has come from a synthesis of biochemical, biophysical, and ultrastructural studies. These include detailed analysis of the digestion of chromatin by several different deoxyribonucleases including an endogenous nuclease (Hewish and Burgoyne, 1973), micrococcal nuclease (Noll, 1974; Sahasrabuddhe and Van Holde, 1974; Sollner-Webb and Felsenfeld, 1975; Shaw et al., 1976; Noll and Kornberg, 1977; Simpson, 1978), DNase I (Noll, 1974), and DNase II (Oosterhof et al., 1975); neutron scattering of chromatin in solution (Baldwin et al., 1975; Lilley and Tatchell, 1977; Hjelm et al., 1977); X-ray diffraction of crystallized monomer nucleosomes (Finch et al., 1977) and studies of histone-histone and histone-DNA interactions (D'Anna and Isenberg, 1974; Hardison et al., 1975; Martinson and McCarthy, 1975;

Thomas and Kornberg, 1975; Levina and Mirzabekov, 1975; Mirzabekov et al., 1977).

Coincident with the use of these indirect probes of chromatin structure, electron microscopic studies of chromatin focused on levels of organization above the nucleosome. These analyses approached questions such as the arrangement of nucleosomes within the basic chromatin fiber of transcriptionally active and inactive chromatin and the manner in which the basic fiber is folded into higher order structures. These studies, in conjunction with chemical and physical analyses, are important in the development of a better understanding of the relationship between chromatin structure and function. The results summarized in this chapter focus on recent information on the structural organization of chromatin that has been obtained using electron microscopy.

2 CHROMATIN STRUCTURE

2.1 The Nucleosome Organization of Chromatin and the Basic Fiber

Historically, electron microscopic investigation of chromatin organization has relied on thin sections of intact cells or whole mount preparations of water-spread material. Although these studies have suggested many aspects of chromatin and chromosome organization (see reviews, Solari, 1974; Ris, 1975; Wolfe, 1979; Hozier, 1979) our present understanding of chromatin structure is strongly influenced by the introduction of an electron microscopic preparative procedure originally developed by Miller and coworkers to visualize active genes (Miller and Beatty, 1969; Miller and Bakken, 1972; Hamkalo and Miller, 1973; Bakken and Hamkalo, 1979). Modifications of this procedure also produce chromatin and chromosome preparations suitable for detailed structural analysis (Olins and Olins, 1974; Rattner and Hamkalo, 1978a, 1978b, 1979). Although chromatin prepared by the "Miller spreading technique" appears relaxed (i.e., the basic fiber appears as a chain of beads), the technique allows the visualization of the primary interactions between DNA-RNA and proteins.

The Miller procedure consists of the chemical or mechanical disruption of cells and nuclei, the dispersion of the nuclear contents in hypotonic media at pH > 7.0, and centrifugation onto a hydrophilic carbon-coated electron microscope grid through a sucrose cushion (0.1 *M* or greater) containing 10% formalin (pH 7.5–8.5). After centrifugation the grid is rinsed in a solution of Photoflo to reduce the surface tension during drying; then it is stained in alcoholic phosphotungstic acid or uranyl acetate. Details of the procedure can be found in Miller and Bakken (1972) and Bakken and Hamkalo (1979). Application of this procedure (Olins and Olins, 1973, 1974; Woodcock, 1973) provided the first direct visualization of the nucleosome particle. In spread preparations (Figure 1) the basic chromatin fiber appears as a series of

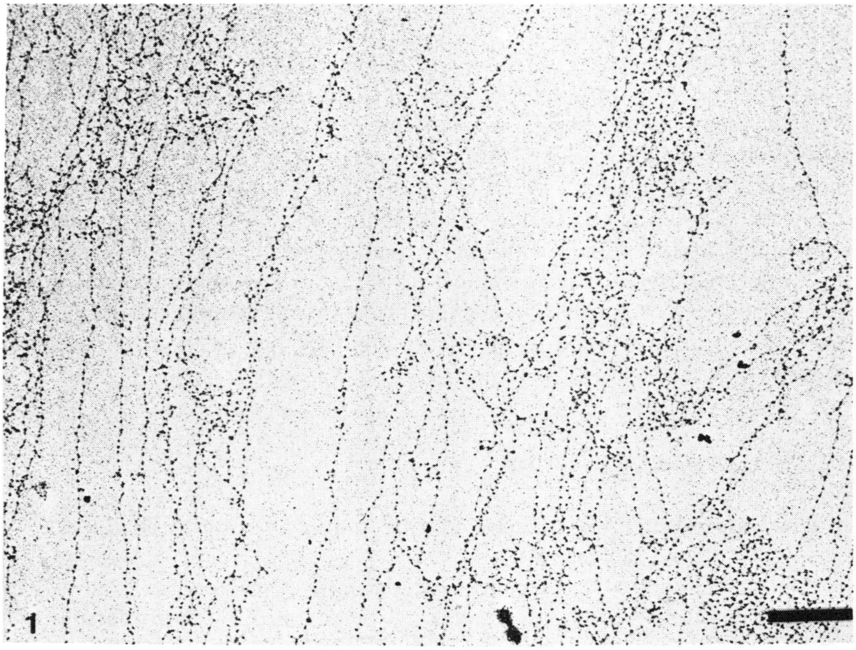

Figure 1 Electron micrograph of chromatin fibers in a beads-on-a-string conformation obtained from a mouse metaphase L929 cell prepared by the Miller spreading procedure and stained with 1% ethanolic phosphotungstic acid. Bar = 1 μm.

spherical particles 60–100 Å in diameter, separated by a linker region of internucleosomal DNA (Olins and Olins, 1974; Olins et al., 1975; Woodcock, 1973; Oudet et al., 1975; Langmore and Wooley, 1974; Kiryanov et al., 1976; Woodcock et al., 1976; Rattner et al., 1975).

It is now generally accepted that the spherical subunit visualized in the electron microscope is analogous to the core particle liberated from chromatin by micrococcal nuclease digestion (Rill and Van Holde et al., 1973; Noll, 1974). These particles are composed of 145 base pairs of DNA superhelically wrapped approximately one and three-quarter times around a protein core that consists of an octomer of histones, H2A, H2B, H3, and H4 (for review, see Rill, 1979). This organization results in a estimated packing ratio of the DNA of 7 : 1 for a simple chain of nucleosomes. This packing ratio has been experimentally documented by Griffith (1975) using Simian virus 40 (SV40) minichromosomes and electron microscopy.

Although the nucleosome appears spherical in the electron microscope, evidence from neutron scattering (Pardon et al., 1977) and X-ray diffraction of crystallized monomers (Finch et al., 1977), has shown that the nucleosome is in fact a flattened disk with dimensions 110 × 110 × 57 Å. In general, an edge-on view of a nucleosome is rarely seen in electron micrographs even when nucleosomes are organized into higher order fibers (Finch and Klug, 1976; Olins, 1977; Rattner and Hamkalo, 1978a, 1978b, 1979). Such observations suggest that nucleosome "faces" may be preferentially adsorbed to the electron microscope support film in dispersed preparations and that they are associated edge to edge as they form higher order fibers (Thoma et al., 1979). Alternate explanations for the inability to distinguish between the two views of the nucleosome in the electron microscope include limitations in the resolution of the instrument and subtle alterations in nucleosome morphology as a result of specimen preparation. These kinds of limitation must be taken into account in evaluating nucleosome arrangement within higher order structures.

The internucleosomal DNA or linker region varies in length from about 30 to 70 base pairs between species and in different tissues within the same organism (see review, Van Holde et al., 1978). Although the diameter of the linker fiber approximates that of naked duplex DNA (20 Å), it is likely that proteins are associated with the linker. In chromatin samples prepared at low ionic strength the linker DNA occasionally appears to enter and exit the nucleosome at about the same point (Thoma et al., 1979). It remains unclear, however, exactly where the linker DNA exits and enters the nucleosome and how this DNA is folded when nucleosomes are closely apposed. Despite this uncertainty, there is evidence that the linker DNA contains sites for the association of histone H1 with the beaded fiber, and that this association is necessary for the formation or maintenance of chromatin packing (Brasch et al., 1971; Brasch, 1976; Thoma and Koller, 1977; Renz et al., 1977a, 1977b).

The extended beads-on-a-string conformation, which allows the visualization of the internucleosomal linker regions in Miller spread preparations, is generally considered an artifact of preparation in that it does not reflect the arrangement of nucleosomes *in vivo*. Rather, it has been suggested that nucleosomes are closely apposed *in vivo,* forming a 100 Å diameter fiber (Finch et al., 1975; Griffith, 1975; Compton et al., 1976; Finch and Klug, 1976), similar to those observed in some thin-section preparations (Frenster, 1974). In general, 100 Å fibers are most readily visualized in chromatin that has been prepared in the presence of chelating agents or under conditions that promote the breakdown of higher order structures (Finch and Klug, 1976; Ris, 1969, 1978; Rattner and Hamkalo, 1978b), suggesting that although such fibers may not exist *in vivo,* they represent a transitional state in the unfolding of higher order fibers. The close packing of nucleosomes visible as 100 Å fibers in some electron micrographs may be generated as a result of incomplete adhesion of the chromatin fiber to the carbon support film during specimen preparation, so that subsequent negative staining results in atypical nucleosome associations (Thoma et al., 1979). It should be noted that isolation of chromatin or nuclei and fixation of tissue in the presence of chelators or metal-binding buffers (e.g., Tris, phosphate buffers) probably causes the relaxation of the native 200–300 Å chromatin fibers into a 100 Å fiber conformation.

Although histone H1 is not required for the formation or maintenance of the nucleosome core particle, it does appear to play a role in nucleosome-nucleosome association and in the formation or maintenance of higher order chromatin structure. Several experiments have suggested the location of the H1 molecules relative to nucleosomes. Nuclear magnetic resonance studies (Baldwin et al., 1975) suggest that H1 is bound to the outside of the nucleosome, whereas micrococcal nuclease digestion studies suggest that it binds to the linker DNA 10–30 bp from the core particle (Varshavsky et al., 1976; Noll and Kornberg, 1977; Simpson, 1978). In addition, experiments utilizing ultraviolet-induced cross-linking are consistent with a model in which histone H1, though anchored to the linker, is bound to the DNA at additional sites. Thus the histone H1 molecule could span the nucleosome clamping together the DNA, which is folded around the histone core (Sperling and Sperling, 1978). Electron microscopy of chromatin prepared under conditions that release histone H1 illustrates the transition from higher order fibers to the beads-on-a-string conformation (Brasch et al., 1971; Brasch, 1976; Thoma and Koller, 1977; Renz et al., 1977a, 1977b). Thus it appears that both histone H1 and cations contribute to stable folding of the chromatin fiber.

A closed chain of nucleosomes has been visualized in SV40 minichromosomes (Griffith, 1975). When prepared in 0.15 *M* NaCl the minichromosome appears as a circular fiber with a diameter of 100 Å and a contour length of 2100 Å. When the salt concentration is reduced tenfold, the fiber relaxes and approximately 21 nucleosomes can be seen. From the known length of the circular DNA molecule in SV40 (14,800 Å) and the contour length of the minichromosome, a packing ratio of 7 : 1 has been calculated, in agreement with estimates for a simple chain of nucleosomes based on X-ray diffraction and nuclease digestion data. Under conditions of physiological ionic strength, and in the presence of H1, the minichromosome appears to be more compact, with globular subunits 190 Å in diameter, which represents a further condensation of the chromatin fiber (Muller et al., 1978). In addition, it has been shown that the interaction between histones and DNA to form nucleosomes results in the introduction of about one superhelical turn per nucleosome in relaxed closed circular SV40 DNA (Germond et al.). Although the minichromosome of SV40, polyoma, and some other DNA viruses provides a valuable model system for the study of the interaction of DNA with histones to form nucleosomes, it is probably too simple a system for the analysis of chromatin folding typical of more complex genomes.

There are few exceptions to the nucleosomal organization of chromatin, but these are notable. In free-living dinoflagellates such as *Prorocentrum micans*, the nucleus contains an enormous amount of DNA (up to 100 pg/cell), which is organized into 50–200 rodlike chromosomes that are permanently condensed throughout the cell cycle. It has been suggested that the dinoflagellates represent an evolutionary transition between procaryotes and eucaryotes. The chromosomes of these organisms appear to lack histones (Rizzo and Nooden, 1973), and they contain large amounts of repeated DNA sequences (Allen et al., 1975). They appear to be composed of smooth chromosomal fibers 40–70 Å in diameter in both thin section and spread preparations (Figure 2 and Dodge, 1963; Haller et al., 1964; Leadbeater and Dodge, 1967; Kubai and Ris, 1969; Soyer and Haapla, 1973; Allen et al., 1975; Hamkalo and Rattner, 1977).

The chromatin of some sperm is also devoid of nucleosomes. During spermiogenesis the haploid genome is sequentially packed into a tight nucleoprotein complex; this packing is coincident with the cessation of RNA transcription and, in some organisms, with the replacement of the somatic histones by small (average molecular weight = 4000 daltons), highly basic, arginine-rich proteins (protamines). In the trout testis, the loss of the nuclease digestion pattern typical of nucleosome-containing chromatin corresponds to the replacement of histones by protamines (Honda et al., 1974). Electron microscopy of chromatin fibers during spermiogenesis in the mouse shows a transition from a beaded chromatin fiber to a smooth fiber of uniform diameter (Kierzenbaum and Tres, 1975) which agrees temporally with protamine-DNA complex formation. In contrast, in sperm that retain his-

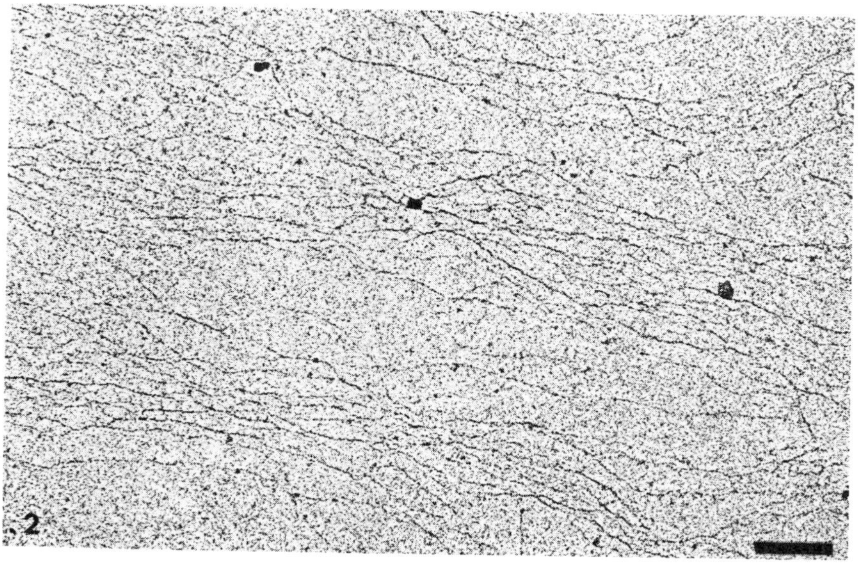

Figure 2 Chromosomal fibers from the dinoflagellate *Prorocentrum micans* prepared by the Miller spreading procedure and stained with ethanolic phosphotungstic acid. The fibers measure 40–70 Å in diameter and show no evidence of a beaded structure. Bar = 0.5 μm.

tones, the nucleosome organization of the chromatin fiber is not altered (Spadafora et al., 1976).

2.2 Chromatin Structure and Transcription

As noted above, the Miller spreading technique was developed to permit direct visualization of the regions of the eucaryotic genome that are active in RNA synthesis. Since its inception in the late 1960s by Miller and Beatty (1969), the technique has been applied successfully by many workers to the ultrastructural analysis of RNA synthesis in a diverse collection of organisms. Several reviews that summarize these many studies have appeared recently (e.g., McKnight et al., 1979). We focus on a few examples of these studies to discuss the interpretation of data and to suggest the potentialities and limitations of this approach in studying the structural features of transcriptionally active chromatin.

Ribosomal Precursor RNA Genes

Historically, the developing amphibian oocyte represents the first system in which active genes were visualized. These cells are atypical for several reasons, including their large size (at maturity > 1 mm in diameter), the presence of an enormous number of selectively replicated extrachromosomal, ribosomal precursor RNA genes all active in RNA synthesis, and the existence of a large number of sites of intense nonribosomal RNA transcription along the paired lateral loops emanating from otherwise condensed lampbrush chromosomes. These unusual features provided Miller and Beatty (1969) the opportunity to readily visualize and identify both ribosomal (Figure 3) and nonribosomal (Figure 4) transcription. Analyses of such preparations led these workers to the conclusion that an active ribosomal RNA precursor gene is only slightly shorter than would be expected if the active gene existed as B conformation DNA. An extended conformation of active ribosomal genes has been documented ultrastructurally in many other systems (for review, see Franke et al., 1979). In light of the observation that the DNA in closely apposed nucleosomes is compacted about sevenfold (Griffith, 1975), these data argue that an active ribosomal gene is not compacted to form nucleosomes along most of its length. However, the packing ratio of dispersed inactive chromatin exhibiting internucleosomal linkers is only about 2.0, regardless of the source of material or the investigator. Therefore active ribosomal genes are approximately twice as long as they would be if in a beads-on-a-string conformation.

The close packing of RNA polymerases on ribosomal genes has made it difficult to visualize the chromatin fiber structure underlying them. Nevertheless, Woodcock et al. (1976) studied slightly stretched ribosomal genes from the newt *Notophthalmus viridescens* and concluded that occasionally one can detect a nucleosome along these active genes in between the RNA polymerases. In agreement with this observation, Osheim et al. (1978)

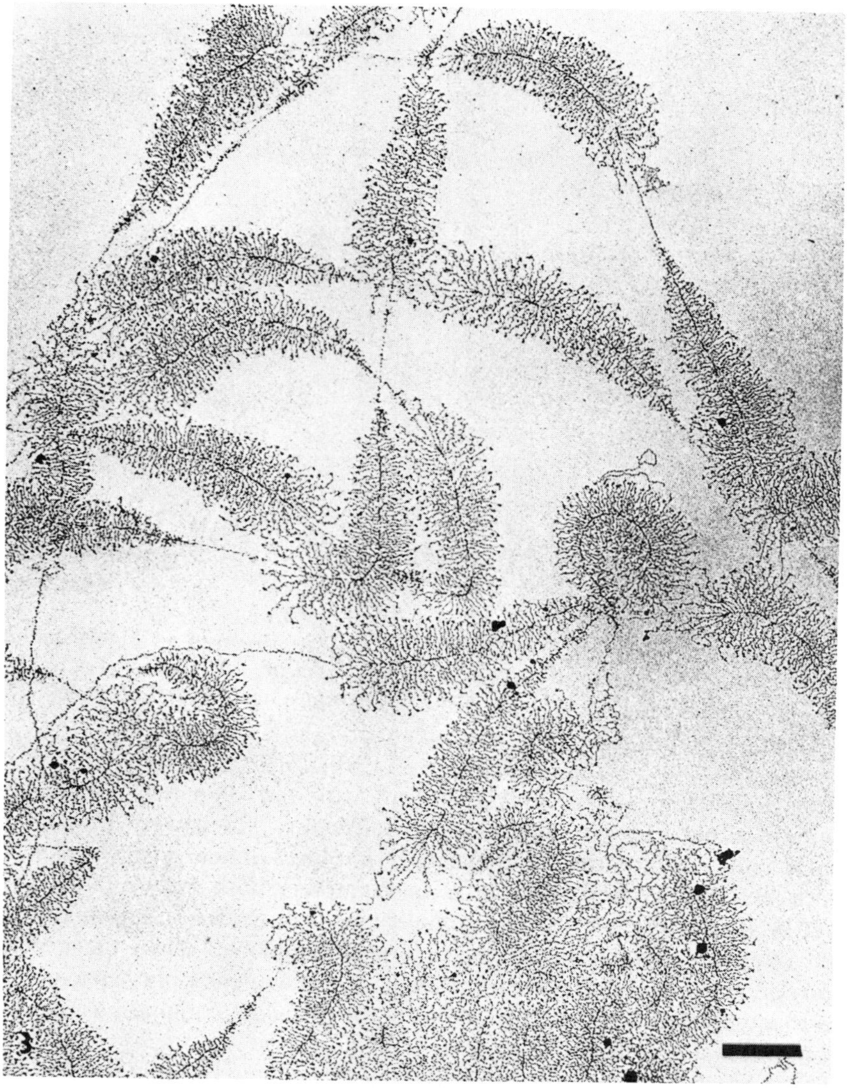

Figure 3 Active ribosomal precursor RNA genes from a developing oocyte of the spotted newt *Notophthalmus viridescens*. Manually isolated nuclear contents dispersed in water were prepared for electron microscopy by the Miller spreading procedure. Bar = 0.5 μm. From Miller and Beatty, *Science* **164:**955, 1969. Copyright 1969 by the American Association for the Advancement of Science.

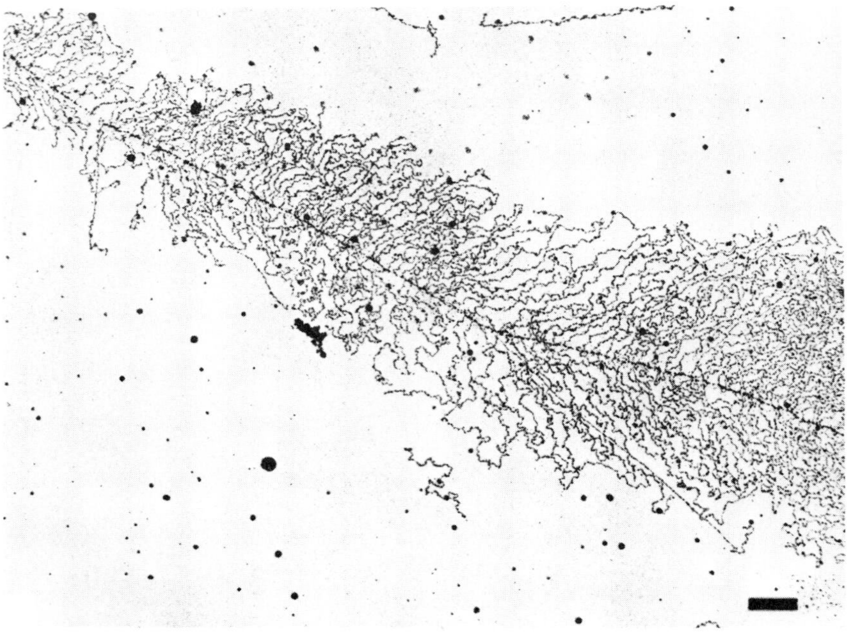

Figure 4 Portion near the thin insertion end of a lateral loop of a lampbrush chromosome, from material as in Figure 3. The high level of transcriptional activity is evident. Bar = 0.2 μm. From Miller et al., 1971; reproduced by permission.

provided electron microscopic evidence for the existence of beaded chromatin along *Xenopus laevis* ribosomal genes, which were transcribed submaximally. In contrast to these observations, Scheer (1978) observed smooth chromatin fibers (30–50 Å in diameter) both between distantly spaced RNA polymerases on ribosomal genes transcribed suboptimally and in the intergene spacer regions in several amphibians. When these genes were repressed, however, all the ribosomal DNA (genes plus spacers) returns to a typical beaded conformation. Some possible explanations for such discrepancies are considered below.

There has been some dispute regarding the definitive identification of a particle along a chromatin strand as an RNA polymerase or a nucleosome. However, after phosphotungstic acid staining, RNA polymerases have a larger diameter than nucleosomes and are noticeably more intensely stained. When preparations are both stained and rotary shadowed, Osheim and Miller (1979) suggest that the polymerase is larger in size than a nucleosome and has a different shape. In addition, Scheer (1978) has used the ionic detergents (Sarkosyl and Joy) to selectively remove nucleosomes from chromatin under conditions that do not extract transcribing RNA polymerases to aid in the identification of enzyme molecules.

In an attempt to study changes in chromatin morphology with transcription, several investigators have taken advantage of the modulation of ribosomal RNA transcription during oogenesis or embryogenesis; some of these investigations were mentioned above. In addition to these studies, two contrasting pictures have emerged from investigations of the activation process using embryos of two insects. Activation of the ribosomal genes in *Drosophila melanogaster* is accompanied by the appearance of closely packed ribonucleoprotein fibril gradients of varying lengths, which grow to the length of complete genes with time. The distal portions of activated genes that are not yet transcribed do not appear to be structurally distinct from inactive chromatin (McKnight and Miller, 1976).

In contrast to this pattern, activation proceeds in a different manner in *Oncopeltus fasciatus*. Foe (1977) studied this species in detail and showed that prior to gene transcription, the ribosomal gene regions appear as smooth fibers about 70 Å in diameter with a DNA packing ratio of about 1.2, the same as it is in the fully active gene. These regions, defined as ρ chromatin, are separated from each other by nucleosome-containing intergene spacer segments. During activation, individual genes show varying levels of transcription but, in contrast to *Drosophila*, the distribution of nascent RNPs at these times appears to be random along the genes. Although the change from nucleosomes to ρ chromatin is not a universal phenomenon, and although the molecular basis of this conversion is not understood, the existence of a distinct structural change that is selective for a locus prior to its transcription implies that in this system the activation step or steps are accompanied by one or more changes in macromolecular interactions that are transmitted along the length of the gene, an idea proposed by Yamamoto and Alberts (1976).

The nearly complete unfolding of active ribosomal chromatin discussed above is in apparent contradiction to results of micrococcal nuclease digestion of these genes, which shows that a substantial amount of active ribosomal chromatin is cleaved into nucleosome-sized DNA fragments (Reeves, 1976; Mathis and Gorovsky, 1977). As noted by Foe (1977), digestion experiments merely indicate that the chromatin is protected in some periodic way, and this is not necessarily equivalent to the occurrence of typical nucleosomes. However, the intense activity of the ribosomal precursor genes and their length when active argue in favor of the idea that these genes, and possibly others of equivalent activity, for the most part do not exist as typical nucleosomes.

If one assumes that nucleosomes do not exist in highly active chromatin, one cannot assume that the histones have been displaced. In fact, evidence derived from the analysis of the proteins associated with regions that are being transcribed favors the persistence of histones in active chromatin (Gottesfeld and Bonner, 1977; Levy-Wilson et al., 1979). A direct approach to this question is the immunological localization of histones in association with active chromatin. This experiment is difficult to perform on ribosomal chromatin because both chromatin and nascent RNPs aggregate in the presence of the high concentration of protein added as immunoglobulins (McKnight et al.; 1977). This technique has been used, however, to show that regions of nonribosomal transcription, where adjacent RNPs are not closely packed, contain histones (McKnight et al., 1977).

Nonribosomal Transcription

If chromatin subunit structure unfolds during RNA synthesis, one might predict a correlation between the level of RNA synthesis and the degree of unfolding. If we accept this hypothesis, we would expect to find that lampbrush chromosome loops, such as that illustrated in Figure 4, exist in a highly extended state. Since the nascent RNPs are about as closely packed as those associated with active ribosomal genes, again one cannot readily visualize the chromatin strand that is being transcribed. In addition, typical lampbrush loop transcription units are so large that it is not feasible to observe an entire unit on a grid square and, thus, one cannot estimate the DNA packing ratio.

Two studies have been performed on very active nonribosomal transcription units whose primary transcript sizes had been estimated biochemically. In the first, McKnight et al. (1976) located the genes actively synthesizing silk fibroin in the silk gland of *Bombyx mori*: the active transcription unit measured about 12% shorter than that predicted from the size of the transcript; this DNA packing is equivalent to that of active ribosomal genes. In a similar type of study Lamb and Daneholt (1979) visualized the intensely active Balbiani ring puffs on chromosome IV of the polytene set of *Chironomus tentans*. These active transcription units are compacted about 1.6:1 as compared to a packing of 1.9:1 for inactive chromatin in the same preparation. Thus the active puff chromatin, although somewhat unfolded, is not as unfolded as the fibroin gene, even though both exhibit high levels of transcription. Hence no simple linear relationship emerges between the level of activity of a region and the reduction in the degree of DNA unfolding.

Because of the difficulty of visualizing chromatin fibers in very actively transcribed regions, several groups have analyzed regions exhibiting lower levels of nonribosomal transcription (e.g., Laird and Chooi, 1976; Foe et al., 1976; Laird et al., 1976; McKnight and Miller, 1976, 1979; Foe et al., 1977; McKnight et al., 1977; Cotton et al., 1978; Busby and Bakken, 1979), a situation typical of the material one observes in the spread contents of almost any cell type (Figure 5). There is uniform agreement that there are nucleosomes along such regions of chromatin and that they are interspersed with RNA polymerases. However, the number of nucleosomes per unit length of chromatin varies considerably at such sites from a packing ratio as low as that for active ribosomal genes [1.1 in preimplantation rabbit embryos (Cotton et al., 1978)] to one very close to that of inactive chromatin [1.9 in sea urchin embryos (Busby and Bakken, 1979)]. Again, although there is a trend of reduced DNA packing with gene expression as assayed by electron

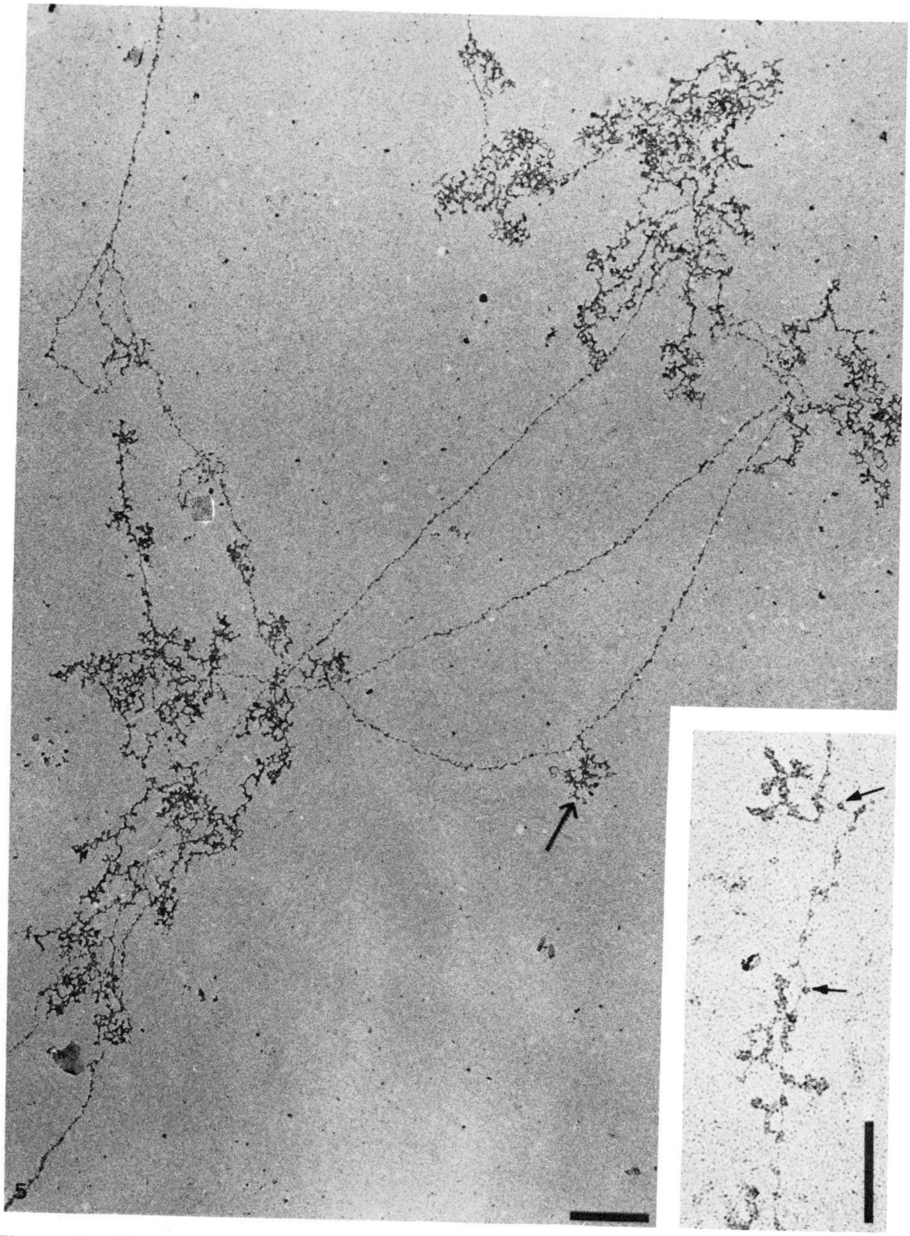

Figure 5 Electron micrograph of dispersed chromatin from a four-cell rabbit embryo. Insert shows chromatin from 16-cell embryo with nucleosomes visible between nascent RNP fibers (arrows). Bar = 0.2 μm. Courtesy of R. W. Cotton (unpublished).

microscopy of spread material, there is no obvious direct correlation between the amount of transcription and the packing ratio. The significance of this variability is not understood. To properly compare the data from different systems, however, one should have available comparable preparations (e.g., the same buffers, pH, etc.), and this is not possible with the information at hand.

In some cases it is possible to analyze the variability inherent in the pattern of RNA synthesis on a single transcription unit. To investigate this question, McKnight and Miller (1979) took advantage of the synchronous divisions of early *Drosophila* embryos and the fact that one can choose embryos at the first G2 after cellularization by measuring the length of the cell membrane that is being deposited. In these specimens, parallel strands of chromatin bearing strikingly similar patterns of nascent transcripts can be located; the authors propose that these strands represent sister chromatids (Figure 6). If one assumes that transcription initiation occurred after a replication fork passed through a promoter, the data of McKnight and Miller suggest that the pattern of activity of a specific transcription unit at a specific time during development is inherent in the promoter or some feature of the region that is involved in the initiation of RNA synthesis. Similar observations have been made in preparations of preimplantation rabbit embryo chromatin (Cotton, unpublished). The implications of these observations for the interaction between replication and transcription are discussed below.

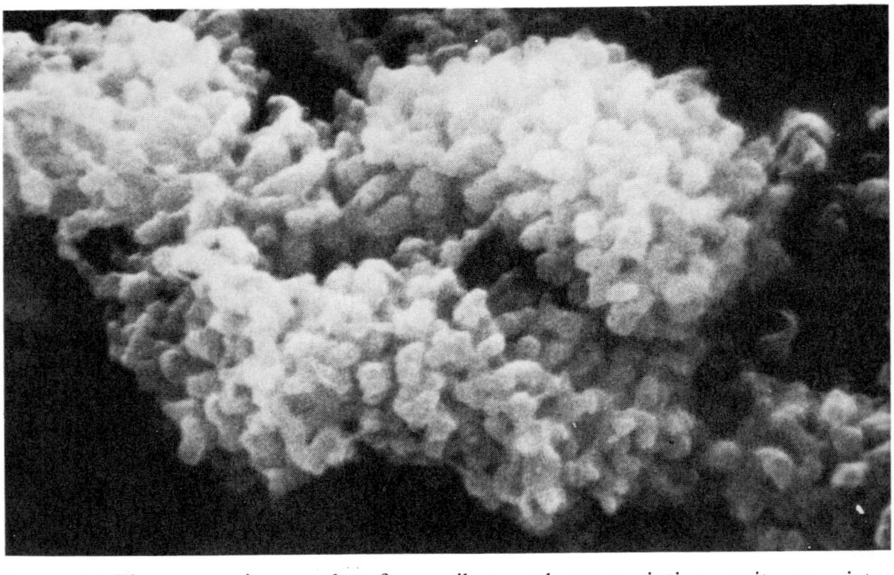

Figure 6 Electron micrograph of nonribosomal transcription unit on sister chromatids of *D. melanogaster* from late cellular blastoderm embryo. Distinct gradients of short-to-long lateral RNP fibers are visible. From McKnight and Miller, 1979, copyright 1979, MIT Press. Reproduced by permission.

Several electron microscopic studies on patterns and levels of RNA synthesis during embryogenesis have provided useful information on the expression of the embryonic genome in such diverse organisms as insects (Laird and Chooi, 1976; McKnight and Miller, 1976, 1979; Foe et al., 1977), sea urchins (Busby and Bakken, 1979) and mammals (Cotton et al., 1978; Hughes et al., 1979). Since only small amounts of the entire genome can be analyzed in spreads [estimated to be about 1.3% by Busby and Bakken (1979)] and this material tends to lie at the periphery of the nucleus, one must ask whether that small sample is representative of synthesis throughout the nucleus. Recently Bakken and Busby (1979) analyzed the distribution of RNA synthesis in sea urchin embryo nuclei by light microscope autoradiography of material after *in vivo* labeling. The equivalent distribution of silver grains over both peripheral and central regions of sectioned nuclei was shown to be statistically significant, suggesting that there are no major differences in the amount of RNA synthesis at various sites in the nucleus. This simple experiment is an important validation that is necessary before general conclusions can be drawn about rates and amounts of transcription from the analysis of only a minuscule portion of the genome.

Interpretation of Ultrastructural Data

There is a large body of evidence that active genes are both compositionally and structurally distinct from inactive genes: they are more rapidly digested by micrococcal nuclease (Reeves and Jones, 1976; Staron et al., 1977; Bloom and Anderson, 1978; Johnson et al., 1978); they are extremely sensitive to solubilization by pancreatic DNase (Weintraub and Groudine, 1976; Garel and Axel, 1976; and many others since 1976); they possess hyperacetylated core histones (Nelson et al., 1978; Simpson, 1978b; Vidali et al., 1978; Levy-Wilson et al., 1979) as well as certain nonhistone chromosomal proteins [among them a subset of the high mobility group proteins (Levy-Wilson and Dixon, 1979; Weisbrod and Weintraub, 1979)]. At least some of these alterations in active chromatin undoubtedly result in the formation of nucleoprotein complexes that are less stable under a given set of conditions than those complexes associated with inactive material. This is significant when one considers the manipulations that precede the visualization of active chromatin by any procedure.

The Miller procedure is described in detail in Section 2.1. Variations in the basic procedure have been introduced by several workers. Although we know very little about the perturbations introduced during preparation, certain observations are worth noting with regard to the effects of some treatments on chromatin structure as visualized in the electron microscope.

To lyse plasma and nuclear membranes that are resistant to hypotonic shock, nonionic detergents such as Triton X-100 and NP40 have been used. A few tenths of a percent of Triton X-100 is sufficient to dissociate nucleosomes from *Leishmania donovani* chromatin when viewed in the electron microscope (Rattner, unpublished observations). This was documented by a detergent titration experiment that showed less dissociation as the concentration of detergent was lowered. The use of ionic detergents to remove all nucleosomes and facilitate the identification of transcribing RNA polymerases (Scheer, 1978) has already been mentioned. The presence of detergents also improves chromatin dispersal but has the additional disadvantage of reducing the affinity of the sample for the grid surface.

The elimination of formalin from the cushion results in the visualization of more highly condensed chromatin fibers. This suggests that the formalin used (10%, pH 8.5) effects chromatin unfolding. Preliminary data from both electron microscopic (Rattner, unpublished data) and biochemical (Chalkley, unpublished data) assays show that this formalin solution both unfolds chromatin and induces the rapid dissociation of H1.

Foe et al. (1976) added transfer RNA to dispersal media to combat endogenous ribonuclease activity; these workers noticed that this addition also improved chromatin dispersal. The data on the rapid removal of H1 from chromatin by tRNA competition (Ilyin et al., 1971) readily explains the effect noted.

Although the steps noted above are essential for the successful visualization of chromatin and nascent transcripts in various cell types, it is obvious that they may also selectively alter labile chromatin structure during the preparative procedure. For example, acetylated histones, prevalent in active nucleosomes (Nelson et al., 1978; Simpson, 1978; Vidali et al., 1978; Levy-Wilson et al., 1979) are less tightly bound to the DNA and thus may be induced to move or dissociate with treatments that do not affect core histone–DNA interactions in inactive material. Such induced instability could result in images that are not faithful replicas of *in vivo* structure but amplify or exaggerate subtle changes. This amplification of a change represents a powerful approach to understanding underlying modifications in structure concomitant with function, provided one is aware of the effects of various treatments on the material.

2.3 Chromatin Structure and Replication

Electron microscopy of replicating material has been performed in very few systems, primarily because of the small number of replicating forks active at any one time, which makes them very difficult to locate. McKnight and Miller (1977) exploited the large number of closely spaced active replicating forks present in the cellular blastoderm stage in *Drosophila* to investigate the structure of replicating chromatin and the relationship between replication and transcription. They noted that in well-dispersed preparations newly replicated daughter chromatin strands possess a nucleosomal configuration and spacing not noticeably different from unreplicated regions. This observation argues that nucleosomes assemble very rapidly at a newly replicated region, in contrast to experiments by Worcel et al. (1978), which imply that chromatin subunit maturation requires about 10 minutes in *Drosophila* tissue culture cells. This difference may be related to the difference in the cell cycle time between the two systems.

Detailed analysis of many replicating regions showed that there were transcription complexes associated with both arms of replicated DNA, and the spacing and density of fibril arrays were homologous on each daughter strand, similar to the pattern seen in Figure 6 (McKnight and Miller, 1977). The occurrence in the same region of both replicative and transcriptive events suggests that these two processes are not mutually exclusive. However, this may be a situation that is unique to rapidly dividing cells. At the time that ribosomal genes are activated in these nuclei, regions containing both active genes and replication bubbles also were evident. Interestingly, sites for replication initiation appear only in the intergenic spacer regions, implying some restriction of the two processes when transcription rates are very high.

The distribution of preexisting nucleosomes to newly replicated daughter DNA strands has been investigated both biochemically and ultrastructurally. In the presence of cycloheximide to inhibit the synthesis of histones, half of the newly replicated DNA is highly sensitive to DNase I; this sensitivity is repaired within a short period of incubation in the absence of the inhibitor (Weintraub, 1973). Weintraub interpreted these and other data (Leffak et al., 1977) to mean that parental nucleosomes segregate preferentially to one strand and that they do not dissociate and reassociate with some complement of newly synthesized histones. A direct confirmation of this mode of segregation was provided by Riley and Weintraub (1979) in spreads of chromatin derived from chicken cells in culture that replicated for various periods of time in the presence of cycloheximide. In such preparations one daughter strand is nucleosome free and the other bears a typical distribution of nucleosomes (Figure 7). At increasing times in the presence of the inhibitor, the lengths of nucleosome-free chromatin increased. Thus the mode of parental nucleosome segregation proposed above is documented by the ultrastructural observations.

3 HIGHER ORDER CHROMATIN STRUCTURE

3.1 Fiber Organization

The nucleosome organization of chromatin is found throughout the cell cycle, in both heterochromatin and euchromatin, and at all stages of development examined (McKnight and Miller, 1976; Woodcock, 1976; Lipchitz and Axel, 1976; Musich et al., 1977; Rattner et al., 1978; Busby and Bakken, 1979; Cotton et al., 1978). Considerable attention is now focused on the manner in which the chain of nucleosomes is folded in the formation of higher order structures. Thin sectioning, freeze-fracturing, and surface spreading of plant and animal cells at both interphase and metaphase have

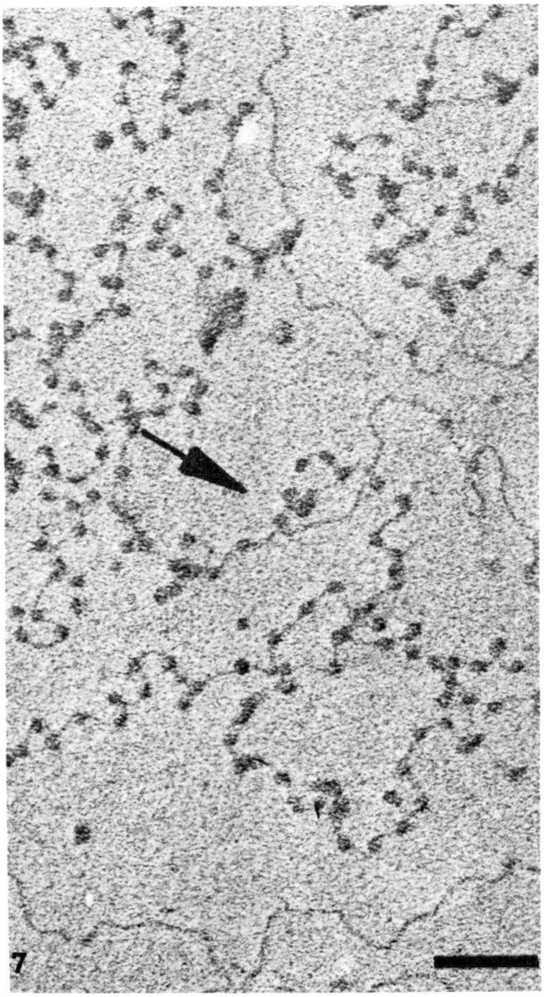

Figure 7 Presumptive asymmetric replication fork generated by replication in the presence of cyclohexamide. Bar = 0.2 μm. From Riley and Weintraub, 1979. Reproduced by permission.

suggested that chromatin exists *in situ* as a knobby fiber 200–300 Å in diameter (Bernstein and Mazia, 1953; Ris, 1956, 1975, 1978; Gall, 1966; Lampert, 1971; Solari, 1971; Filip et al., 1975; Zentgraf et al., 1975; Davies and Haynes, 1976; Rattner and Hamkalo, 1978a, 1979). A simple helical model for the folding of the nucleosome containing fiber within the 200–300 Å fiber has been proposed by several workers (Finch and Klug, 1976; Alberts et al., 1977; Carpenter et al., 1976; Worcel and Benyajati, 1977; Suau et al., 1979).

This proposal was also made in the prenucleosomal literature (Dupraw, 1965; Gall, 1966; Lampert, 1971).

Finch and Klug (1976) observed that chromatin fragments of up to 40 nucleosomes in length, generated by brief micrococcal nuclease digestion of rat liver chromatin and then brought to physiological concentrations of Mg^{2+}, appear in the electron microscope as a supercoil or solenoid structure 250–300 Å in diameter with a pitch of 100 Å. It was estimated that such an arrangement with six to seven nucleosomes per turn of the helix would give a DNA packing ratio of 40:1. Thoma et al. (1979) recently illustrated that with increasing ionic strength, chromatin fragments fold progressively from a chain of nucleosomes at about 1 m*M* monovalent salt through intermediate higher order structures to a thick fiber 250 Å in diameter at 60 m*M* monovalent salt or 0.3 m*M* Mg^{2+}. At intermediate ionic strengths the fiber observed exhibits a constant pitch and shows an increasing number of nucleosomes per turn as the salt concentration is increased. The formation of a 250 Å fiber on these studies requires the presence of histone H1 and has the form of a flexible but definite superhelix with an estimated six nucleosomes per turn.

Helical structures were visualized by Ris and Korenberg (1977) and Ris (1978) in amphibian erythrocyte chromatin prepared for the high-voltage electron microscope in the presence of mono- or divalent cations. In such preparations right- and left-handed coils could be detected within the fiber, suggesting a random coiling. In addition, Filip et al. (1975) observed helical structures in water-spread, acute-angle shadowed preparations of metaphase chromosomes. It should be noted, however, that in none of these studies is it possible to visualize individual nucleosomes within the condensed fiber, and in many cases helicity is seen only along short regions of the fiber. Thus these observations do not preclude the possibilities that helical packing of the nucleosome-containing fiber is confined to only short regions of the higher order fiber, or that variation may occur in the packing of adjacent helical turns along the fiber.

When interphase nuclei or metaphase cells are mechanically lysed with glass beads and the contents prepared for electron microscopy by the Miller procedure in the absence of formalin, 200–300 Å fibers also can be visualized (Figures 8 and 9). In negatively stained preparations the nucleosome particles are identifiable within the fiber (Rattner and Hamkalo, 1978a, 1979). Analysis of the packing patterns within these fibers suggests that nucleosomes are closely apposed along the fiber length and that they are arranged into several distinct packing arrangements. Discontinuities in the fiber present in these preparations may represent transitions between packing patterns and may account for the knobby morphology frequently reported in thin-section and water-spread preparations. Similar packing patterns have also been observed in meiotic chromatin (Rattner, unpublished).

Although some of the patterns observed can be accounted for by a helical folding of the basic nucleosome-containing fiber, others suggest alternate forms of nucleosome packing. It is possible that where helical arrangements

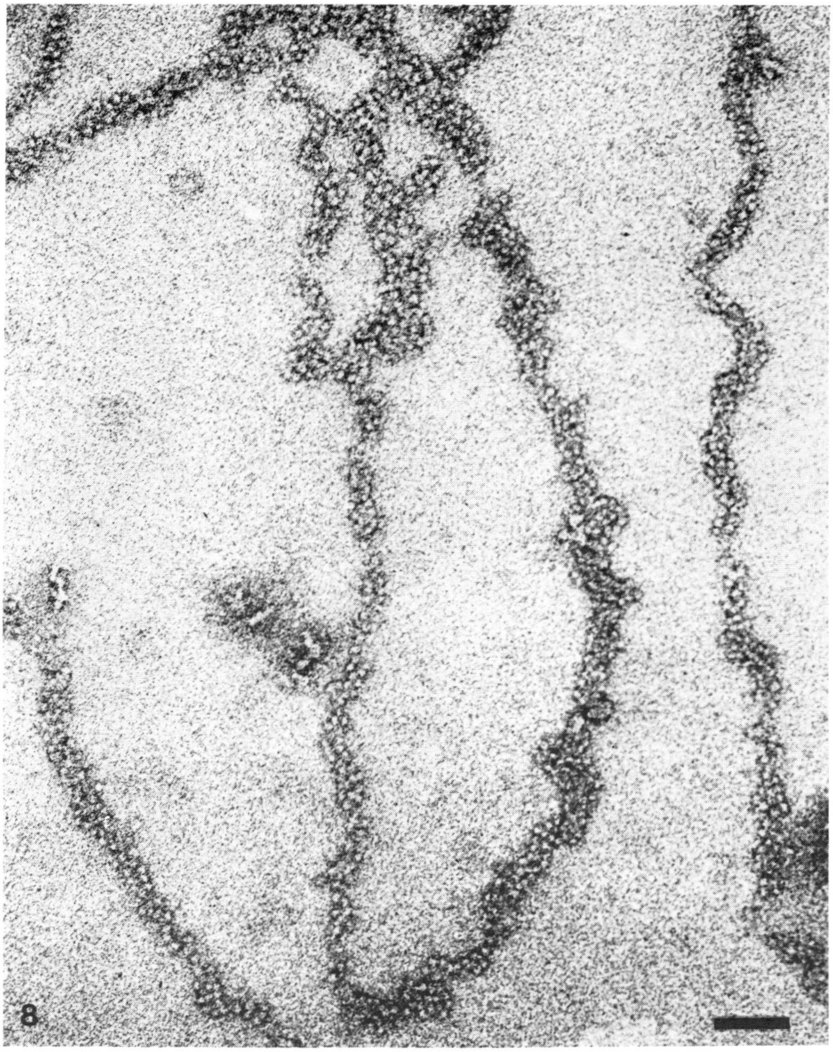

Figure 8 Loops (200–300 Å in diameter) of interphase chromatin from mechanically lysed mouse L939 cell prepared by the Miller spreading procedure in the absence of 10% formalin and negatively stained in 0.005 *M* methanolic uranyl acetate. The fibers are composed of closely apposed arrays of nucleosomes, two to three nucleosomes span the width of the fiber. Bar = 0.2 μm. Reproduced from Rattner and Hamkalo, *J. Cell Biol.* **81**:456, 1979, by permission of the Rockefeller University Press.

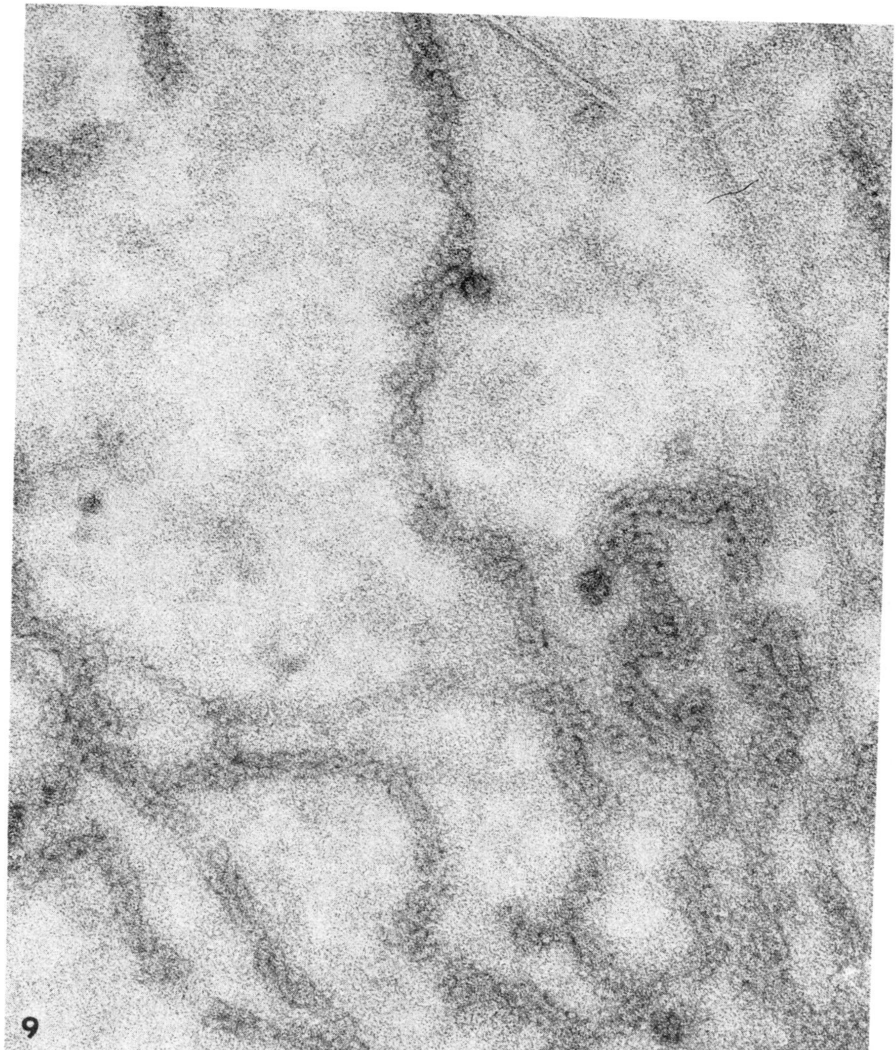

Figure 9 Fibers (200–300 Å in diameter) from a mouse metaphase cell prepared by the Miller spreading procedure in the absence of 10% formalin and negatively stained with 0.005 *M* methanolic uranyl acetate. The fiber appears to be composed of closely apposed nucleosomes arranged in a variety of packing patterns along the length of the fiber. Bar = 0.2 μm. From Rattner and Hamkalo, 1978. Reproduced by permission.

exist, the number of nucleosomes per turn may vary (Worcel, 1977; Rattner and Hamkalo, 1978a), giving rise to different packing patterns. Variation in packing along the length of the fiber suggested by these studies may correspond to variations in the distribution of NHCP and/or histone modifications or variants that may reflect functional differences along the chromatin fiber. Agarose gels of DNA that was isolated from DNase I–cleaved nuclei of *Drosophila melanogaster*, blotted by the Southern transfer technique, and probed with specific cloned fragments, exhibit a unique pattern of higher order domains of chromatin organization for each fragment. Such observations suggest that although these enzymes do not permit detection of regular supranucleosomal structures throughout the genome, a given locus exists as a discrete domain (Wu et al., 1979).

A different view of higher order structure was presented by Hozier et al. (1977), who reported that in the presence of mono- or divalent cations, erythrocyte chromatin prepared for electron microscopy by the Ficoll drying procedure displays a discontinuous 250 Å fiber within which nucleosomes are clustered into knobs or "superbeads." It is estimated that such an arrangement would provide a DNA packing ratio of at least 25:1. Similar chromatin fiber structures were reported by Kiryanov et al. (1976) and Franke et al. (1976). The "superbead" appearance of the fiber is present at 40 m*M* NaCl but is lost at lower ionic strength (Hozier et al., 1977). Superbeads can be released from nuclei by brief digestion with micrococcal nuclease and can be separated as a 40S peak in sucrose gradients. In the absence of histone H1, the 40S peak is lost and digestion produces a peak in the 11S monomer nucleosome position (Hozier et al., 1977). Recent studies of "superbead" monomers suggest that they may be very heterogeneous in size, containing 8–24 nucleosomes (Renz, 1979; Meyer and Renz, 1979). The composition of the 40S peak has been disputed, since in at least some systems the bulk of the material sedimenting at 40S can be accounted for by nuclease-trimmed RNP complexes (Walker et al., 1979). Although the packing arrangement of the nucleosomes within these superbeads is difficult to identify from micrographs, Worcel (1977) has suggested that they may contain limited domains of nucleosome supercoiling. In this way a "superbead" structure is not necessarily in conflict with a helical structure if discontinuities occasionally exist between gyres of the helix.

In an effort to resolve the relationship between the discontinuous "superbead" fiber and the more continuous fiber reported for chromatin fragments and mechanically dispersed, mouse L929 metaphase chromosomes have been prepared under conditions (100 m*M* PIPES, 0.1 m*M* MgCl and 1 m*M* EGTA) in which the transition from a 200–300 Å fiber to a beads-on-a-string conformation can be followed (Rattner and Hamkalo, 1978b). During this transition (which is complete within 20 minutes at room temperature) "superbead"-like arrays of nucleosomes are generated, suggesting that this conformation may represent a metastable intermediate in the unfolding of the higher order fiber (Figures 10 and 11).

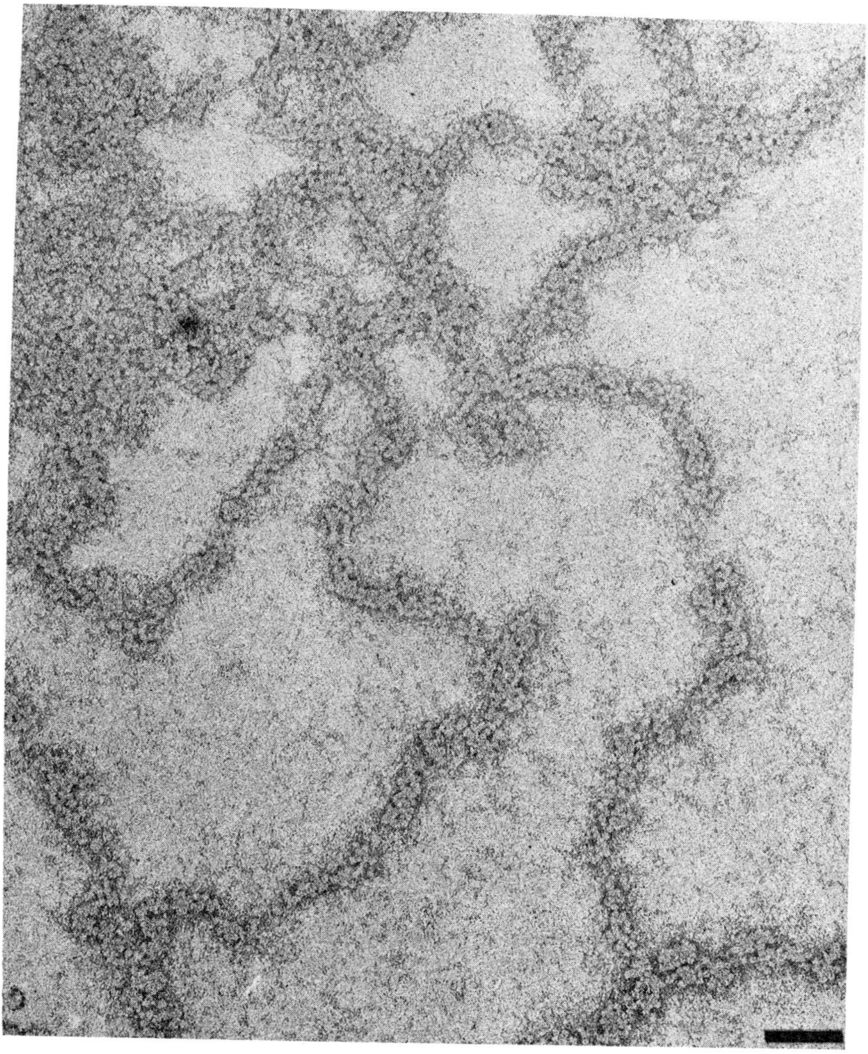

Figure 10 Mouse L929 metaphase chromatin prepared in the presence of PIPES, MgCl, and EGTA and prepared for electron microscopy by the Miller spreading procedure in the absence of 10% formalin. These fibers show occasional discontinuities, and the relationship between adjacent nucleosomes appears more relaxed than those observed in Figure 9. Bar = 0.2 μm. From Rattner and Hamkalo, 1978b. Reproduced by permission.

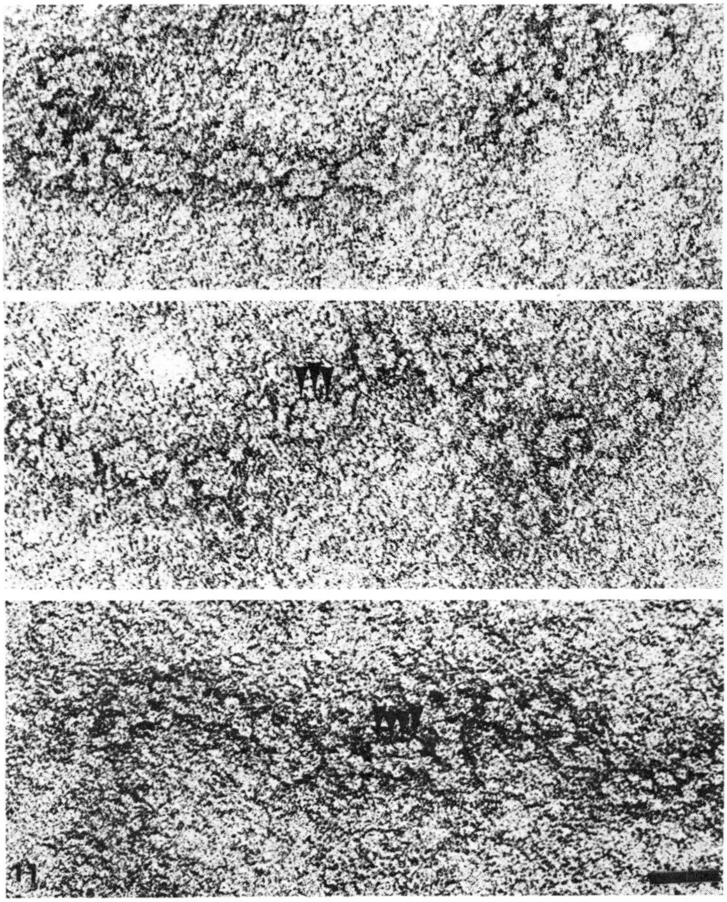

Figure 11 High magnification of partially relaxed fiber in which the helicity of nucleosome packing is suggested. Bar = 0.2 μm. From Rattner and Hamkalo, 1978b. Reproduced by permission.

It now appears that *in vivo* chromatin exists as a fiber approximately 200–300 Å in diameter that is formed by the folding of a single nucleosome-containing fiber (Ris, 1967; Rattner and Hamkalo, 1978B). This fiber is unstable at low ionic strength (Ris, 1969; Thoma et al., 1979), its existence is dependent on the presence of histone H1 (Finch and Klug, 1976; Brasch, 1976; Renz et al., 1977a, 1977b; Hozier, 1979), and it is stabilized by mono- or divalent cations (Ris, 1975; Rattner and Hamkalo, 1978, 1979). The precise way in which the nucleosome-containing fiber is folded within the higher order fiber requires further investigation, and existing information should be evaluated with the understanding that the three-dimensional structure that exists *in vivo* may be perturbed by isolation and electron microscopic preparative procedures. It seems reasonable, however, to expect that variations in nucleosome arrangements may be detected, and that different arrangements may reflect variations in the distribution of histone variants, histone modifications, and/or nonhistone chromosomal proteins. Such structural domains may correspond to specific structural and functional units (Alberts, 1977; Rattner and Hamkalo, 1978a, 1978b).

3.2 Chromosome Organization

Metaphase chromosomes represent a highly compacted state of the nucleosome-containing 200–300 Å fiber (Figure 12). The nature of this compaction has been the subject of numerous investigations. However, these studies have been complicated by the necessity of relaxing the chromosome to visualize its internal organization. This relaxation tends to destroy the higher order organization that is under study, often producing a wide array of artifacts. In addition, the formation of metaphase chromosomes undoubtedly results from several hierarchies of folding of the 200–300 Å fiber. Several investigators have proposed that a single mode of folding accomplishes all levels of higher order organization. The data now available are in-

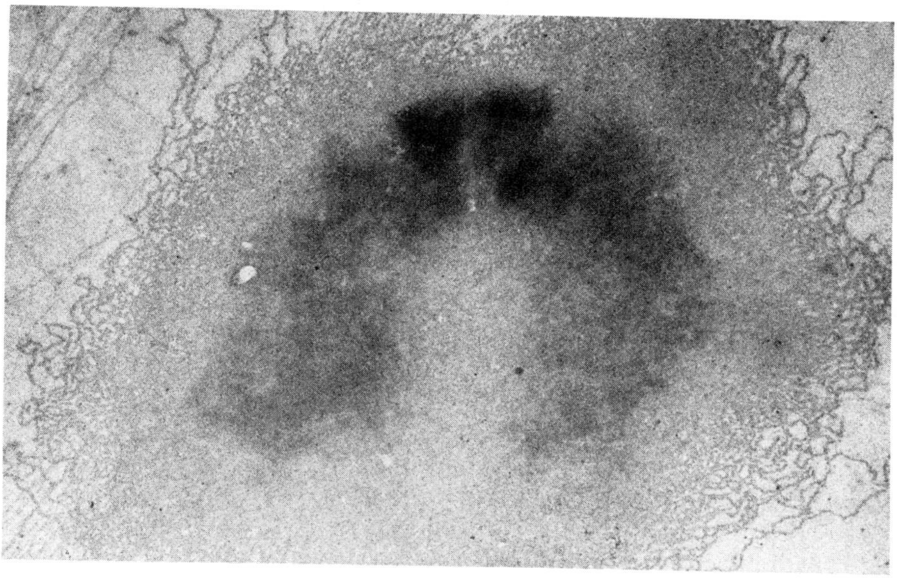

Figure 12 Mouse L929 cell metaphase chromosome obtained by mechanical cell lysis. Centromeric heterochromatin, kinetochore, and euchromatic arms are illustrated. Fibers 200–300 Å in diameter extend as a series of loops from the central, less dispersed region of the chromatid. From Rattner and Hamkalo, 1978b. Reproduced by permission.

sufficient to form firm conclusions, and many studies suggest that different modes of folding operate at different levels of organization. The superimposition of such differences complicates the interpretation of ultrastructural studies of chromosome organization. Nonetheless, three alternatives for the folding of the 200–300 Å fiber have been proposed based on light and electron microscopic data: kinking or localized folding, helical coiling, and looping. These observations form the basis for several different models of chromatin organization.

DuPraw (1965) proposed the "folded fiber model" in which the thick fiber is folded repeatedly transversely and longitudinally to form a chromatid. The evidence for this model is derived from electron microscopic studies on metaphase chromosomes prepared by standard surface-spreading procedures. In these preparations chromosomes retain their basic shape but are somewhat expanded and appear to be composed of a mass of chromatin fibers 200–300 Å in diameter folded in an indeterminate manner.

In contrast, several groups have proposed a model based on successive coiling of a 100 Å fiber to form tubes of increasing diameter (Bak and Zeuthen, 1977; Bak et al., 1977; Sedat and Manuelidis, 1977; Ris et al., 1978). This model of chromatin organization is based on a variety of light and electron microscopic observations. In some instances, a helix with a diameter of 0.2 μm can be visualized in living *Haemanthus* endosperm cells (Bajer, 1966; Holm and Bajer, 1966). In general, however, such coils are best visualized after pretreatment with hypertonic salt (Ohnuki, 1968) or exposure to 0.7 *M* KCl or to 0.04 *M* phosphate buffer. A hierarchy of coiling has been seen in high-voltage electron micrographs of whole Chinese hamster ovary metaphase chromosomes (Ris, 1976). After treatment of cells with hypotonic phosphate buffer (0.04 *M*), 200 Å fibers appear to be organized into a thicker unit, 1000–2000 Å in diameter. In stereoscopic photographs this unit appears to be helically coiled. Frequently, at the edge of these chromosomes 200 Å fibers appear to be coiled to form a fiber 500–700 Å in diameter, which in turn is helically coiled to form the 1000–2000 Å fiber (Ris, 1978).

Light and electron microscopic studies by Sedat and Manuelidis (1977) provide evidence for helical structures in interphase nuclei and metaphase chromosomes of tissue culture cells and in polytene chromosomes gently released from *Drosophila melanogaster* salivary gland cells. From these observations the investigators suggest that metaphase chromosomes are formed from a basic fiber 100 Å in diameter that forms successive coils of 300–500 and 6000 Å, the latter representing a chromatid.

Recently, Laemmli and co-workers (Adolph et al., 1977; Laemmli et al., 1977; Paulson and Laemmli, 1977) proposed a "radial loop model" in which loops of 200–300 Å fibers are arranged radially around a central protein-containing element or scaffold that exists along the length of the chromatid. This model is based partly on their observations of histone-depleted metaphase chromosomes prepared by competition with the polyanions dextran sulfate and heparin or high salt, after isolation by the procedure of Wray and Stubblefield (1970) utilizing hexylene glycol and Ca^{2+}. In the electron microscope these chromosomes display a loose network or scaffolding that extends along most of the length of the chromatid. This scaffold appears to consist of a subset of nonhistone chromosomal proteins, and DNA loops (10–30 μm long) can be seen emanating from the central element. The scaffolding structure is maintained after exhaustive digestion of histone-depleted chromosomes with micrococcal nuclease, suggesting that the shape of the chromosome is retained by a DNase-resistant structure. These observations are consistent with earlier observations of Stubblefield and Wray (1971), which showed a protein core in chromosomes isolated in the same manner. Loops of chromatin fibers 200–300 Å in diameter are also seen in electron micrographs of mouse metaphase chromosomes prepared by the Miller spreading procedure, although a scaffolding structure is not identifiable (Rattner and Hamkalo, 1978a, 1978b).

Additional support for the radial loop model is drawn from thin sections of isolated metaphase chromosomes swollen by EDTA chelation of divalent cations. Electron micrographs illustrate a series of loops of chromatin in a beads-on-a-string conformation radiating from a central dense core. In the same study chromosomes fixed in the presence of 1 m*M* MgCl display loops of 200–300 Å fibers radiating from the less dispersed central axis of the chromatid (Marsden and Laemmli, 1979) (Figure 13). An additional level of

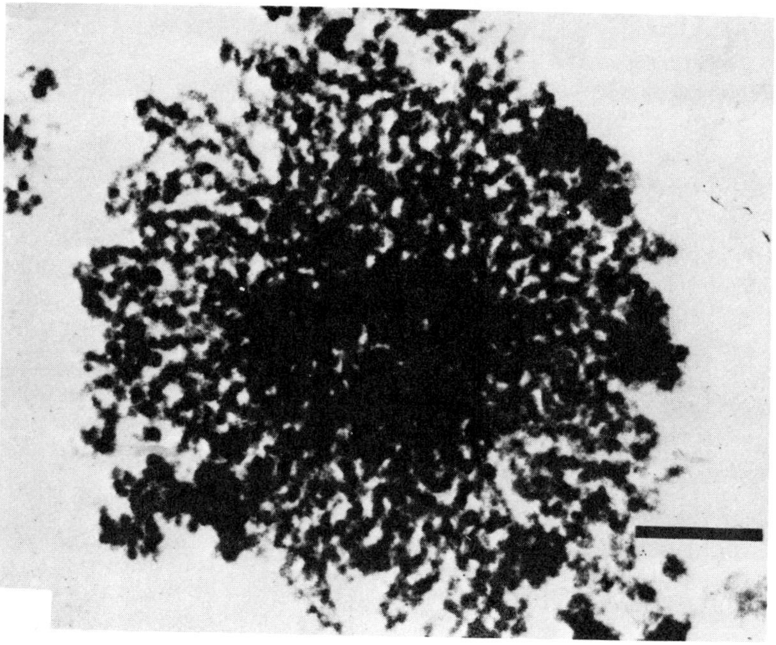

Figure 13 Cross section of a HeLa S_3 chromatid fixed in a buffet containing 1.0 *mM* MgCl. The 200–300 Å fibers appear as radial loops that extend from the dense central region of the chromatid. Bar = 0.2 μm. From Marsden and Laemmli, 1979, copyright 1979 by MIT Press. Reproduced by permission.

folding was observed in metaphase chromosomes isolated by the Wray and Stubblefield procedure and prepared for scanning electron microscopy. In these preparations the chromatid appears as a series of projections approximately 500 Å in diameter that may represent supercoiling or folding of the radially arranged 200–300 Å fibers (Figure 14). (Similar observations have also been reported by Daskel et al., 1976).

Many of the studies noted above utilize hexylene glycol and Ca^{2+} for chromosome isolation (Wray and Stubblefield, 1970) because chromosomes prepared in this manner can be purified by sucrose or metrizamide gradients free of other cellular components (Stubblefield and Wray, 1973). However, this preparative procedure is known to result in highly contracted chromosomes that are difficult to disperse after isolation (Krystal, unpublished observations; Marsden and Laemmli, 1979). A scaffold structure has not been visualized in chromosomes prepared in aqueous buffers and deposited on electron microscope grids by centrifugation through solutions that remove histones (Krystal and Hamkalo, unpublished observations). In addition, treatment of mitotic cells or chromosomes in aqueous buffers with mi-

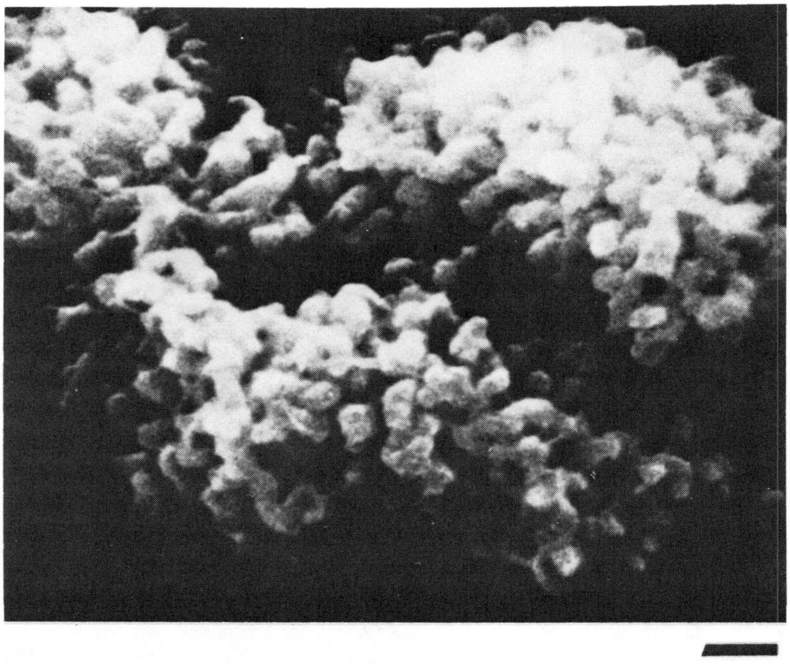

Figure 14 Scanning electron micrograph of a HeLa S_3 chromosome isolated with hexylene glycol. Compact projections are thought to represent further condensation of the loops of the 200–300 Å diameter fiber. Bar = 0.2 μm. From Marsden and Laemmli, 1979, copyright 1979 by MIT Press. Reproduced by permission.

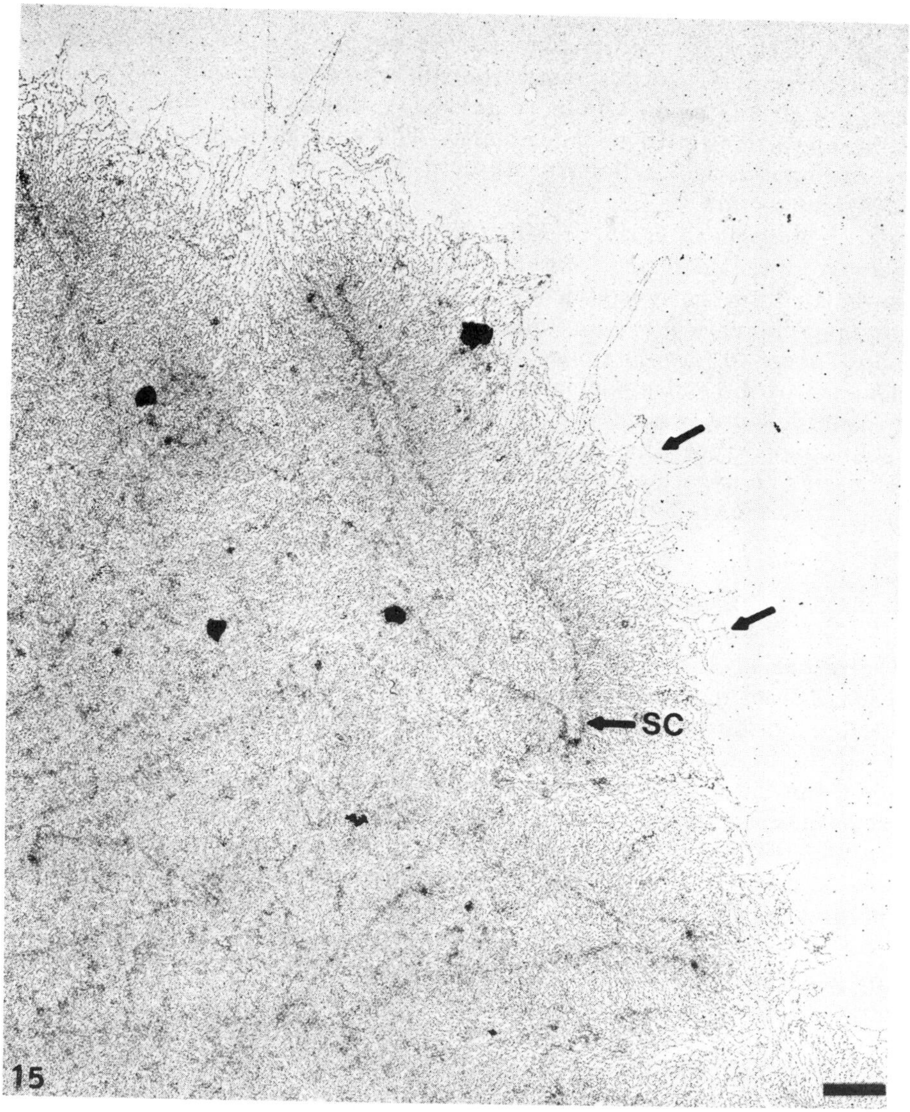

Figure 15 Male meiotic prophase chromosome of *Bombyx mori*, prepared by the Miller spreading procedure and stained with ethanolic uranyl acetate. The chromatin fibers appear as a series of loops that extend from the synaptonemal complex (SC). Bar = 0.5 μm.

crococcal nuclease results in the rapid solubilization of all the chromosomal DNA, and electron micrographs of these samples do not show scaffolds. These observations suggest that chromatin aggregation may be occurring in hexylene glycol–treated chromosomes and that, as a consequence, normally labile interactions involved in maintaining chromosome shape are stabilized. In addition, this aggregation may obscure certain subtle features of chromosome architecture.

Recent application of the Miller procedure to meiotic prophase spermatocytes of male *Bombyx mori* clearly demonstrates that loops of chromatin radiate from the synaptonemal complex, a protein structure that is present along the pairing axis of homologous chromosomes (Figure 15). These loops appear to form just prior to or at the time of formation of the axial elements of the synaptonemal complexes. These loops are 200–300 Å in diameter and are composed of closely apposed nucleosomes. These observations support the concept that a looplike organization of chromatin fibers represents an important aspect in the organization of chromatin into higher order structures in both mitotic and meiotic chromosomes.

4 CONCLUDING REMARKS

The recent advances in our understanding of chromosome structure have resulted from the significant increase in information on the interaction of DNA and histones at the level of the basic chromatin fiber. Above this level of organization, however, our knowledge is less sound because of uncertainties in the degree of preservation of higher order organization and because the material is less amenable to the detailed chemical and physical analyses that have been so important in determining nucleosome substructure. In addition, a large number of variations in preparative procedures, only some of which have been noted above, make it difficult to compare and reconcile inconsistencies in the literature. As with all biological structure problems, knowledge of the structural arrangement of chromatin and chromosomes *in situ* is crucial to a correct interpretation of *in vitro* data. This problem is obviously unsolved, and it is only beginning to be approached experimentally. Clearly its solution is essential for a complete understanding of the relationship between chromosome structure and function.

ACKNOWLEDGMENTS

The authors thank N. Hutchison for her critical reading of the manuscript and R. Cotton, T. P. Yang, L. Lica, and A. Demetrescu for helpful discussion. During the preparation of this manuscript B.A.H. was supported by grant GM-002-33-04 and J.B.R. and B.A.H. by grant PCM-78-08-930.

REFERENCES

Adolph, K. W., Cheng, S. M., and Laemmli, U. K. (1977) *Cell* **12:**805.

Alberts, B., Worcel, A., and Weintraub, H. (1977) In *The Organization and Expression of the Eukaryotic Genome* (E. M. Bradbury and K. Javaherian, Eds.), Academic Press, New York, pp. 165–208.

Allen, J. R., Roberts, T. M., Loeblich, A. R., and Klotz, L. (1975) *Cell* **6:**161.

Bajer, A. (1966) In *Probleme der biologischen Reduplikation* (P. Sutte, Ed.), Springer Verlag, New York, pp. 99ff.

Bak, A. L. and Zeuthen, J. (1977) *Cold Spring Harbor Symp. Quant. Biol.* **42:**319.

Bak, A. L., Zeuthen, J., and Crick, F. H. C. (1977) *Proc. Natl. Acad. Sci. U.S.A.* **74:**1595.

Bakken, A. H. and Busby, S. J. (1979) *J. Cell Biol.* **83:**164a.

Bakken, A. H. and Hamkalo, B. A. (1978) In *Principles and Techniques of Electron Microscopy*, Vol. 9 (M. A. Hayat, Ed.), Van Nostrand Reinhold, New York, pp. 84–104.

Baldwin, J. P., Boseley, P. G., Bradbury, E. M., and Ibel, K. (1975) *Nature (London)* **253:**245.

Bernstein, M. H. and Mazia, D. (1953) *Biochim. Biophys. Acta* **10:**59.

Bloom, K. S. and Anderson, J. N. (1978) *Cell* **15:**141.

Brasch, K. (1976) *Exp. Cell Res.* **101:**396.

Brasch, K., Setterfield, G., and Neeline, J. M. (1971) *Exp. Cell Res.* **101:**396.

Busby, S. J. and Bakken, A. H. (1979) *Chromosoma (Berlin)* **71:**249.

Carpenter, B. G., Baldwin, J. P., Bradbury, E. M., and Ibel, K. (1976) *Nucleic Acids Res.* **3:**1739.

Compton, J. L., Bellard, M., and Chambon, P. (1976) *Proc. Natl. Acad. Sci. U.S.A.* **73:**4382.

Cotton, R. W., Manes, C., and Hamkalo, B. A. (1978) *J. Cell Biol.* **79:**160a.

D'Anna, J. A., Jr. and Isenberg, I. (1974) *Biochemistry* **13:**2098.

Daskel, Y., Mace, M. L., Wray, W., and Busch, H. (1976) *Exp. Cell Res.* **100:**204.

Davies, H. G. and Haynes, M. E. (1976) *J. Cell Sci.* **21:**315.

Dodge, J. D. (1963) *Arch. Mikrobiol.* **45:**46.

DuPraw, E. J. (1965) *Nature (London)* **206:**338.

Filip, D. A., Gilly, C., and Mouriquand, C. (1975) *Humangenetik* **30:**155.

Finch, J. T. and Klug, A. (1976) *Proc. Natl. Acad. Sci. U.S.A.* **73:**1897.

Finch, J. T. and Klug, A. (1977) *Cold Spring Harbor Symp. Quant. Biol.* **42:**1.

Finch, J. T., Noll, M., and Kornberg, R. D. (1975) *Proc. Natl. Acad. Sci. U.S.A.* **72:**3320.

Finch, J. T., Lutter, L. C., Rodes, D., Brown, R. S., Rushton, B., Levitt, M., and Klug, A. (1977) *Nature (London)* **269:**29.

Foe, V. E. (1977) *Cold Spring Harbor Symp. Quant. Biol.* **42:**723.

Foe, V. E., Wilkinson, L. E., and Laird, C. D. (1976) *Cell* **9:**131.

Franke, W. W., Scheer, M. F., Trendelenberg, H., and Zentgraf, H. (1976) *Cytobiologie* **13:**401.

Franke, W., Scheer, U., Spring, H., Trendelenberg, M., and Zentgraf, H. (1979) In *The Cell Nucleus,* Vol. 7, *Chromatin,* Part D (H. Busch, Ed.), Academic Press, New York, p. 49.

Frenster, J. H. (1974) In *The Cell Nucleus*, Vol. 1 (H. Busch, Ed.) Academic Press, New York, pp. 220–238.

Gall, J. (1966) *Chromosoma (Berlin)* **20:**221.

Garel, A. and Axel, R. (1976) *Proc. Natl. Acad. Sci. U.S.A.* **73:**3966.

Germond, J. E., Hirt, B., Oudet, P., Gross-Bellard, M., and Chambon, P. (1975) *Proc. Natl. Acad. Sci. U.S.A.* **27:**1843.

Gottesfeld, J. M. and Bonner, J. (1977) In *The Molecular Biology of the Mammalian Genetic Apparatus* ((P. O. P. Ts'o, Ed.), Elsevier/North-Holland, Amsterdam, p. 381.

Griffith, J. (1975) *Science* **187:**1202.

Haller, C. D., Kellenberger, E., and Rouiller, C. J. (1964) *J. Microsc.* **3:**627.

Hamkalo, B. A. and Miller, O. J., Jr. (1973) *Annu. Rev. Biochem.* **42:**379.

Hamkalo, B. A. and Rattner, J. B. (1977) *Chromosoma (Berlin)* **60:**39.

Hardison, R. M., Eichner, E., and Chalkley, R. (1975) *Nucleic Acid Res.* **2:**1751.

Hjelm, P. P., Kneale, G. G., Suau, P., Baldwin, J. P., and Bradbury, E. M. (1977) *Cell* **10:**139.

Hewish, D. R. and Burgoyne, L. A. (1973) *Biochem. Biophys. Res. Commun.* **52:**504.

Holm, G. and Bajer, A. (1966) *Hereditas* **54:**356.

Honda, B. M., Baillie, D. L., and Candido, E. P. M. (1974) *FEBS Lett.* **48:**156.

Hozier, J. C. (1979) In *Molecular Genetics,* Part 3, *Chromosome Structure* (J. H. Taylor, Ed.), Academic Press, New York, pp. 315–380.

Hozier, J. C., Nehls, P., and Renz, N. (1977) *Chromosoma (Berlin)* **61:**301.

Hughes, M. E., Burki, K., and Fakan, S. (1979) *Chromosoma (Berlin)* **73:**179.

Ilyin, Y. V., Varshavsky, A. Ya., Mickelsaar, U. N., and Georgiev, G. P. (1971) *Eur. J. Biochem.* **22:**235.

Johnson, E. M., Allfrey, V. G., Bradbury, E. M., and Matthews, H. (1978) *Proc. Natl. Acad. Sci. U.S.A.* **75:**1116.

Kierzenbaum, A. L. and Tres, L. L. (1975) *J. Cell Biol.* **65:**258.

Kiryanov, G. I., Manamshjan, T. A., Polyakov, V. Y., Fais, D., and Chentsor, J. S. (1976) *FEBS Lett.* **67:**323.

Kornberg, R. D. (1974) *Science* **184:**868.

Kubai, D. F. and Ris, H. (1969) *J. Cell Biol.* **40:**508.

Laemmli, U. K., Cheng, S. M., Adolph, K. W., Paulson, J. R., Brown, J. A., and Baumback, W. R. (1977) *Cold Spring Harbor Symp. Quant. Biol.* **42:**351.

Laird, C. D. and Chooi, W. Y. (1976) *Chromosoma (Berlin)* **58:**193.

Laird, C. D., Wilkinson, L. E., Foe, V. E., and Chooi, W. Y. (1976) *Chromosoma (Berlin)* **58:**169.

Lamb, M. M. and Daneholt, B. (1979) *Cell* **17:**835.

Lampert, F. (1971) *Humangenetik* **13:**285.

Langmore, J. P. and Wooley, J. C. (1975) *Proc. Natl. Acad. Sci. U.S.A.* **72:**2691.

Leadbeater, B. and Dodge, J. D. (1967) *Arch. Microbiol.* **57:**239.

Leffak, I. M., Grainer, R., and Weintraub, H. (1977) *Cell* **12:**837.

Levina, E. S. and Mirzabekov, A. K. (1975) *Dokl. Akad. Nauk SSSR* **221:**1222.

Levy-Wilson, B. and Dixon, G. H. (1979) *Proc. Natl. Acad. Sci. U.S.A.* **76:**1682.

Levy-Wilson, B., Connor, W., and Dixon, G. H. (1979) *J. Biol. Chem.* **254:**609.

Levy-Wilson, B., Watson, D. C., and Dixon, G. H. (1979) *Nucleic Acids Res.* **6:**259.

Lilley, D. M. J. and Tatchell, K. (1977) *Nucleic Acids Res.* **4:**2039.

Lipchitz, L. and Axel, R. (1976) *Cell* **9:**355.

Marsden, M. P. F. and Laemmli, U. K. (1979) *Cell* **17:**849.

Martinson, H. G. and McCarthy, B. J. (1975) *Biochemistry* **14:**1073.

Mathis, D. J. and Gorovsky, M. A. (1977) *Cold Spring Harbor Symp. Quant. Biol.* **42:**773.

McKnight, S. L. and Miller, O. L., Jr. (1976) *Cell* **8:**305.

McKnight, S. L. and Miller, O. L., Jr. (1977) *Cell* **12:**795.

McKnight, S. L. and Miller, O. L., Jr. (1979) *Cell* **17**:551.

McKnight, S. L., Sullivan, N. L., and Miller, O. L., Jr. (1976) *Prog. Nucleic Acids Res. Mol. Biol.* **19**:313.

McKnight, S. L., Bustin, M., and Miller, O. L., Jr. (1977) *Cold Spring Harbor Symp. Quant. Biol.* **42**:741.

McKnight, S. L., Martin, K. A., Beyer, A. L., and Miller, O. L., Jr. (1979) In *The Cell Nucleus,* Vol. 7, *Chromatin,* Part D (H. Busch, Ed.) Academic Press, New York, p. 97.

Meyer, G. F. and Renz, M. (1979) *Chromosoma (Berlin)* **75**:177.

Miller, O. L., Jr. and Beatty, B. R. (1969) *Science* **164**:955.

Miller, O. L., Jr. and Bakken, A. H. (1972) *Acta Endocrinol. (Copenhagen)* **168**:155.

Miller, O. L., Jr., Beatty, B. R., Hamkalo, B. A. and Thomas, C. A., Jr. (1970) *Cold Spring Harbor Symp. Quant. Biol.* **35**:508.

Mirzabekov, A. D., Schick, V. V., Belyavsky, A. V., Karpor, V. L., and Bavykin, S. G. (1977) *Cold Spring Harbor Symp. Quant. Biol.* **149**:155.

Muller, V., Zentgraf, H., Eicken, I., and Keller, W. (1978) *Science* **201**:406.

Musich, P. R., Brown, F. L., and Maio, J. J. (1977) *Proc. Natl. Acad. Sci. U.S.A.* **74**:3297.

Nelson, D. A., Perry, W. M., and Chalkley, R. (1978) *Biochem. Biophys. Res. Commun.* **82**:356.

Noll, M. (1974) *Nature (London)* **251**:249.

Noll, M. and Kornberg, R. D. (1977) *J. Mol. Biol.* **109**:393.

Ohnuki, Y. (1968) *Chromosoma (Berlin)* **25**:402.

Olins, A. L. (1977) *Cold Spring Harbor Symp. Quant. Biol.* **42**:325.

Olins, A. L. and Olins, D. E. (1973) *J. Cell. Biol.* **59**:252a.

Olins, A. L. and Olins, D. E. (1974) *Science* **183**:330.

Olins, A. L., Carlson, R. D., and Olins, D. E. (1975) *J. Cell Biology* **64**:528.

Oosterhof, D. K., Hozier, J. C., and Rill, R. L. (1975) *Proc. Natl. Acad. Sci. U.S.A.* **72**:633.

Osheim, Y. N. and Miller, O. L., Jr. (1979) *J. Cell Biol.* **83**:166a.

Osheim, Y. N., Martin, K., and Miller, O. L., Jr. (1978) *J. Cell Biol.* **79**:126a.

Oudet, P., Gross-Bellard, M., and Chambon, P. (1975) *Cell* **4**:281.

Pardon, J. F., Worcester, D. L., Wooley, J. C., Cotter, R. I., Liley, D. M. J., and Richards, B. M. (1979) *Nucleic Acids Res.* **4**:3199.

Paulson, J. R. and Laemmli, U. R. (1977) *Cell* **12**:817.

Rattner, J. B., Branch, A., and Hamkalo, B. A. (1975) *Chromosoma (Berlin)* **52**:319.

Rattner, J. B. and Hamkalo, B. A. (1978a) *Chromosoma (Berlin)* **69**:363.

Rattner, J. B. and Hamkalo, B. A. (1978b) *Chromosoma (Berlin)* **69**:373.

Rattner, J. B. and Hamkalo, B. A. (1979) *J. Cell Biol.* **81**:453.

Rattner, J. B., Krystal, G., and Hamkalo, B. A. (1978) *Chromosoma (Berlin)* **66**:259.

Reeves, R. (1976) *Science* **194**:529.

Reeves, R. and Jones, D. (1976) *Nature (London)* **260**:495.

Renz, M. (1979) *Nucleic Acids Res.* **6**:2761.

Renz, M., Nehls, P., and Hozier, J. C. (1977a) *Cold Spring Harbor Symp. Quant. Biol.* **42**:245.

Renz, M., Nehls, P., and Hozier, J. C. (1977b) *Proc. Natl. Acad. Sci. U.S.A.* **74**:1879.

Riley, D. and Weintraub, H. (1979) *Proc. Natl. Acad. Sci. U.S.A.* **76**:328.

Rill, R. L. (1979) In *Molecular Genetics,* Part III, *Chromosome Structure* (J. H. Taylor, Ed.), Academic Press, New York, pp. 247–306.

Rill, R. L. and Van Holde, K. E. (1973) *J. Biol. Chem.* **248:**1080.

Ris, H. (1969) In *Handbook of Molecular Cytology* (A. Lima-de-Faria, Ed.), American Elsevier, New York, pp. 221–250.

Ris, H. (1975) In *CIBA Foundation Symposium on Structure and Function of Chromatin* (D. W. Fitzsimmons and G. E. W. Wolstenholme, Eds.), Amsterdam, North Holland, New York, pp. 7–28.

Ris, H. and Korenberg, J. (1977) *Abstracts,* Helsinki Chromosome Conference, Painovalmiste, Helsinki, pp. 54.

Ris, H. and Korenberg, J. (1978) In *Cell Biology*, Vol. 2 (L. Goldstein and D. Prescott, Eds.), Academic Press, New York, 268–327.

Rizzo, P. J. and Nooden, L. P. (1973) *J. Protozool.* **20:**666.

Sahasrabuddhe, C. G. and Van Holde, K. E. (1974) *J. Biol. Chem.* **249:**152.

Scheer, U. (1978) *Cell* **13:**535.

Sedat, J. and Manuelidis, L. (1977) *Cold Spring Harbor Symp. Quant. Biol.* **42:**331.

Shaw, B. R., Herman, T. M., Kovacic, R. T., Beaudreau, G. S., and Van Holde, K. E. (1976) *Proc. Natl. Acad. Sci. U.S.A.* **73:**505.

Simpson, R. T. (1978) *Biochemistry* **17:**5523.

Solari, A. J. (1971) *Exp. Cell Res.* **67:**161.

Solari, A. J. (1974) In *The Cell Nucleus*, Vol. 1 (H. Busch, Ed.), Academic Press, New York, pp. 493–535.

Sollner-Webb, B. and Felsenfeld, G. (1975) *Biochemistry* **14:**2915.

Soyer, M. O. and Haapla, O. K. (1973) *J. Microsc.* **18:**267.

Spadafora, C., Bellard, M., Compton, J. L., and Chambon, P. (1976) *FEBS Lett.* **69:**281.

Spadafora, C., Noviello, L., and Geraci, G. (1976) *Cell Diff.* **5:**225.

Sperling, J. and Sperling, R. (1978) *Nucleic Acids Res.* **5:**2755.

Staron, K., Jerzmanowski, A., Tyniec, B., Urbanska, A., and Toczo, K. (1977) *Biochim. Biophys. Acta* **475:**131.

Stubblefield, E. and Wray, W. (1971) *Chromosoma (Berlin)* **32:**262.

Stubblefield, E. and Wray, W. (1973) *Cold Spring Harbor Symp. Quant. Biol.* **38:**835.

Suau, P., Bradbury, E. M., and Baldwin, J. P. (1979) *Eu. J. Biochem.* **97:**593.

Thoma, F. and Koller, T. (1977) *Cell* **12:**101.

Thoma, F., Koller, T., and Klug, A. (1979) *J. Cell Biol.* **83:**403.

Thomas, J. O. and Kornberg, R. D. (1975) *FEBS Lett.* **58:**353.

Van Holde, K. E., Allen, J. R., Gorden, J., Lohr, D., Tatchell, K., and Weischet, W. O. (1978) In *Chromatin Structure and Function. Levels of Organization and Cell Function,* Part B (C. A. Nicolini, Ed.), Plenum Press, New York–London, pp. 389–412.

Van Holde, K. E., Sahasrabuddhe, C. G., and Shaw, B. R. (1974) *Nuclei Acids Res.* **1:**1597.

Varshavsky, A. J., Bakayer, V. V., and Georgiev, G. P. (1976) *Nucleic Acids Res.* **3,** 477–492.

Vidali, G., Boffa, L. C., Bradbury, E. M., and Allfrey, V. G. (1978) *Proc. Natl. Acad. Sci. U.S.A.* **75:**2239.

Walker, B. W., Lothstein, L. K., and LeStergen, W. M. (1979) *J. Cell Biol.* **83:**777.

Weintraub, H. M. (1973) *Cold Spring Harbor Symp. Quant. Biol.* **38:**247.

Weintraub, H. and Groudine, M. (1976) *Science* **193:**847.

Weisbrod, S. and Weintraub, H. (1979) *Proc. Natl. Acad. Sci. U.S.A.* **76:**630.

Wolfe, S. L. (1979) In *The Biological Basis of Medicine* (E. E. Bittar, Ed.), Academic Press, New York, pp. 3–43.

Woodcock, C. L. F. (1973) *J. Cell Biol.* **59:**368a.

Woodcock, C. L. F., Safer, J. P., and Stanchfield, J. E. (1976) *Exp. Cell Res.* **97:**101.

Worcel, A. and Benyajati, C. (1977) *Cell* **12:**83.

Worcel, A., Han, S., and Wong, M. L. (1978) *Cell* **15:**969.

Wray, W. and Stubblefield, E. (1970) *Exp. Cell Res.* **59:**469.

Wu, C., Bringham, P., Livak, K., Holmgren, R., and Elgin, S. C. R. (1979) *Cell* **16:**797.

Yamamoto, K. R. and Alberts, B. M. (1976) *Annu. Rev. Biochem.* **45:**721.

Zentgraf, H., Falk, H., and Franke, W. W. (1975) *Cytobiologie* **11:**10.

THE ELECTRON MICROSCOPY OF RIBONUCLEIC ACIDS

3

Claire L. Moore
Cancer Research Center and Department of Bacteriology and Immunology
University of North Carolina
Chapel Hill, North Carolina

CONTENTS

1 Introduction
2 The basic formamide-spreading technique
3 Reagents and supplies
  3.1 Reagents
  3.2 Ribonuclease contamination
  3.3 Other contaminants
  3.4 Stain preparation
  3.5 Grid preparation
4 Modifications of formamide spreading useful for RNA electron microscopy
  4.1 The high-formamide method
  4.2 The urea-formamide method
  4.3 Glyoxal treatment
  4.4 Formaldehyde denaturation
  4.5 Other techniques
5 Length measurements
6 The study of secondary structure
7 The visualization of proteins on RNA
Appendix
References

1 INTRODUCTION

During the past decade, electron microscopy has become an important and routine analytical technique for probing nucleic acid structure and nucleic acid–protein interactions. It has been used repeatedly to complement and verify biochemical observations. Electron microscopy is particularly useful when only limited amounts of sample are available or when nucleic acids must be located and characterized, but cannot be labeled with radioactivity or followed by other methods such as ultraviolet absorption.

DNA electron microscopy is now a standard tool in many laboratories and several good reviews are available (44, 73–76). Techniques first applied to DNA are also proving useful for investigations of ribonucleic acids. However, RNA has been difficult to work with because of its susceptibility to nucleases and the ubiquity of these enzymes. As an added complication, nonspecific interactions between bases are stronger in single-stranded RNA than in single-stranded DNA. This often gives the RNA molecule a kinky appearance in the electron microscope with methods commonly used to visualize single-stranded DNA. These factors have previously limited the precision of molecular weight determinations for RNA species by electron microscopy. Recently several techniques have been developed to disrupt the secondary structure seen in RNA. The most useful of these involve treatment with high concentrations of formamide, urea-formamide, or glyoxal. These methods, and special applications for the study of secondary structure and visualization of proteins, are discussed after a description of the basic surface spreading procedure developed originally by Kleinschmidt (1) and modified by the addition of formamide for single-stranded DNA (2).

2 THE BASIC FORMAMIDE–SPREADING TECHNIQUE

Single-stranded RNA, like DNA, requires denaturing conditions if it is to be seen as anything more than a knotted structure in the electron microscope. Therefore, most spreading solutions include formamide in addition to cytochrome *c*. As this cytochrome *c* and nucleic acid solution is spread onto a liquid surface, called the hypophase, the cytochrome *c* denatures and coats the nucleic acid in a sheath of protein. This thickens the nucleic acid's diameter 10–20 times, and also traps it in a floating protein film, which can be collected on a coated grid. Ingredients vary, but the manipulations are the same for the different spreading techniques, and are described in the next paragraphs.

In our laboratory, the final concentrations of reagents in the spreading solution are 50% formamide, 0.1 M Tris, pH 8.5, 0.01 *M* EDTA, 40–60 μg/ml cytochrome *c*, and 0.25–1 μg/ml nucleic acid. For the best results, this solution should be used within an hour after mixing. The hypophase contains 20% formamide, 10 mm Tris, pH 8.5, and 1 mm EDTA, in distilled water, and because of the breakdown of the formamide should be used within 15 minutes of preparation. The hypophase is poured into a trough such that a slight meniscus forms. Petri dishes 9 cm square or round make convenient troughs, and, if placed on a dark background, permit the spreading solution to be seen swirling on the surface. A glass slide, supported by a Plexiglas bar about halfway across the dish, is placed in position as a ramp into the hypophase (Figure 1*a*). Most slides can be used directly from the box without further treatment. (A good criterion for cleanliness is the formation of a sharp line where the slide enters the hypophase. If the hypophase begins to creep up the slide, however, the slide should be discarded.) Just above the slide-hypophase boundary, 50 μl of the spreading solution is layered onto the slide, and a Parlodion-coated grid, film side down, is touched to the surface in a straight down-and-up motion. Care should be taken not to slant the grid, since this could cause the molecules to stretch in one direction, affecting length measurements. The best grids are often obtained by picking up the protein film approximately one grid width away from the slide-hypophase boundary immediately after the spreading solution has been applied. However, this choice will vary with the sample, shape of the trough, and labora-

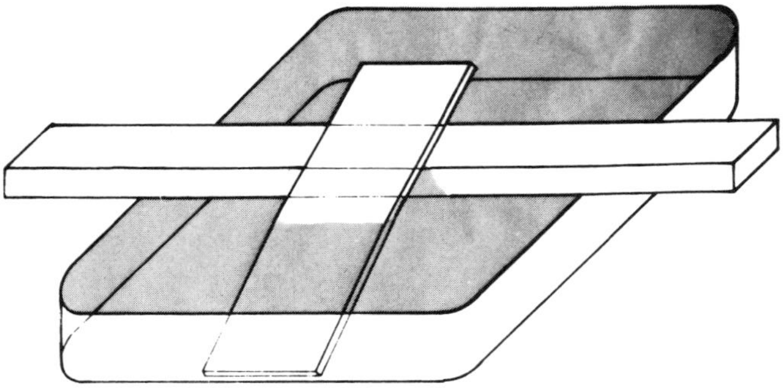

(a)

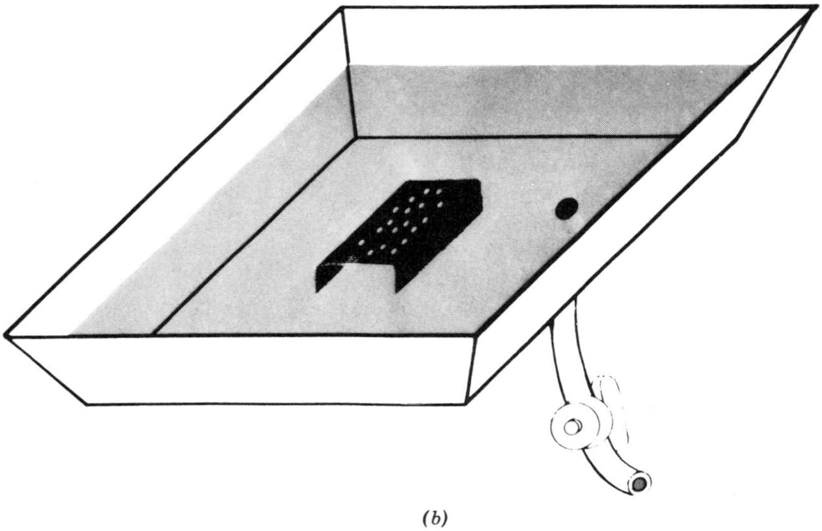

(b)

Figure 1 (a) Diagram of spreading apparatus. (b) Tray for preparation of Parlodion-coated grids. (Courtesy of J. Glass.)

tory, and an hour spent varying the time and location on the trough at which the grid is touched to the surface is well worthwhile. The coated side of the grid should be completely covered with a small droplet. If this has pulled away from the grid's edge, it is probable that no film has been picked up, and the grid is usually discarded. The coated grid is dipped in a stain of uranyl acetate in 75% ethanol (details of stain preparation are described later) for 15–30 seconds and rinsed in 75% ethanol for the same amount of time. A second grid can sometimes be picked up from the same spread. The hypophase can also be reused once, after cleaning the surface by sliding the

Plexiglas bar over it. After air-drying on filter paper, the grids are rotary-shadowed at a 10° angle with platinum-palladium (80:20). Since the Parlodion film will stretch when exposed to the electron beam, causing variations in molecular length, a thin carbon film is evaporated onto the grids to give the Parlodion stability.

A variation of this method is the drop-spreading technique described by Inman and Schños (3). A Pasteur pipette serves as a ramp into an 0.9 ml droplet of hypophase on Teflon or parafilm, and 5 μl of spreading solution is used each time. The advantage of this technique is that only small quantities of RNA (1–5 ng) are used for each spreading. Another simple method is to place 50–100 μl drops of spreading solution (same reagent concentrations as described previously) onto parafilm and allow the nucleic acid molecules to absorb to the surface for a minute or more before touching a grid to the surface (77).

3 REAGENTS AND SUPPLIES

3.1 Reagents

Reproducibility in the formamide-spreading technique is often dependent on the purity of the reagents, and a few additional notes on these ingredients will be helpful. First, the formamide, particularly in the spreading solution, must be of high quality. We have had most success with formamide from Matheson, Coleman and Bell; if care is taken to get fresh lots from the manufacturer, and stocks are aliquoted and stored frozen at −4°C, recrystallization is not necessary. The conductivity of a 10% formamide solution should be less than 10 times that of deionized water. The formamide can be deionized by recrystallization (4) or with ion-exchange resins (80). The cytochrome *c* (Calbiochem, equine heart, salt free, A grade) is made up as a stock of 2 mg/ml in water and stored at 4°C. For a given sample and specified spreading conditions, the amount of cytochrome *c* giving the best appearance of the RNA will have to be titrated; this should be near 40–60 μg/ml final concentration. Water used to make solutions should be double-distilled, made fresh at least every 2 weeks, and stored in glass containers, not plastic. (Components of the plastics often leach out into the water and interfere with the spreads.)

3.2 Ribonuclease Contamination

When preparing RNA for electron microscopy, extreme care must be taken to prevent ribonuclease contamination. Detergents such as SDS may be used in the initial purification of the RNA sample, but if they are not carefully removed from the final stock used for electron microscopy, poor spreads will be obtained. Buffers, pipettes, and vials should be sterilized by autoclaving or treatment with diethyl pyrocarbonate. Since exposure to the formamide and cytochrome *c* is brief, these are not usually sterilized in any way, though precaution is taken by using sterile pipettes to eliminate subsequent contamination. If it is necessary to treat the RNA with deoxyribonuclease, commercial preparations of this enzyme can be rendered ribonuclease free by a simple affinity chromatography method using Agarose-UMP (78). Ribonucleoside-vanadyl complexes have been used successfully to inhibit ribonucleases in the isolation of intact RNA from animal cells (79). Treatments such as those described above during the purification steps should give full-length RNA for electron microscopy.

3.3 Other Contaminants

Depending on base composition, RNA prepared by formamide spreading will be more or less extended, with few crossover points, as in Figure 2. The contrast should be high and the background "stipple" from the cytochrome *c* should be low. As well as the factors mentioned previously, certain contaminants can cause poor spreads. Detergents, phenol and other organic compounds, and free protein should be removed from the sample as thoroughly as possible. High concentrations of carrier RNAs will produce excessive background. Phosphate, borate, or nitrate will interact with the cytochrome *c*, and high concentrations in the spreading solution should be avoided. High magnesium concentrations can cause aggregation and condensation of the molecules, but this can usually be offset by adding EDTA.

3.4 Stain Preparation

To prepare the uranyl stain, a saturated solution of uranyl acetate, good for about 6 hours, is made by mixing 0.1 g of uranyl acetate in 2 ml of 75% ethanol, shaking for 1 minute, and letting settle for another minute. This stock is then diluted one part per 1000 in 75% ethanol to give a stain useful for about 1 hour. Preparing the stain in this manner eliminates the deposition of dark uranyl granules on the grid. Commercial preparations of ethanol sometimes contain contaminants that interfere with specimen contrast and also cause poor recovery of sample after ethanol precipitation. If this is a problem, the ethanol should be redistilled.

3.5 Grid Preparation

When the denatured cytochrome *c* film is formed, hydrophobic residues on the protein probably orient toward the air, while hydrophilic ones favor the liquid hypophase. For this reason, grids coated with a hydrophobic surface give more reproducible results in picking up the protein monolayer. Several different support films (carbon, carbon-Formvar, Formvar, and Parlodion) have been used with the surface-spreading technique, but we find that Par-

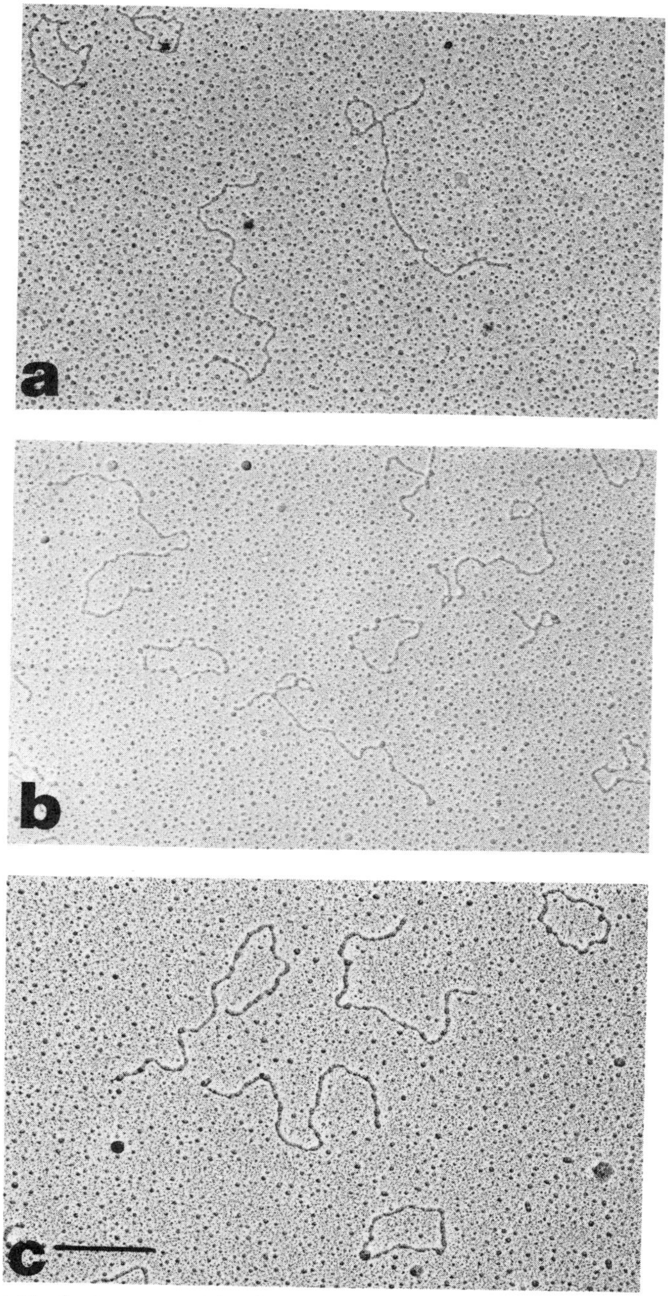

Figure 2 (*a*) Vesicular stomatitis virus (VSV) RNA spread from 80% formamide. (*b*) VSV RNA prepared by the urea-formamide spreading method. The hypophase was 30% formamide. (*c*) VSV RNA incubated in 0.5 M glyoxal, 10 mM sodium phosphate, pH 6.9, 1 mM EDTA, 73% formamide at 37°C for 60 minutes and spread from 80% formamide onto a 50% formamide hypophase. Bar = 0.5 μm. Micrographs courtesy of J. Glass.

lodion films are the most hydrophobic, therefore the most effective. Parlodion strips are dried overnight in a 60°C vacuum oven and then dissolved in amyl acetate to give a solution of 3.5% (weight/volume) Parlodion. Water contamination of this stock which will produce holes in the film, should be avoided. Figure 1*b* illustrates the placement of 200–400 mesh copper grids on a wire screen submerged in distilled water. Next 20–50 μl of the Parlodion solution is dropped onto the water surface, and the amyl acetate is allowed to evaporate. The first film is used to clean the surface, then removed. Another film is formed in the same manner, and when dry, is lowered onto the grids by draining the water out of the container. The grids are dried before use and are good for at least 2 weeks.

4 MODIFICATIONS OF FORMAMIDE SPREADING USEFUL FOR RNA ELECTRON MICROSCOPY

Three methods have been effective in giving better extension of RNA molecules. The first two, high-formamide spreading and urea-formamide spreading, use stronger denaturing conditions in the spreading solutions to remove secondary structure. The third uses pretreatment with glyoxal, which by binding almost irreversibly to the guanosine residues, prevents secondary structure from forming under spreading conditions.

4.1 The High-Formamide Method

High-formamide spreading simply means using higher formamide concentrations, up to 90%, in the spreading solution. The concentrations of the other reagents are the same as before. The hypophase, to remain isodenaturing with the spreading solution, usually contains a formamide concentration 30 percentage points less than that used in the spreading solution. Figure 2*a* gives an example of vesicular stomatitis virus (VSV) RNA spread from 80% formamide.

4.2 The Urea-Formamide Method

Depending on the RNA, high-formamide spreading may not be sufficient to disrupt all the nonspecific base interactions, and it may be necessary to use the urea-formamide technique. This method, introduced by Robberson et al. (4) to study mitochondria ribosomal RNA, has been useful in investigations of many other RNA species (see Appendix 1). A urea-formamide stock solution is made that consists of 5 *M* urea (Schwarz-Mann, Ultrapure grade), freshly dissolved in 100% formamide. The RNA sample is mixed in a solution that is 80% urea-formamide stock, 0.1 *M* Tris, pH 8.5, 0.01 *M* EDTA, and 30–60 μg/ml cytochrome *c*. This is spread onto a hypophase of 30% formamide with one-tenth as much Tris-EDTA as the spreading solution, or in some cases onto distilled water. Figure 2*b* shows VSV RNA spread with urea-formamide.

4.3 Glyoxal Treatment

The first application of glyoxal for RNA electron microscopy was by Hsu et al. (5), and this method has been used in numerous studies since it was reported in 1973 (see Appendix). Glyoxal adds to the adenosine, cytodine, and guanosine residues of nucleic acids and blocks the hydrogen-bonding abilities of these bases. The reaction with guanosine has a slow rate of dissociation and can be considered essentially irreversible. Therefore, under mild conditions, glyoxal will disrupt weaker secondary structure, but it will not denature long duplex regions. The RNA sample is incubated in a solution of 0.5–1 *M* glyoxal (Eastman, 40% in water) in 10 m*M* sodium phosphate, pH 7. (Technical grade glyoxal, which contains impurities that will degrade RNA, should be deionized with an ion-exchange resin such as BioRad Ag 501 × 8 (D), 20–50 mesh.) The glyoxal is then dialyzed away or the mixture simply diluted at least tenfold into a formamide-spreading solution. One hour at 37°C will remove most guanosine: cytosine base pairing. Decreasing the time or temperature or increasing the salt concentration results in less denaturation. With glyoxal treatment in combination with formamide, which is effective at denaturing adenosine:thymidine regions, most secondary structure is eliminated. Kung et al. (6) used methylmercuric hydroxide, a powerful denaturant even at room temperature, to completely disrupt any base pairing and thus expose all the guanosines to glyoxal fixation. Secondary structure observed in urea-formamide spreads or with glyoxal treatment alone was entirely removed with this treatment, though the length measurement did not change significantly. Figure 2*c* shows VSV RNA treated with glyoxal.

4.4 Formaldehyde Denaturation

Formaldehyde has also been used to denature RNA for electron microscopy. This is accomplished by heating the RNA for 10–15 minutes at 60–65°C in the presence of formaldehyde (7, 8) or at only 50°C in a formaldehyde-formamide mixture (9), and then diluting into a formamide-spreading solution.

4.5 Other Techniques

Heteroduplex mapping has been a powerful tool in DNA electron microscopy (10) for identifying regions of nonhomology between two different nucleic acids. Noncomplementary stretches do not base pair to form a duplex; they appear as single-strand loops or tails in the electron microscope. This technique has been used extensively with the retravirus system, in which heteroduplexes were formed between viral RNA and complementary DNA transcripts from a different virus (11–17); it also has been applied to the double-stranded viruses of yeast (18) and replicative intermediates of polio virus (19, 92). Lundquist et al. (19) characterized defective interfering particles of poliovirus, a single-stranded RNA virus, by heteroduplex analysis using complementary strands obtained from the double-stranded replicative form (Figure 3). In some cases, this type of electron microscopic analysis has advantages over biochemical methods such as T1 RNAse finger-print analysis because of the small quantities of sample required and the accuracy possible with this technique.

A similar hybridization procedure was used to map the position of the mRNAs of vesicular stomatitis virus, a negative-strand RNA virus, and to demonstrate the presence of intervening polyadenylate sequences in the RNA transcripts (20).

Partial denaturation mapping, in which heat treatment with formaldehyde

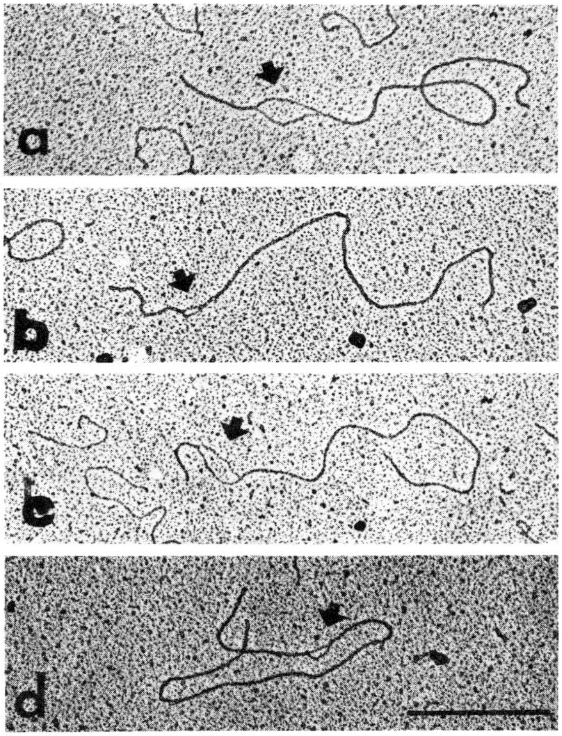

Figure 3 Heteroduplexes formed from the RNA of two different poliovirus defective interfering particles by the method of Lundquist et al. (19). Bar = 0.5 μm. Courtesy of P. Lundquist; micrograph by M. Sullivan.

or high formamide concentrations specifically melts out A-U-rich regions, can be applied to double-stranded RNAs such as replicative intermediates or double-stranded RNA viruses (73, 75, 91, 92). In conjunction with the formamide denaturation, glyoxal can be used to fix the denatured regions (93) and to give a more reproducible denaturation pattern. A denaturing temperature should be determined for a given RNA that produces small but easily detectable "bubbles" in the RNA strand

In formamide spreads, the single-strand regions should appear thinner and less smooth than double-strand portions. Often, when analyzing molecules that contain both double and single-stranded regions but lack forks or loops, it is difficult to tell where the single strand begins. One way to enhance the difference is to coat the single strand with T4 phage gene 32 protein or *E. coli* single-strand binding protein (21–23, 88). This gives a single-strand diameter almost twice that of the double strand. However, the resulting condensation in the case of the *E. coli* protein (35%) or extension with gene 32 protein should be taken into account when calculating lengths. The amount of condensation or extension of the single-strand length can vary with spreading conditions, and a suitable single stranded standard should be included in the reaction.

Other compounds [e.g., BAC (45) and Anthrabis (81)], discussed later, have also been used to extend RNA; but for routine work not requiring protein visualization, the cytochrome *c* method is probably more convenient. R-Looping, a valuable technique for mapping mRNAs, is described in Chapter 5.

5 LENGTH MEASUREMENTS

If electron microscopy is to be used for determinations of molecular weight, it is important to include molecules of known molecular weight in the spreading preparations. This eliminates error due to magnification changes or the slight variations that can occur from spread to spread, and facilitates comparisons with other spreading conditions. Lately many RNAs have been sequenced and their genome size determined exactly. These are listed in Table 1 and will provide useful markers. Some of these nucleic acids can be obtained commercially (Miles Laboratories), but purity should be checked before use. ϕX174 DNA has also been sequenced and used as a length calibration. This is a single-stranded circular molecule and therefore easily recognized when mixed with other nucleic acids. However, there was no evidence that the DNA chain should behave identically to RNA under spreading conditions. Recently Glass and Wertz (82) completed a thorough study comparing the base-to-length ratios of ϕX174 DNA and the sequenced RNAs mentioned above, using a variety of spreading techniques. The average difference in base-to-length ratios for ϕX174 DNA compared to the

Table 1 Base-to-length ratios of sequenced nucleic acids with different spreading conditions: FA = formamide, Gly = glyoxal

		Length (μm)			
Nucleic Acid	Bases	80% FA[a]	80% FA + Gly[b]	4 *M* Urea[c]	4 *M* Urea + Gly[d]
ϕX174	5386	1.61 ± 0.07	1.64 ± 0.07	1.61 ± 0.07	1.64 ± 0.07
MS2	3569	0.84 ± 0.06	0.88 ± 0.06	0.86 ± 0.06	0.91 ± 0.07
Qβ	4214	0.97 ± 0.06	1.03 ± 0.06	0.99 ± 0.07	1.03 ± 0.06
E. coli 23s rRNA	2904	0.64 ± 0.06	0.69 ± 0.06	0.67 ± 0.06	0.70 ± 0.06

		Bases per Micron			
Spreading Condition	Average % Difference[e]	ϕX174 DNA	MS2 RNA	Qβ RNA	23s RNA
80% FA	29%	3350	4420	4300	4500
80% FA + gly	26%	3280	4100	4100	4200
4 *M* Urea	27%	3350	4200	4300	4300
4 *M* Urea + gly	23%	3280	3900	4100	4100

Courtesy of J. Glass and G. Wertz.

[a] Samples were spread from 80% formamide onto a hypophase of 50% formamide.

[b] Samples were prepared as described in Figure 2*c*.

[c] Samples were prepared as described in Section 4.2.

[d] Samples were pretreated with glyoxal as before, and then spread by the urea-formamide technique.

[e] [(RNA − DNA)/DNA] × 100.

RNAs was about 26%. The results are summarized in Table 1. With this correction factor, ϕX174 can be a convenient internal standard for RNA length measurements.

Several digitizer systems for length measurements are on the market. We use a Hewlett-Packard system with a fully smoothed length calculation program, and trace the molecules from negatives projected onto the digitizer surface. In applying such tracing systems and programs, it is very important to select values for "jitter" and "smoothing" appropriate for the stiffness of each type of molecule. These are best determined by trial and error with known line lengths.

6 THE STUDY OF SECONDARY STRUCTURE

Secondary structure features can serve as handy reference points to orient an RNA molecule. However, there is increasing evidence that the secondary structure of RNA may have an important biological function as enzyme recognition sites in processing of precursor molecules [splicing (24, 96, 97), tRNAs (25)], in the efficiency of translation initiation (90), or in the initiation of replication in RNA viruses. The binding of replicase to the Qβ phage genome (26) and the initiation of reverse transcription in retraviruses (27) may involve specific secondary structure. The familiar cloverleaf structure of tRNAs is crucial for their role in translation (28, 29). In some systems such as the RNA bacteriophages, by screening ribosome-binding sites, secondary structure serves to regulate the appearance of specific proteins or to prevent internal intiations (30). Hairpins on nascent mRNA chains have been postulated as a factor in the termination of transcription in procaryotes (31, 32). Often the same RNA sequence has the potential for several types of secondary structure. RNA isolated from Moloney sarcoma virus has a different secondary structure after it has been denatured and allowed to reanneal (33). This versatile aspect of secondary structure may have implications in the maturation of RNA species or as a mechanism to discriminate between functions as a replicative template or as a message. Many of the known secondary features mentioned above are too small to be detected by electron microscopy. However it is likely that the prominent ones visible in the electron microscope have a biological function. Knowledge of secondary structure has also been useful in mapping the evolution of rRNA species, as in the detailed work of Schibler et al. (34) and in determining the processing pathways of rRNAs (35–38).

All the methods described previously have been employed in secondary structure studies. Often a combination of different denaturing conditions, from mild to strongly denaturing, helps determine the most stable features. In a study by Pettersson and Hewlett (39) on Uukuniemi virus, under mild conditions the RNA exists as a single-stranded circular molecule with reproducible hairpin features (Figure 4*a*). Under more stringent conditions, linear molecules were the predominant species, indicating closure of the circles by base pairing at the ends. In another example, a strong denaturing treatment showed that certain mitochondrial RNAs are covalently closed circles and might represent remnants derived from a splicing event or a method for control of mRNA expression (94, 95). A similar study by Kung et al. (6) on RD-114 (a feline retravirus) demonstrated that the virion RNA consisted of two subunits linked together by a characteristic T-shaped structure.

Several useful modifications are available to orient the RNA molecules and stabilize the secondary features. One involves hybridizing the 3′ poly(A) ends found on some RNA molecules to either poly (dT) (5, 40) or the more visible poly(dT)- or poly(dBrU)-labeled circular DNA molecule (41, 42). Other RNAs can be polyadenylated in vitro using calf thymus poly(A) polymerase. (98) This tagging provides rapid orientation of the molecule under investigation. Examples are the micrographs of poliovirus RNA (Fig. 6*b, c*) and of Moloney sarcoma virus RNA (Figure 4*b*), which also demonstrates both the dimer linkage structure and the loop features often seen in the retravirus

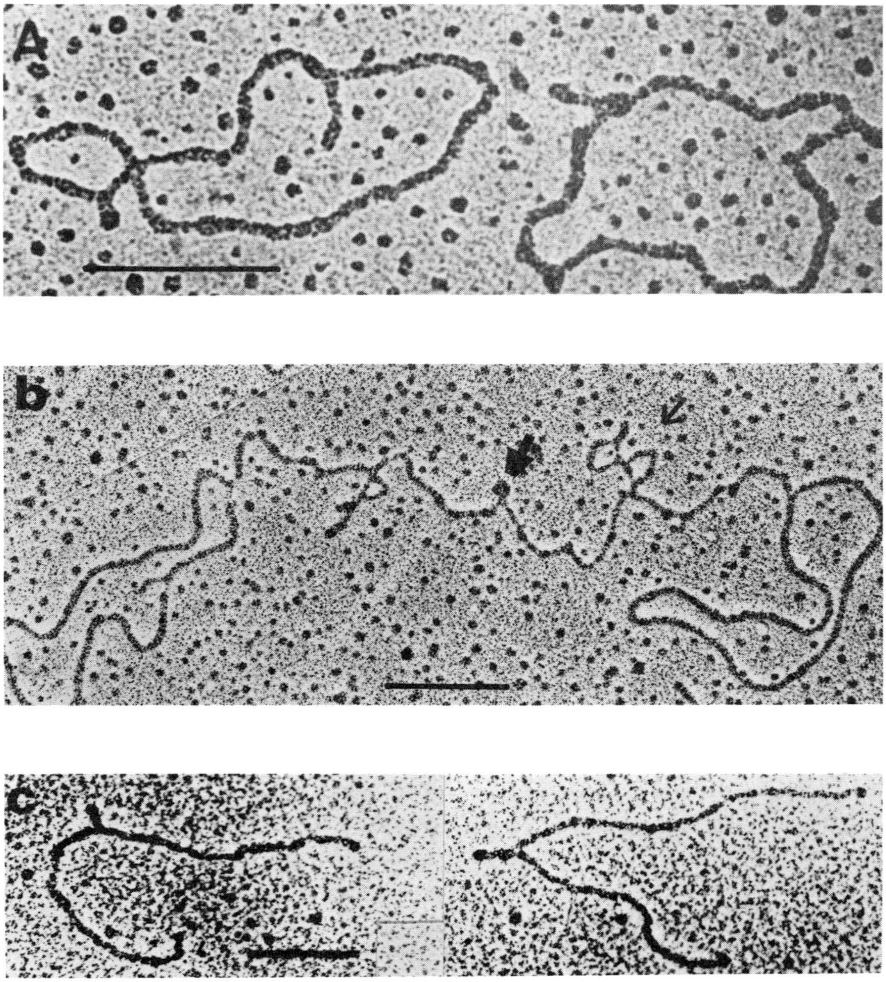

Figure 4 (*a*) Duplex "hairpins" on Uukuniemi virus RNA spread from 70% formamide (39). Bar = 0.25 μm. Micrograph courtesy of R. Pettersson. (*b*) Moloney sarcoma virus RNA, glyoxal-treated and spread from 40% formamide, with the poly (A) ends hybridized to poly (dT) tails on SV40 DNA (large circles). Note the dimer linkage structure (bold arrow) and the characteristic large loops (light arrow) (33). Bar = 0.25 μm. Micrograph courtesy of W. Bender, by S. Hu. (*c*) The hairpin structure of 26S rRNA from *Drosophila melanogaster*, stabilized by psoralen cross-linking. Samples were then denatured by formaldehyde and spread from 80% formamide (9). Bar = 0.2 μm. Micrograph courtesy of P. Wollenzien.

RNA. It is also possible to bind ferritin, an iron-containing and thus electron-dense molecule, to the 3′ end of RNA. This approach has been used in mapping tRNA genes on heteroduplexes (43), but it has not been applied to electron microscopic studies of single-stranded RNA molecules.

Psoralen will cross-link double-stranded regions of nucleic acid and has been used by Wollenzien et al. (9) to stabilize hairpin structures in rRNA and to give more reproducible maps of these features (Figure 4*c*). This is necessary to localize hairpins rich in A + T (U) and most likely to melt out in partially denaturing conditions.

7 THE VISUALIZATION OF PROTEINS ON RNA

RNA in living cells must rarely be in a free state not complexed with protein—enzymes use it as a template or modify its structure; proteins bind and change its conformation. Proteins found in message ribonucleoprotein particles may influence the RNA's ability to become associated with ribosomes and thus expressed through translation (83, 84). The future of RNA electron microscopy will lie in the contribution it can make toward understanding these interactions.

Negative staining with uranyl acetate or phosphotungstic acid, which enhances structural detail by depositing a thin film of electron-dense material around a particle, can be used to study the structure of RNA highly complexed with proteins, such as viral nucleocapsids (85, 86). Figure 5 shows negatively stained Sendai virus nucleocapsids (Figure 5*a*) and tobacco mosaic virus particles (Figure 5*b*). These stains are very acidic and can damage some protein structures unless these have been fixed. Phosphotungstic acid solutions, but not uranyl acetate solutions, can also be neutralized to pH 7. Negative staining has been used to demonstrate variations due to ionic strength in the conformation of the helical nucleocapsids of paramyxoviruses and rhabdoviruses (85), and also to examine the *in vitro* assembly of tobacco mosaic virus (87). In another method, ribonucleoproteins extracted from avian myeloblastosis virus have been direct mounted and visualized by platinum-palladium rotary shadowing (89). However, free nucleic acid is poorly extended and hard to follow in these techniques; therefore they have not been useful in the localization of a few or single proteins on RNA.

Although many technqiues have been developed recently for directly visualizing proteins on extended DNA molecules (44, 76), only one of these, the BAC-spreading technique, has been successfully applied to the electron microscopy of ribonucleoprotein complexes. It has been difficult to prepare single-stranded RNA for electron microscopy without first complexing the molecules with a basic compound. Cytochrome *c*, the small protein routinely used in surface spreading, is still large enough (13,000 daltons) to obscure the presence of other proteins on the RNA molecule. Benzyldimethylalkylam-

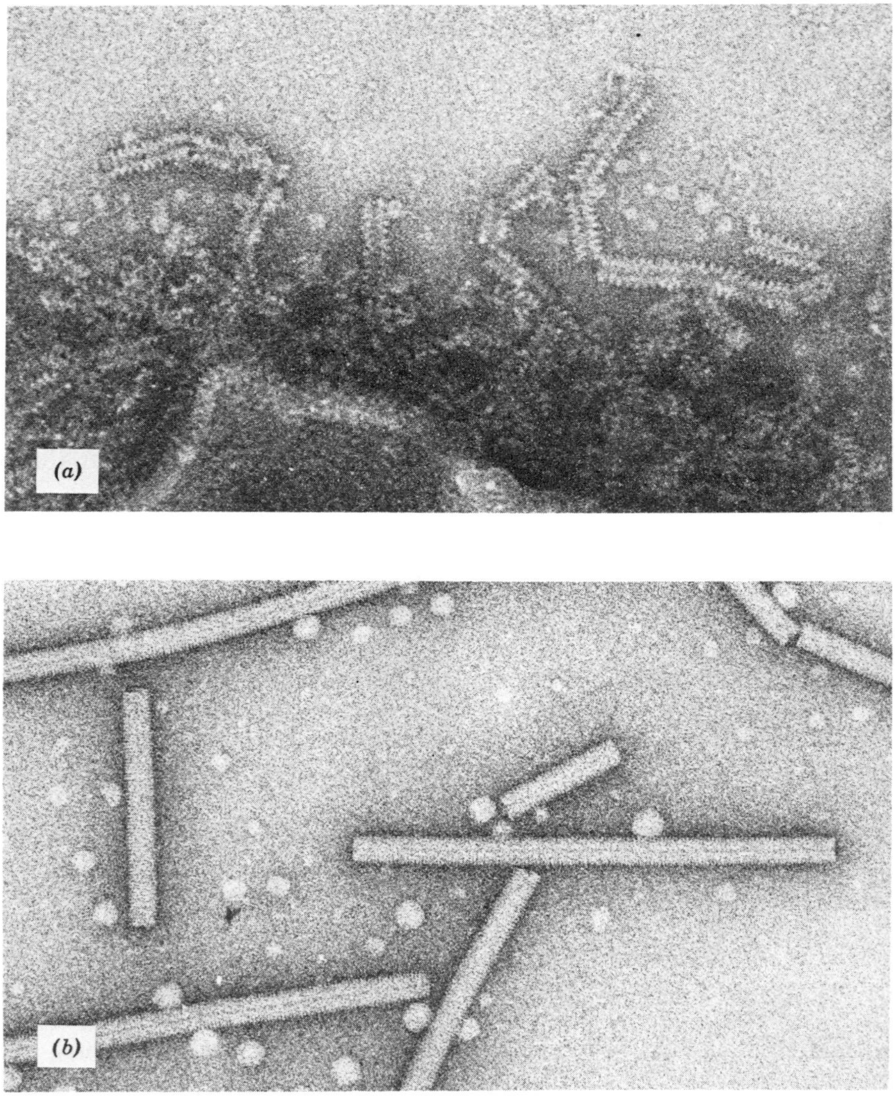

Figure 5 (*a*) Sendai virus nucleocapsids negatively stained with phosphotungstic acid. (*b*) Tobacco mosaic virus negatively stained with uranyl acetate. Micrographs courtesy of J. Griffith.

monium chloride (BAC) is a basic compound weighing only 350 daltons, and a BAC-spreading method has been used to see Qβ replicase bound to Qβ RNA (Figure 6*a*) (45) and to map the binding sites of ribosomal proteins on rRNA (46). Some of the factors affecting the quality of these spreads are discussed by Coetzee and Pretorius (47). Anthrabis (81) is another small molecule with potential for the visualization of protein complexes.

Good fixation is often critical in high-resolution direct visualization of nucleic acid protein complexes. We find that a two-step fixation (44) gives the best preservation. The sample (in 0.01 *M* NaCl, 0.01 *M* sodium phosphate pH 7.5) is treated with 1% formaldehyde (diluted from a 37% stock solution and heated to 90° for 5 minutes) for 15 minutes on ice to cross-link

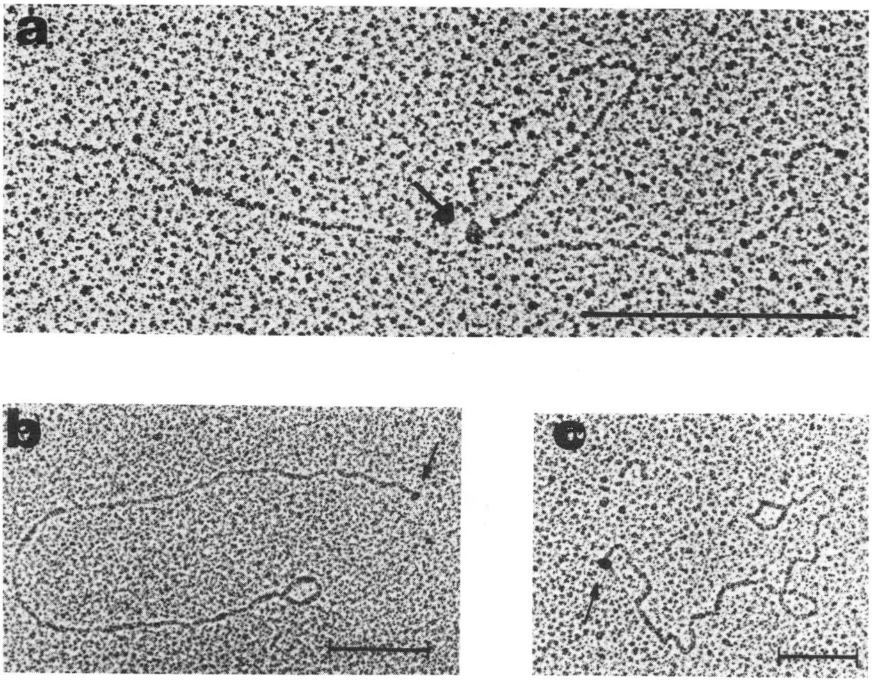

Figure 6 (*a*) Complex of Qβ replicase (arrow) bound to Qβ RNA, spread by the BAC technique (45). Micrograph courtesy of H. J. Vollenweider and T. Koller. Bar = 0.25 μm. (*b*) Electron micrograph of poliovirion RNA spread by the BAC technique after treatment with DNFB and rabbit anti-DNP IgG by the method of Wu and Davidson (52), and with the 3′-poly(A) end labeled with trypanosome circles tailed with poly(dBrU). There is a terminal protein dot at the 5′ end (arrow). (*c*) A cytochrome *c* spread of poliovirion RNA treated with DNFB, anti-DNP, and the FAb fragment of goat antirabbit IgG, and with the 3′-poly(A) end labeled by a trypanosome circle. An arrow points to the terminal protein dots (52). From reference 52, courtesy of the publishers of *Nucleic Acids Research*. Bar = 1 kb.

the protein to the nucleic acid. It is essential to use a buffer without a primary amine that would react with formaldehyde. Phosphate in the solution may destabilize some RNA-protein complexes, and a sulfonate buffer such as HEPES or PIPES can be substituted. The addition of glutaraldehyde to 0.6% for another 15 minutes stabilizes the protein into a globular structure that should be both more representative and more visible.

Nascent RNA in transcribing regions of chromatin can be beautifully demonstrated with the method of Miller and Beatty (48), which is reviewed in Chapter 2.

Several procedures permit the location of proteins on nucleic acid molecules using the cytochrome *c* spreading technique. These involve reacting chemicals or antibodies with the bound protein to effectively make it larger. Broker and Chow (49) developed a ferritin-avidin-biotin conjugate to label the 55,000 dalton protein bound to the ends of adenovirus DNA. Richards et al. (50) also used the biotin-avidin reaction to couple plastic spheres to a covalently bound protein on the ends of poliovirus double-stranded RF RNA. Ferritin-labeled antibodies have been used to identify the T-antigen on SV40 DNA (51). In this case, as often happens with this approach, nonspecific binding to the DNA of proteins in the antibody preparation gave a high background. Recently Wu and Davidson (53) introduced a promising method for mapping proteins on nucleic acids, which has the advantage of not requiring antibodies specific for the protein being investigated. This technique utilizes the covalent attachment of dinitrophenyl (DNP) groups to proteins, followed by binding of rabbit anti-DNP IgG molecules, and spreading with either cytochrome *c* or BAC. If needed, a second antibody (the FAb fragment of goat antirabbit IgG) can be added to further enlarge the complex. This sucessfully labeled the protein at the 5′ end of polio RNA (Figure 6*b*, *c*) (52) and a protein at the 3′ end of avian sarcoma virus RNA (53). Noncovalently bound proteins must be cross-linked to the nucleic acid with glutaraldehyde. If this is used at low concentration (0.2%), the DNP reaction is not blocked. It is likely that the same techniques can be productively applied to the study of other ribonucleoprotein complexes.

ACKNOWLEDGMENTS

Portions of this work were supported by a grant from the National Science Foundation and the National Institute of Health. I also thank Dr. Jack Griffith for his encouragement and helpful criticism.

APPENDIX: RNA ELECTRON MICROSCOPIC STUDIES

This reference list, organized by subject, is by no means comprehensive. It is, however, a starting point for further research.

1 mRNA 40, 72

2 rRNA 4, 9, 34–38, 46, 62–64, 69

3 Viral RNA 74

- **A** Bacteriophage 8, 45, 54, 56, 71
- **B** Eucaryotic viruses
  - **1** Single-stranded RNA
    - **a** New Castle's disease 7, 8
    - **b** Polio 19, 50, 52, 61, 68
    - **c** Retravirus 6, 11–17, 21, 33, 41, 53, 57–59, 67, 70, 89
    - **d** Sendai 7
    - **e** Sindbis 5
    - **f** Uukuneimi 39, 86
    - **g** Vesicular stomatitis virus 20, 85
  - **2** Double-stranded RNA
    - **a** Yeast viruses 18
    - **b** Reovirus 55
  - **3** Viroids 60, 65, 66

REFERENCES

1 Kleinschmidt, A. K. (1968) In *Methods in Enzymology*, Vol. 40B (S. P. Colowick and N. O. Kaplan, Eds.), Academic Press, New York, pp. 361–377.

2 Westmoreland, B. C., Szybalski, W., and Ris, H. (1969) *Science* **163:**1343–1348.

3 Inman, R. B. and Schnös, M. (1970) *J. Mol. Biol.* **49:**93–98.

4 Robberson, D., Aloni, Y., Attardi, G., and Davidson, N. (1971) *J. Mol. Biol.* **60:**473–484.

5 Hsu, M., Kung, H., and Davidson, H. (1973) *Cold Spring Harbor Symp. Quant. Biol.* **38:**943–950.

6 Kung, H. J., Bailey, J. M., Davidson, N., Nicholson, M. O., and McAllister, R. M. (1975) *J. Virol.* **16:**397–411.

7 Kolakofsky, D., Boy de la Tour, E., and Deliss, H. (1974) *J. Virol.*, **13:**261.

8 Chi, Y. Y. and Bassel, A. R. (1974) *J. Virol.* **13:**1194.

9 Wollenzien, P. L., Youvan, D. C., and Hearst, J. E. (1978) *Proc. Natl. Acad. Sci. U.S.A.* **75:**1642–1646.

10 Davis, R., Simon M., and Davidson, N. (1971) In *Methods in Enzymology*, Vol. 21 (L. Grossman and K. Moldave, Eds.), Academic Press, New York, pp. 413–428.

11 Chien, Y. H., Lai, M., Shih, T. Y., Verna, I. M., Scolnick, E. M., Roy-Burman, P., and Davidson N. (1979) *J. Virol.* **31:**752–760.

12 Donoghue, D. J., Rothenberg, E., Hopkins, N., Baltimore, D., and Sharp, P. A. (1978) *Cell* **14:**959–970.

13 Donoghue, D. J., Sharp, P. A., and Weinberg, R. A. (1979) *Cell* **17:**53–63.

14 Hu, S., Davidson, N., and Verma, I. (1977) *Cell* **110:**469–477.

15 Hu, S., Lai, M. M., and Vogt, P. K. (1979) *Proc. Natl. Acad. Sci. U.S.A.* **76:**1265–1268.

16 Lai, M. C., Hu, S., and Vogt, P. (1979) *Virology* **97:**366–377.

17 Rothenberg, E., Donoghue, D. J., and Baltimore, D. (1978) *Cell* **13:**435–451.

18 Fried, H. M. and Fink, G. R. (1978) *Proc. Natl. Acad. Sci. U.S.A.* **75:**4224–4228.

19 Lundquist, R. E., Sullivan, M., and Maizel, J. Jr. (1979) *Cell* **18:**759–769.

20 Herman, R., Adler, S., Lazzarini, R., Colonno, R., Banerjee, A., and Westphal, H. (1978) *Cell* **15:**587–596.

21 Delius, H., Duesberg, P. H., and Mengel, W. F. (1974) *Cold Spring Harbor Symp. Quant. Biol.* **39:**835–843.

22 Siegel, N., Delius, H., Kornberg, T., Geffer, M., and Alberts, B. (1972) *Proc. Natl. Acad. Sci. U.S.A.* **69:**3537–3541.

23 Wu, M. and Davidson, N. (1975) *Proc. Natl. Acad. Sci. U.S.A.* **72:**4506–4510.

24 Khoury, G., Gruss, P., Dhar, R., and Lai, C. (1979) *Cell* **18:**85–92.

25 Lewin, B. (1978) *Gene Expression*, Vol. 1, *Bacterial Genomes*, Wiley, New York, pp. 169–187.

26 Mills, D. R., Nishihara, T., Doblein, C., Kramer, F., Cole, P., and Spiegelman, S. (1977) In *Nucleic Acid–Protein Recognition*, P & S Biomedical Sciences Symposium Series (Henry Vogel, Ed.), Academic Press, New York, pp. 533–547.

27 Darlix, J., Spahr, P., and Bromley, P. (1978) *Virology* **90:**317–329.

28 Rich, A. and RajBhandary, U. (1976) *Annu. Rev. Biochem.* **45:**805–860.

29 Rich, A. (1977) In *Nucleic Acid–Protein Recognition,* P & S Biomedical Sciences Symposium Series (Henry Vogel, Ed.), Academic Press, New York, pp. 281–291.

30 Steiz, J. (1979) In *Gene Expression—Biological Regulation and Development*, Vol. 1 (R. F. Goldberger, Ed.), Plenum Press, New York, pp. 281–291.

31 Adhya, S., Sarker, P., Valenzuela, D., and Maita, U. (1979) *Proc. Natl. Acad. Sci. U.S.A.* **76:**1613–1617.

32 Pribnow, D. (1979) In *Gene Expression—Biological Regulation and Development*, Vol. 1 (R. F. Goldberger, Ed.), Plenum Press, New York, pp. 219–277.

33 Maisel, J., Bender, W., Hu, S., Duesberg, P. H., and Davidson N. (1978) *J. Virol.* **25:**384–394.

34 Schibler, U., Wyler, T., and Hagenbüchle, D. (1975) *J. Mol. Biol.* **94:**503–517.

35 Wellauer, P. K. and Dawid, J. B. (1973) *Cold Spring Harbor Symp. Quant. Biol.* **38:**525–535.

36 Wellauer, P. K. and Dawid, I. B. (1973) *Proc. Natl. Acad. Sci. U.S.A.* **70:**2827–2831.

37 Wellauer, P. K. and Dawid, I. B. (1974) *J. Mol. Biol.* **89:**379–395.

38 Wellauer, P. K. and Dawid, I. B. (1974) *J. Mol. Biol.* **89:**397–407.

39 Pettersson, R. F. and Hewlett, M. J. (1976) *Anim. Virol.* **4:**515–527.

40 Lizardi, P. M., Williamson, R., and Brown, P. D. (1975) *Cell* **4:**199–205.

41 Bender, W. and Davidson, N. (1976) *Cell* **7:**595–607.

42 Bender, W., Davidson, N., Kindle, K., Taylor, W., Silverman, M., and Firtel, R. (1976) *Cell* **15:**779–788.

43 Wu, M. and Davidson, N. (1973) *J. Mol. Biol.* **78:**1–21.

44 Griffith, J. D. and Christiansen, G. (1978) *Annu. Rev. Biophys. Bioeng.* **7:**19–35.

45 Vollenweider, H. (1976) *J. Mol. Biol.* **101:**367–377.

46 Cole, M. D., Beer, M., Koller, T., Strycharz, W. A., and Nomura, M. (1978) *Proc. Natl. Acad. Sci. U.S.A.* **75:**270–274.

47 Coetzee, W. F. and Pretorius, G. H. (1979) *J. Ultrastruct. Res.* **67:**33–39.

48 Miller, O. L., Jr. and Bakken, A. (1972) *Acta Endocrinol. (Copenhagen) Suppl.* **168:**155–177.

49 Broker, T. R. and Chow, L. T. *Abstract*, Tumor Virus Meeting, Cold Spring Harbor Laboratories, New York, 1976.

50 Richards, O. C., Enrenfeld, E., and Manning, J. (1979) *Proc. Natl. Acad. Sci. U.S.A.* **76:**676–680.

51 Reed, S. I., Ferguson, J., Davis, R. W., and Stark, G. R. (1975) *Proc. Natl. Acad. Sci U.S.A.* **72:**1605–1609.

52 Wu, M., Davidson, N., and Wimmer, E. (1979) *Nucleic Acid Res.* **5:**4711–4723.

53 Wu, M. and Davidson, N. (1978) *Nucleic Acid. Res.* **4:**2825–2846.

54 Granboulan, N. and Franklin, R. (1968) *J. Virol.* **2:**129–148.

55 Isaacs, S. T., Shen, C. K., Hearst, J. E., and Rapoport, H. (1977) *Biochemistry* **16:**1058–1064.

56 Jacobson, A. B. (1976) *Proc. Natl. Acad. Sci. U.S.A.* **73:**307–311.

57 Jacobson, A. B. and Bromley, P. A. (1974) *Cold Spring Harbor Symp. Quant. Biol.* **37:**845–846.

58 Kung, H. J., Bailey, J. M., Davidson, N., Vogt, P. K., Nicholson, M. O., and McAllister, R. M. (1974) *Cold Spring Harbor Symp. Quant. Biol.* **39:**827–834.

59 Kung, H., Hu, S., Bender, W., Bailey, J. M., and Davidson, N. (1976) *Cell* **7:**609–620.

60 McClements, W. and Kaesberg, P. (1977) *Virology* **76:**477.

61 Meyer, T., Lundquist, R. E., and Maizel, J. V. Jr. (1978) *Virology* **85:**445–455.

62 Granboulan, N. and Scherrer, K. (1969) *Eur. J. Biochem.* **9:**1–20.

63 Nanninga, N., Garrett, R. A., Stöffler, G., and Klotz, G. (1972) *Mol. Gen. Genet.* **119:**175–184.

64 Nanninga, N., Meyer, M., Sloof, P., and Reijnders, L. (1972) *J. Mol. Biol.* **72:**807–810.

65 Owens, R. A., Erbe, F., Hadidi, A., Steere, R. L., and Diener, T. O. (1977) *Proc. Natl. Acad. Sci. U.S.A.* **74:**3859–3863.

66 Sanger, H. L., Glotz, G., Presner, P., Gross, H. J., and Kleinschmidt, A. K. (1976) *Proc. Natl. Acad. Sci. U.S.A.* **73:**3852–3856.

67 Sarker, N. H. and Moore, D. H. (1970) *J. Virol.* **5:**230.

68 Savage, T., Granboulan, N., and Girard, M. (1971) *Biochimie* **53:**533–543.

69 Verma, I., Edelman, M., Herzberg, M., and Littauer, U. (1970) *J. Mol. Biol.* **52:**137–140.

70 Weber, G. H., Herne, U., Cottler-Fox, M., and Beaudreu, G. S. (1974) *Proc. Natl. Acad. Sci. U.S.A.* **71:**887.

71 Thach, S. S. and Thach, R. E. (1973) *J. Mol. Biol.* **81:**367–380.

72 Naora, H. and Fry, K. (1977) *Biochim. Biophys. Acta*, **478:**350–363.

73 Ferguson, J. and Davis, R. W. (1978) "Quantitative Electron Microscopy of Nucleic Acids," in *Advanced Techniques in Biological Electron Microscopy*, Vol. 2 (J. K. Koehler, Ed.), Springer-Verlag, Berlin, pp. 123.

74 Evenson, D. P. (1977) *Methods Virol.* **6:**219.

75 Brack, C. B. "DNA Electron Microscopy," in *CRC Critical Reviews*, in press.

76 Fisher, H. W. and Williams, R. C. (1979) *Annu. Rev. Biochem.* **48:**649–679.

77 Delain, E. and Brack, C. (1974) *J. Microscop.* **21:**217.

78 Brison, O. and Chambon, P. (1976) *Anal. Biochem.* **75:**402–409.

79 Berger, S. L. and Birkenmeir, C. S. (1979) *Biochemistry* **18:**5143–5149.

80 Maniatis, T., Jeffrey, A., and van de Sande, H. (1975) *Biochemistry* **14:**3787.

81 Thomas, J. O., Kolb, A., and Szer, W. (1978) *J. Mol. Biol.* **123:**163.

82 Glass, J. and Wertz, G. *Nucleic Acids Res.*, Dec., 1980.

83 Preobrazhensky, A. A. and Spirin, A. S. (1978) *Res. Mol. Biol.* **21:**1–38.

84 Princer, H., Eekelen, C., Asselbergs, F., and Venrooij, W. (1979) *Mol. Biol. Rep.* **5:**59–64.

85 Heggeness, M., Scheid, A., and Choppin, P. (1980) *Proc. Natl. Acad. Sci. U.S.A.* **77:**2631–2635.

86 Pettersson, R. F., and von Bonsdorff, C. (1975) *J. Virol.* **15:**386–392.

87 Butler, P., Finch, J., and Zimmerman, D. (1977) *Nature (London)* **265:**217–219.

88 Botchan, P. (1976) *J. Mol. Biol.* **105:**161–176.

89 Chen, M., Garon, C., and Papas, T. (1980) *Proc. Natl. Acad. Sci. U.S.A.* **77:**1296–1300.

90 Iserentant, D. and Fiers, W. (1980) *Gene* **9:**1–12.

91 Senkevich, T. G., Cumakov, I. M., Lipskaya, G. Y., and Agol, V. (1980) *Virology* **102:**339–349.

92 Cumakov, I. M., Lipskaya, G., and Agol, V. I. (1979) *Virology* **92:**259–270.

93 Johnson, D. (1975) *Nucleic Acid. Res.* **2:**2049.

94 Arnberg, A. C., Van Ommen, G. J., Grivell, L. A., Van Bruggen, E. F., and Borst, P. (1980) *Cell* **19:**313–319.

95 Halbreich, A., Pajot, P., and Foucher, M. (1980) *Cell* **19:**321–329.

96 Trapnell, B. C., Tolstoshew, P., and Crystal, R. G. (1980) *Nucleic Acid Res.* **8:**3659–3672.

97 Reymond, C., Agliano, A., and Spohr, G. (1980) *Proc. Natl. Acad. Sci. U.S.A.* **77:**4683–4687.

98 Ackerman, S., Keshgegian, A. A., Henner, D., and Furth, J. J. (1979) *Biochem.* **18:**3232–3242.

4

EXPERIMENTAL STUDIES ON RESOLUTION IN HIGH-VOLTAGE TRANSMISSION ELECTRON MICROSCOPY

Mircea Fotino
Department of Molecular, Cellular, and Developmental Biology
University of Colorado
Boulder, Colorado

CONTENTS

1 Introduction
2 Rationale for high-voltage electron microscopy in biology
3 Resolution measurements
4 Resolution in thin specimens
  4.1 Aperture dependence
  4.2 Focusing
  4.3 Astigmatism
5 Resolution in thick specimens
  5.1 Contrast mechanism
  5.2 Aperture dependence
  5.3 Depth of field
  5.4 Orientation in the beam
  5.5 Thickness dependence
6 Specimen damage
Acknowledgments
References

1 INTRODUCTION

During the last decade the use of transmission electron microscopy at higher accelerating voltages—commonly referred to as high-voltage electron microscopy (HVEM)—has grown steadily in spite of the emergence of and competition from other imaging modes. Although the bulk of HVEM applications is still in materials science and metallurgy, biological applications are progressively increasing in a more massive and systematic way. Established for the purpose of facilitating access by biologists to unusually costly and sophisticated technology, dedicated HVEM installations for biological applications currently operate at 1 MV as national resources sponsored by the National Institutes of Health at the University of Colorado at Boulder and at the University of Wisconsin at Madison. A third and similar HVEM installation engaged primarily in biological research operates at the New York State Department of Health in Albany.

This chapter focuses on aspects of resolution in amorphous specimens that are relevant or of interest in biological applications. Thus the pursuit of the ultimate in resolution is not the aim of this endeavor. Nevertheless, the evidence presented here points clearly to a satisfactory performance level. An experimental point of view was adopted primarily for conveying to the reader a sense of what is accessible in practice and what can emerge from real situations. Theoretical considerations are included only occasionally and mostly for elucidating the background.

Section 2 presents succinctly the theoretical foundation for extending transmission electron microscopy to high accelerating voltages and its main benefits in biological research.

Section 3 describes the resolution concept and a method of general applicability used throughout this chapter for determining experimentally the resolution in various configurations. In presenting the results of these measurements, a distinction is made between thin and thick specimens not only because the latter involve multiple scattering, hence a different mechanism in image formation, but also because it approximates more closely the experimental situations encountered in the biological applications of HVEM.

Included in the discussion of resolution in thin specimens in Section 4 are critical factors such as correction of astigmatism, focusing dependence, and aperture dependence.

With thick specimens, covered in Section 5, factors other than aperture dependence play a role in determining the resolution, including orientation in the beam, depth of field, and thickness dependence. The influence of accelerating voltage on the resolution attainable in thick specimens is illustrated by data collected near the 3 MV limit of presently available accelerating voltages in HVEM at the Laboratoire d'Optique Electronique in Toulouse, France.

Section 6 briefly discusses beam-induced specimen damage.

2 RATIONALE FOR HIGH-VOLTAGE ELECTRON MICROSCOPY IN BIOLOGY

As is well known, the resolving power δ_0 of an electron microscope is much poorer than the associated electron wavelength λ because of the limitation introduced by the spherical aberrations of the objective lens characterized by a constant C_s. Thus the resolving power is described by the expression:

$$\delta_0 = A\ C_s^{1/4}\ \lambda^{3/4}$$

where A is a constant reflecting a combination of the diffraction and spherical aberration limits. This constant is generally taken as $A = 0.43$, but more detailed wave mechanical considerations lead to the value $A = 0.56$, obtained by Glaser (1), with an optimum acceptance angle $\alpha_{\text{opt}} = 1.13\lambda^{1/4} C_s^{-1/4}$. Clearly, improvement in resolving power (i.e., reduction of δ_0) can be obtained either by reducing the spherical aberration constant C_s of the objective lens at a given excitation or by increasing the accelerating voltage to reduce the wavelength λ. The latter approach has led to installations of increasing dimensions and complexity and operating at higher and higher accelerating voltages. The approximate relationship between theoretical resolving power δ_0 and accelerating voltage of contemporary installations in the world is summarized in Figure 1. If one keeps in mind that an increase in accelerating voltage is accompanied by a more-than-proportional increase in complexity because of the usual requirements of stability and monochromaticity essential to electron microscopy, it becomes clear that a practical cutoff must be reached when one determines the real price, both in expenditure and in effort, of benefiting from improvements associated with higher operating voltages. It may seem surprising, then, that high-voltage installations maintain their attractiveness and materialize to the extent they have in the last decade.

One reason for the continued interest in high-voltage installations is simply that technological progress has reduced to manageable levels many of the erstwhile difficulties. The other, more compelling, reason is that there is more than resolution to be gained from higher accelerating voltages.

Indeed, the electron loses energy as it interacts with matter. This energy loss per unit thickness is a function of the accelerating voltage, or equivalently of the kinetic energy of the electron, and it goes through a broad minimum around 1 MeV, as illustrated in Figure 2 for several substances of biological interest from values calculated by Berger and Seltzer (2). Without going into the details of this mechanism, the decrease in characteristic energy loss corresponds qualitatively to the fact that as its energy increases, hence its velocity as well, the electron's interaction time with matter is reduced, and thus the transfer of energy from the electron becomes smaller. That an *increase* in energy loss should appear at even higher electron energies is the result of relativistic effects, a simple discussion of which, however, is beyond the scope of this book.

Of the various consequences of this decrease in energy loss, three main features stand out as beneficial for biological electron microscopy.

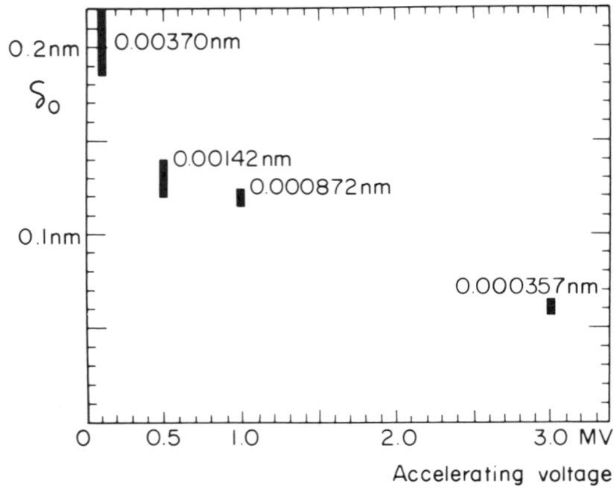

Figure 1 Dependence of the theoretical resolving power δ_0 of current instruments on accelerating voltage. Data in nanometers represent electron wavelengths.

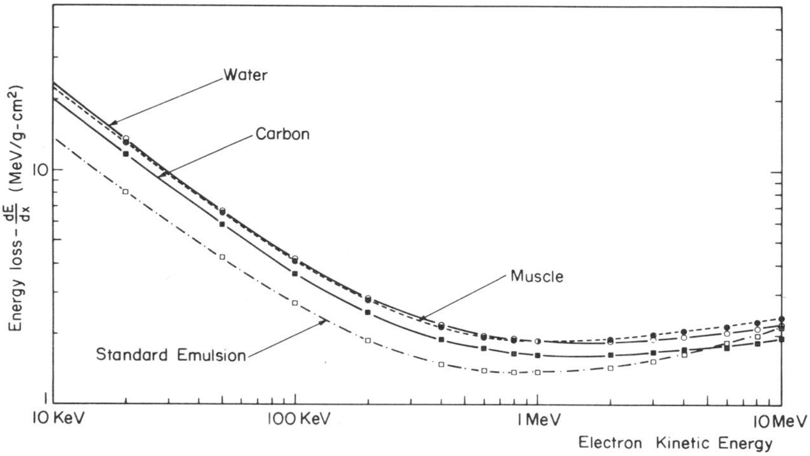

Figure 2 Dependence of characteristic energy loss $-dE/dx$ on kinetic energy for electrons in various substances. Adapted from values calculated by Berger and Seltzer (2).

First, on one hand the energy loss per unit thickness is smaller and on the other hand the total available energy of the electron is larger; therefore interaction with matter as described above can take place for longer intervals or equivalently over longer trajectories before the total energy is reduced to a certain extent or completely dissipated. The overall result is an increase in "range," which is the distance traveled from the point of impact to the point where the electron comes to rest (with no energy left). Range-energy curves for several substances are shown in Figure 3, assembled from values calculated by Berger and Seltzer (2). It should be noted that the increase in range is more than proportional to an increase in energy by a factor of about 3–4. It is thus clear that operating at higher voltages makes thick specimens more transparent to the beam than by simply prorating from lower voltages. This improvement is often and vaguely referred to as increased "penetration," a term that commonly designates only a small fraction of the total range through which some visibility or recording capability is still present, without consideration of transmitted intensities or detection nonlinearities. In reality and in a stricter sense, all specimen thicknesses smaller than the range are in fact "penetrated," since their images could be obtained by adequately adjusted exposures.

This feature is possibly the most significant and beneficial for biological applications because (1) a new, three-dimensional view of global structures within thick tissue slices and whole cells becomes readily available, (2) larger specimen volumes can be viewed in shorter times, and (3) extended structures are viewed directly in their natural habitat without the artifacts and disruptions inherent to extraction or serial sectioning. For discussions of

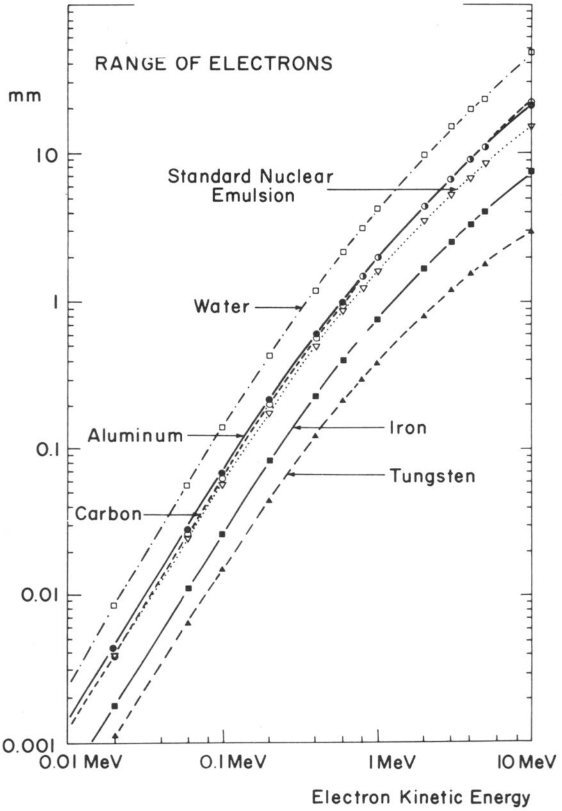

Figure 3 Dependence of electron range on kinetic energy in various substances. Adapted from values calculated by Berger and Seltzer (2).

such biological applications and for general accounts of research accomplishments by HVEM, the reader is referred to several reviews (3–8).

Second, the resolution attainable in thick preparations is substantially better than could be obtained at lower voltages because the scattering of imaging electrons, taking place at higher energies and with less energy loss, is more strongly peaked forward. As an indication of how an improvement in resolution is linked to the energy-dependent scattering of electrons, a comparison between scattering cross sections in carbon (in the form of the ratio between inelastic σ_i and elastic σ_e) at 100 and 1000 kV is illustrated in Figure 4 on the basis of data given by Dupouy (9). The implication of these curves is that at 1 MV there is a preponderance both of elastically scattered electrons at angles larger than about 0.5 mrad and of inelastically scattered electrons in the forward direction. The latter, however, having lost less energy than at lower accelerating voltages, bring about reduced chromatic aberrations. The net result thus is less deterioration in resolution. Some experimental evi-

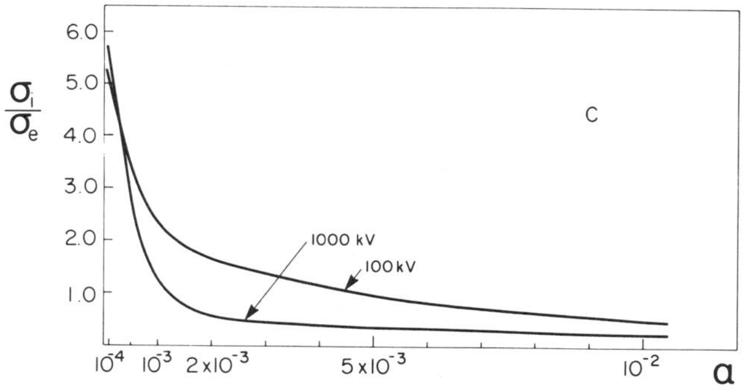

Figure 4 Relationship between the inelastic and elastic cross sections σ_i and σ_e, respectively, in carbon, in terms of scattering angle α and accelerating voltage. Adapted from Dupouy (9).

dence for this feature is presented below. More detailed determinations of cross sections (σ_e and σ_i) as well as of penetration, beam broadening, and transmission coefficients in several elements based on Monte Carlo calculations for multiple scattering and on the Landau formalism for energy loss have been given recently by Soum et al. (10).

Third, lower energy loss by imaging electrons means less energy transferred to the specimen, hence less radiation-induced damage to the specimen's structures, provided adequate attention is paid to the viewing or recording procedures. This point is discussed briefly at the end of this chapter.

Since the physical arrangement of a high-voltage electron microscope is rather unusual by the common standards of traditional electron microscopy, it seems appropriate to include in this presentation an example of an existing HVEM installation dedicated to biological investigations. Figure 5 shows diagrammatically the installation of the 1 MV electron microscope (model JEM-1000, built by JEOL Co. of Tokyo, Japan) housed in the Department of Molecular, Cellular and Developmental Biology of the University of Colorado at Boulder, where most data for this study were collected over the past few years. To reach the resolution level of 0.3 nm currently accessible on a routine basis, certain precautions have been necessary against disruptive factors such as mechanical vibrations and electrical or magnetic disturbances, as well as long-range improvements of critical factors such as high-voltage and lens-current stability, and drift. Details of the installation have been described by Fotino (11).

3 RESOLUTION MEASUREMENTS

To determine accurately and reliably the level of resolution actually attainable in any given configuration is not a straightforward experimental task.

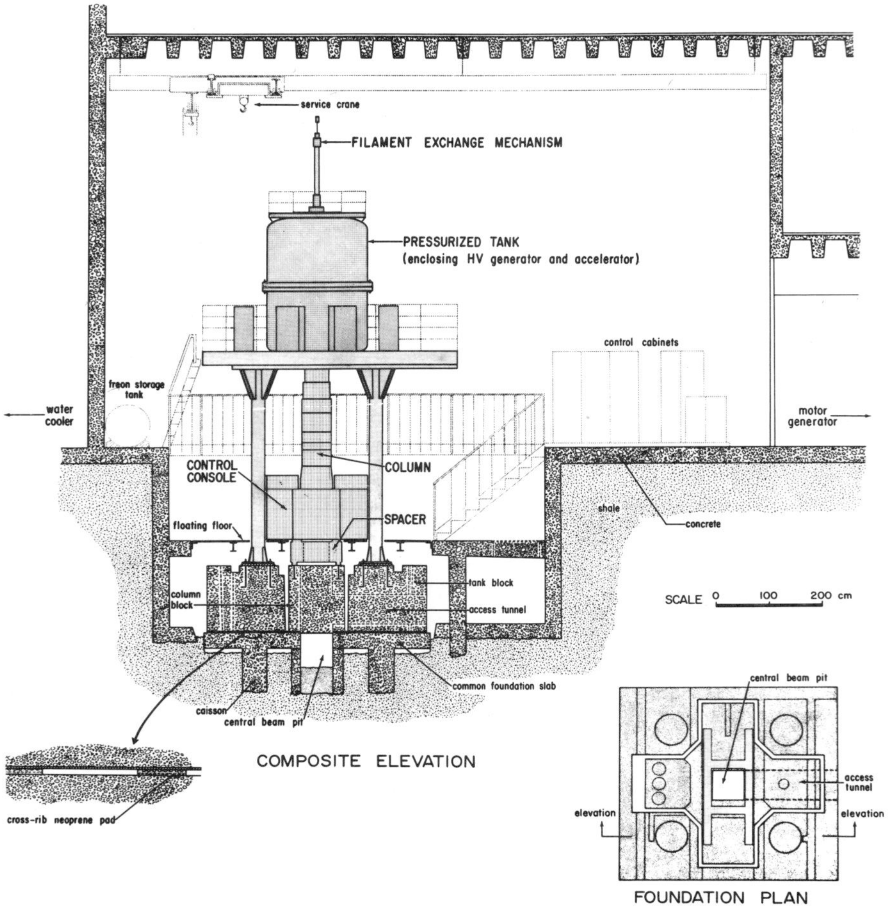

Figure 5 Schematic diagram of the JEM-1000 1 MV electron microscope installation at the Department of Molecular, Cellular and Developmental Biology of the University of Colorado, Boulder.

Conceivably one could use a wide set of standards of discrete and well-known dimensions such as microcrystals or stable macromolecules to incorporate in the desired specimen and thus determine the resolution by a visibility cutoff. The difficult and unsatisfactory requirements of this approach are: (1) the orientation of these standards in the specimen must be known accurately, (2) their visibility must be unambiguous in the field of view, and (3) the distinction between what is and what is not resolvable must be made objectively and reproducibly.

The situation can be illustrated qualitatively by the schematic configuration shown in Figure 6. When a well-defined standard is lying on a thin

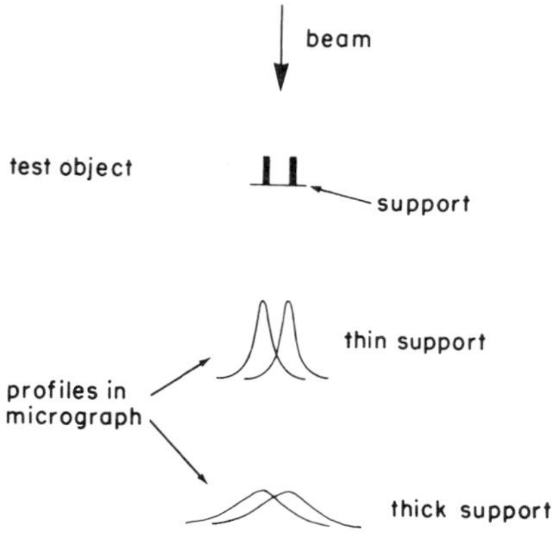

Figure 6 Schematic outline of the deterioration of resolution in thick specimens.

support, its image in the micrographs remains well separated. However, when the support is thicker, individual points in the standard appear spread out in the micrograph, and the resolution deteriorates so that it can no longer be appreciated unambiguously.

Instead of following the disappearance of a known separation between two point elements of a given standard, an alternate method has been defined and used (12, 13) in which resolution is equivalently defined in a simple and unambiguous way. It is an extension of the well-known Rayleigh criterion for the separation of spectral lines. Since the separation of the two components is determined by the extent of their overlap, it is assumed to a good approximation in this method that the profiles of steplike components are linear along the "edge width" δ_e, which is defined as the distance between 10 and 90% of maximum, so that their separation may be determined by the superposition of two identical edges, rather than of two points, as outlined in Figure 7. The limits of 10 and 90% of maximum, between which the edge width δ_e is defined, have been chosen in keeping with a generalized convention applied to many other physical signals of similar appearance, such as single pulses (their "rise time") from photomultipliers and pulse-height distributions from quasimonochromatic particles in an energy detector. Depending on the extent of adopted overlap, the resolution is:

$\delta_R = 1.58\ \delta_e$ if central minimum is 73.5% of maximum (as in the Rayleigh criterion)

$\delta_{1/2} = 1.87\ \delta_e$ if central minimum is 50% of maximum

$\delta_{sep} = 2.50\ \delta_e$ if edges are completely separated

The width of the actual edge needed in this method is conveniently approximated by the peak of the statistical distribution of many edge widths δ_e obtained by scanning normally to their contours the Au islands deposited by

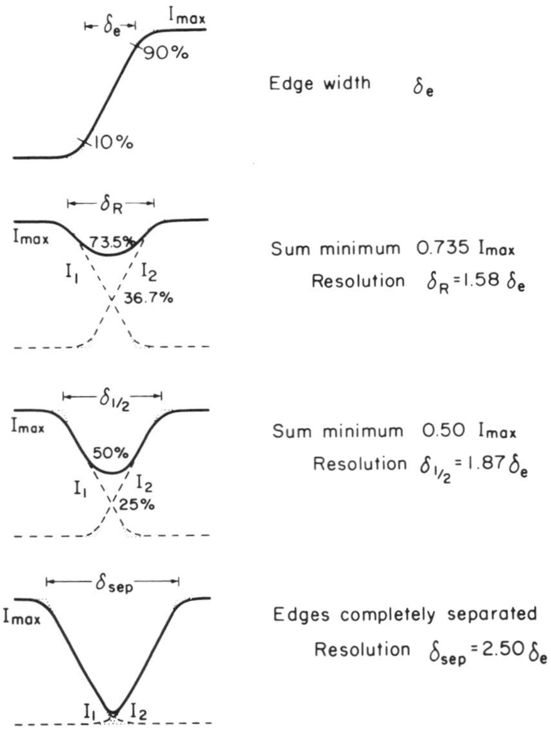

Figure 7 Resolution criteria for identical edges.

evaporation on an amorphous carbon film. The Au islands are estimated to be approximately 5–20 nm thick. Scanning is done in several directions to ensure isotropy. Such distributions are reproducible and their deterioration by any imaging factor (aperture, defocus, thickness, etc.) gives a quantitative determination of changes in resolution. Furthermore, the Au-island contours remain stable under prolonged irradiation and are readily visible under most conditions in spite of the poor screen visibility that one has to contend with in high-voltage electron microscopy. The specimens used in this study accordingly had one of the configurations illustrated in Figure 8. When comparing several specimens, the materials were taken from the same batch, to ensure uniformity of preparation. Specimen dimensions were determined in a straightforward manner. With latex spheres, low-magnification maps of the sphere imaged at high magnification were usually taken before and after a particular sequence so that if a relative change occurred, it could be easily estimated. In this case, since the specimen has spherical symmetry, it is assumed, as evidenced by many observations on transmission and scanning electron micrographs (TEM and SEM), that a beam-induced modification is isotropic, hence its thickness can reasonably be approximated by its

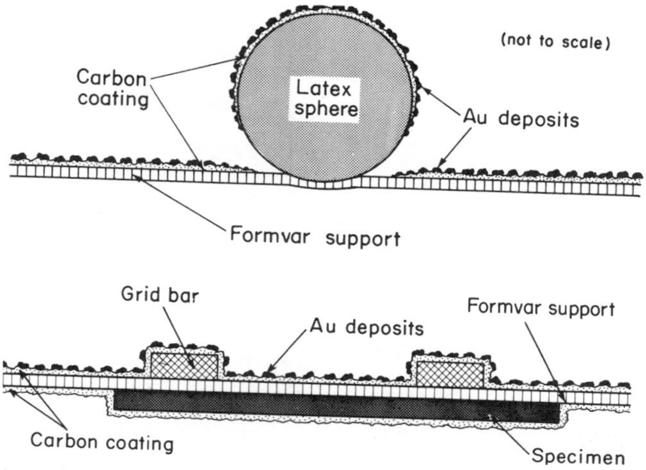

Figure 8 Configuration of preparations with Au islands and specimens of variable thickness used for studying experimentally various aspects of resolution by HVEM.

diameter measured on the map. An example is shown in Section 6 (Figure 38). Furthermore, since imaging and tracing were restricted to a small area around the sphere center, its thickness could be considered constant. With sectioned material, the point of interest was usually selected at the edge of a slice, and low-magnification maps were taken to locate the surroundings, usually with a pointer identifying the exact area imaged at high magnification. The same area was subsequently identified in an SEM and the slice thickness unambiguously determined from an image taken almost edge-on at a known angle, as can be seen on the examples shown in Figure 9. Examples of microdensitometer tracings of Au-island profiles imaged in upstream orientation through epon layers of increasing thickness t are shown in Figure 10.

A similar method, based on the broadening of an edge and using the half-edge width (defined as the distance between 25 and 75% of maximum) of individual indium crystals on a carbon support film was described by Gentsch et al. (14, 15) for measuring the top-bottom effect in thick amorphous specimens by scanning transmission electron microscopy (STEM). No statistical distribution of the edge width is taken, however; thus the extent to which the uncertainty in crystal orientation on the support film has been eliminated remains unclear.

4 RESOLUTION IN THIN SPECIMENS

The trend toward better resolution by means of higher voltages became irrepressible soon after the first electron microscope built in 1931 by Ruska

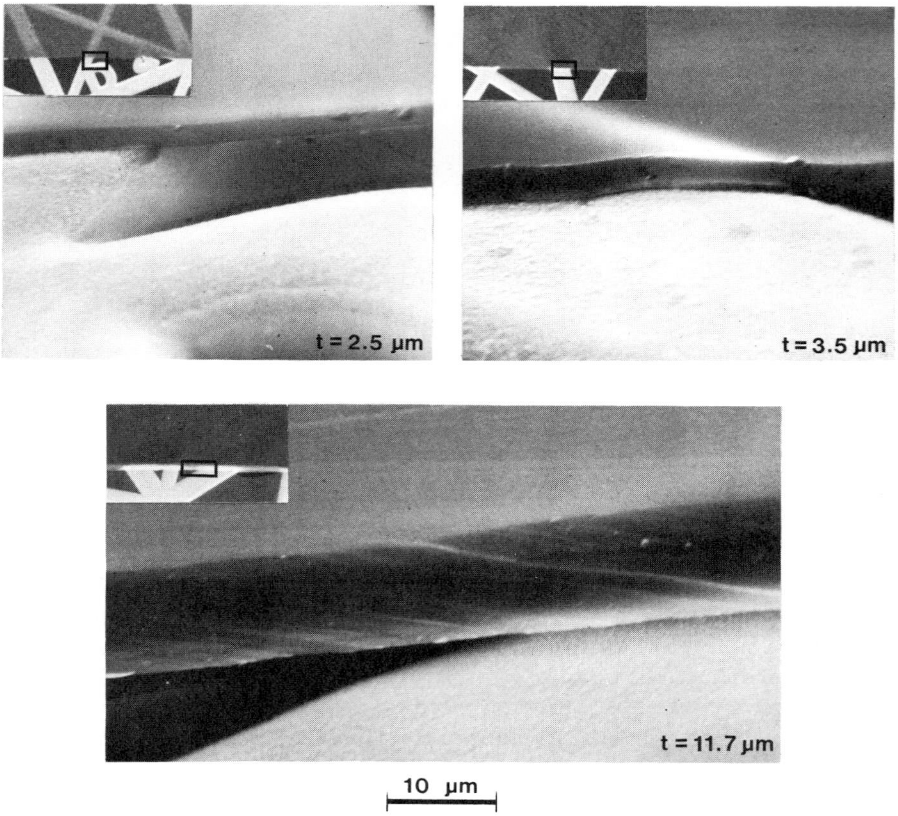

Figure 9 Examples of thickness determination by SEM. The edges of epon slices of increasing thickness are viewed at 60° inclination with respect to the electron beam. Inserts represent maps at normal beam incidence and low magnification (60×) by which the area around each point of interest is identified. Framed area in each inset is shown magnified in the micrograph. Note the reversal of contrast seen through the slice that accompanies the change in the number and energy of electrons backscattered by the grid bar as the specimen thickness increases. Measured slice thickness shown on each micrograph. Magnification: 1,470×.

and Knoll (16) started operating at 50–80 kV. The adopted operating voltage kept steadily increasing: from the "universal electron microscope" at 200 kV of von Ardenne (17) in 1941, to the experimental microscope at 400 kV of Van Dorsten et al. (18) in 1947, to the 750 kV instrument of Smith et al. (19) in 1966 at the Cavendish Laboratory in Cambridge, and finally to the megavolt range heralded successfully by the first installation at 1.5 MV by Dupouy and Perrier (20) in 1960. The most powerful microscopes in the world today operate at up to 3 MV in Toulouse, France (21), and in Osaka, Japan (22).

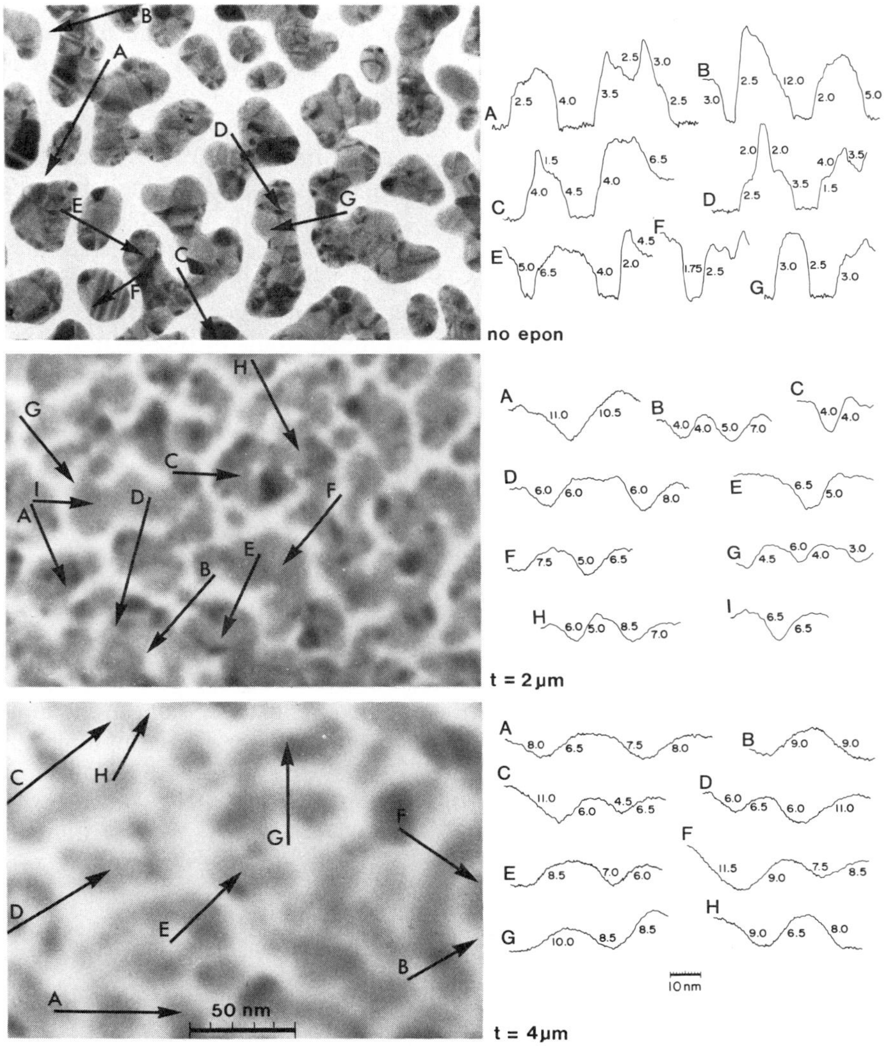

Figure 10 Examples of microdensitometer tracings of Au-island profiles imaged at fixed aperture (45 μm) in upstream orientation through epon layers of increasing thickness t. Numbers represent edge width in nanometers. The identifying letter marks the beginning of each tracing taken along the marked line in the direction of the arrow. Micrograph magnification: 280,000×, 1 MV.

Several microscopes combining specifically high resolution and high voltage have been realized during the last decade (23–25). The first high-quality results at resolution levels around 0.1 nm obtained at Cambridge University with a 600 kV electron microscope designed for atomic resolution have been recently described by Cosslett et al. (26).

The only limitation in this steadfast endeavor has come from the stringent requirements of high mechanical and electrical stabilities, which become more difficult to establish and maintain as the operating voltage increases. When these requirements are rigorously fulfilled, the microscope performance can be usually optimized and maintained at a highly satisfactory level on an almost routine basis. Elements of such performance from the Boulder HVEM installation are presented below as illustrations of the current standard of operations and of what expectations biological microscopists can justifiably entertain for HVEM applications.

4.1 Aperture Dependence

The optimum performance of a high-voltage transmission electron microscope, like that of any conventional instrument, occurs for a value of the acceptance angle α (half-angle of the cone subtended by the objective-lens aperture) imposed by both the diffraction and the spherical aberration limits. The theoretical performance curve of the JEM-1000 instrument (C_s = 5.5 mm, f = 7.2 mm) with universal top-entry gonio stage (ALG) in place is given in Figure 11, in which the vertical arrows indicate the optimum angle α_{opt} for best resolution determined by the various modes of combining the two effects. The value of α_{opt} calculated by Glaser (1) on the basis of more complete wave mechanical considerations is also indicated. A similar curve may be obtained for the high-voltage electron microscope in Toulouse, operating at 2.5 MV (C_s = 6.1 mm, f = 11 mm, λ = 0.00043 nm).

Experimental evidence for the aperture dependence of resolution at 1 MV was obtained with the Au-island substrate of the preparations described above. A given area of this substrate was imaged by fine focal series ($\Delta f \sim$ 130 nm/step) with several apertures, the diameter of which ranged from 8 μm to 3 mm (no aperture). The sequence of micrographs thus obtained at optimum defocus and processed under identical conditions is shown in Figure 12. It can be easily seen that the image appearance follows qualitatively the theoretical description of resolution of Figure 11 in that it deteriorates visibly at small and large angles, with an optimum in between. At large angles, however, the image quality is much better than described by an α^3 dependence. To evaluate quantitatively this behavior, microdensitometer tracings were taken with a rectangular scanning spot (corresponding to about 3 × 0.05 nm in object space) in a direction perpendicular to the long side; this combination of shape and size was chosen to reduce the noise from the carbon film while integrating substantial rectilinear portions of the island contour. The resulting histograms of Figure 13, although based on fairly

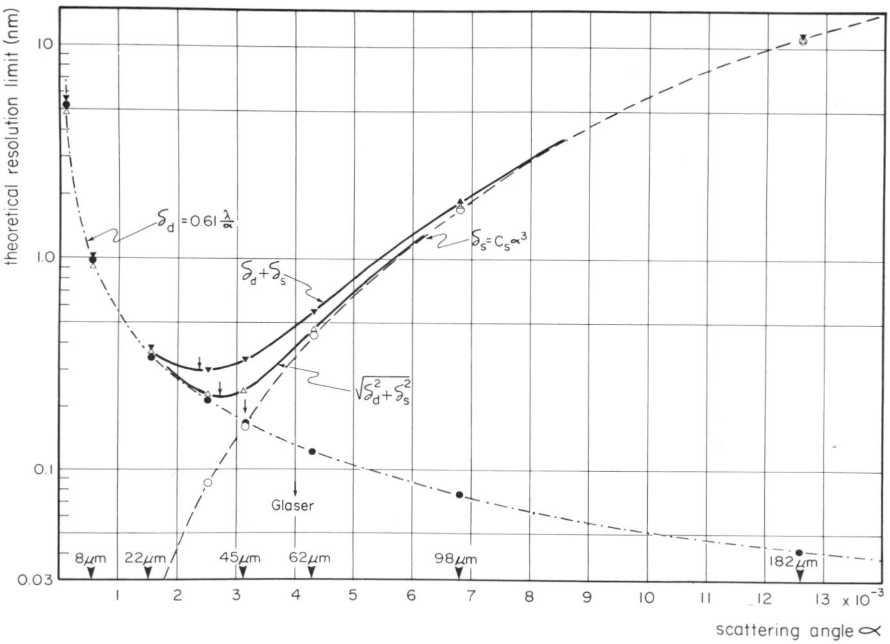

Figure 11 Theoretical performance of the 1 MV electron microscope JEM-1000 as determined by the diffraction and spherical aberration limits: C_s = 5.5 mm, f = 7.2 mm. Vertical arrows locate the optimum angle α_{opt} for best performance in the customary modes of combining the two effects as well as the value of α_{opt} according to detailed wave mechanical considerations made by Glaser (1). The actual diameters of several apertures used in various experimental configurations are also indicated along the abscissa.

limited profile samplings, and, more clearly, the curve of Figure 14 assembled with the peaks of edge-width distributions, show an optimum resolution approximately located as predicted theoretically. The deterioration in resolution at small and large angles, though visible, is smaller than anticipated. Beyond acceptance angles of about 7 or 8 mrad, the resolution becomes practically constant, which is consistent with the fact that the entire beam is accepted within this angle and thus no other electrons that could be affected by spherical aberrations are added at larger angles.

4.2 Focusing

A study of various aspects of resolution reveals that one of the central requirements concerns the adequate focusing that must be obtained in any given circumstance. The adequacy of micrographs used for illustrating an experimental situation is thus ascertained by the rigorous selection of "best

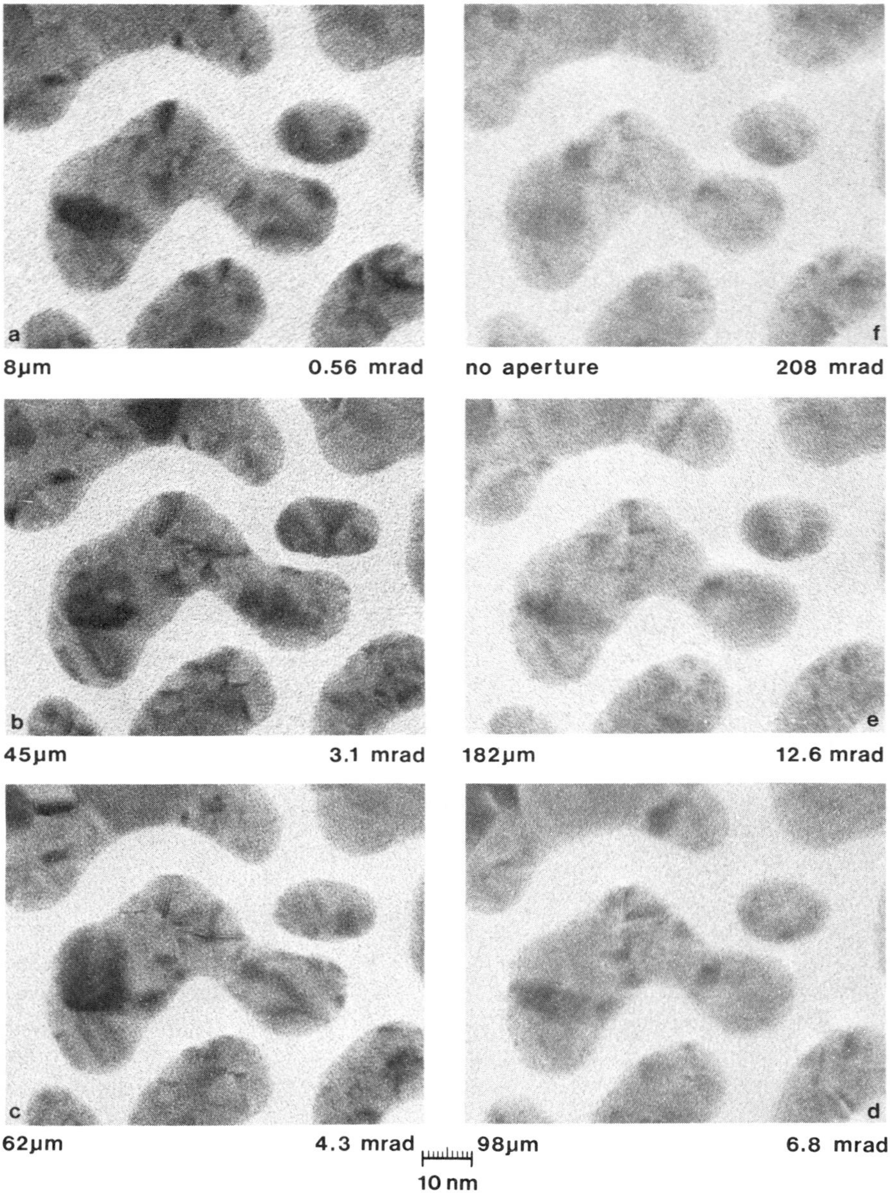

Figure 12 Aperture dependence of resolution in thin specimens illustrated by micrographs of the same area of Au-island substrate at optimum defocus obtained by fine focal series with JEM-1000 at 1 MV. As predicted by theory, the image quality deteriorates at small or large scattering angles, but appears very much better than expected at large angles. Aperture diameter (*left*) and the corresponding acceptance angle α (*right*) given under each image. Magnification: 660,000×.

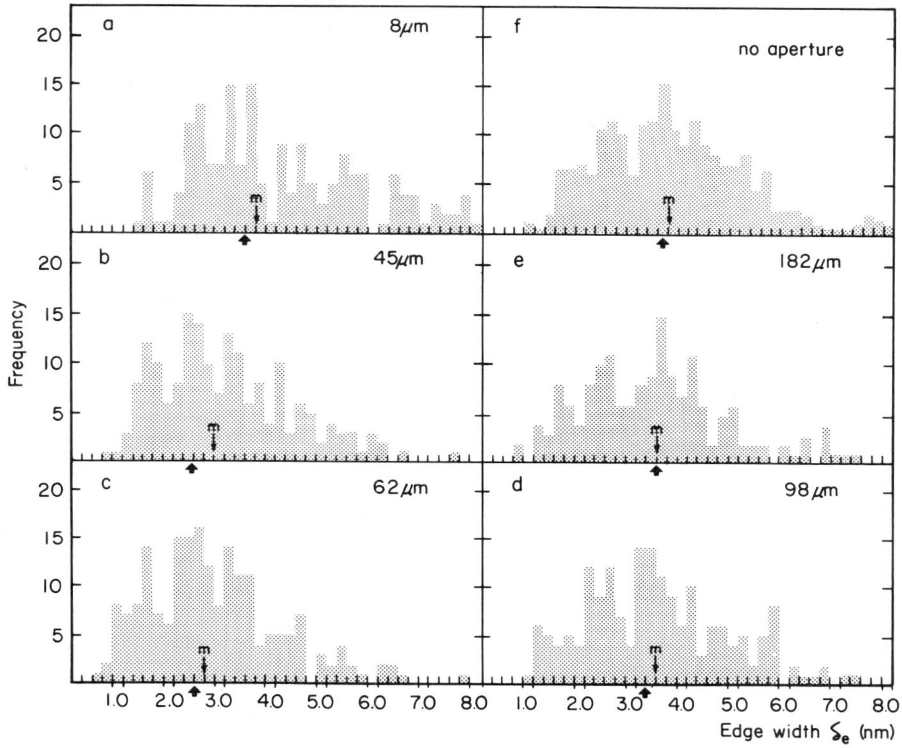

Figure 13 Histograms of the edge width δ_e measured on micrographs in the corresponding aperture-dependence sequence of Figure 12. The heavy arrow under each histogram locates the approximate peak and the light arrow (m) indicates the median value for each distribution.

focus'' from a focal series. This requirement is particularly stringent in HVEM because screen visibility is poor and focusing uncertain, especially at relatively high instrumental magnifications.

When applied to a thin and well-defined specimen such as the Au-island substrate at high magnification, the method for resolution measurement described above can be used both to assess its sensitivity and to correlate a measured (i.e. objective) performance level with what appears to the eye as a best-focus micrograph in a focal series with fixed aperture. Such a series of eight micrographs is illustrated at high magnification for better visibility in Figure 15. These micrographs were taken under the conditions mentioned above (magnification 100,000×, aperture 62 μm) and with the finest focusing range routinely used in the JEM-1000 installation ($\Delta f \sim 130$ nm/step). An adjustment of one such step changes the width of an overfocus Fresnel fringe by about 0.12–0.14 nm at or near focus. A finer range is also available for more refined focusing ($\Delta f \sim 35$ nm/step), but it is difficult and time-

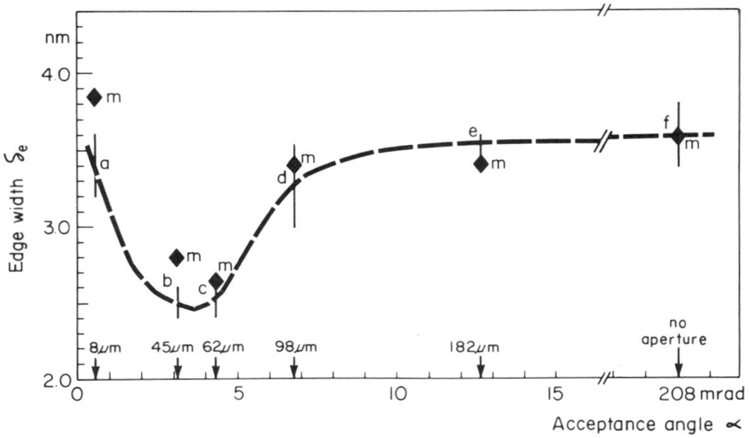

Figure 14 Dependence of the edge width δ_e on acceptance angle α at 1 MV resulting from the approximate peaks of the edge-width histograms of Figure 13. The median value (m) of each distribution is also indicated. Note that the minimum is located approximately as indicated by the performance curve of Figure 11, but the range of deterioration in resolution is much smaller than was anticipated theoretically.

consuming to use. Given its limited extent of only 10 steps, it is highly uncertain that the poor visibility conditions will permit the investigator to bracket the best focus with thin specimens.

The best focus in this sequence is easy to identify visually both by the edge sharpness of the Au islands and by the characteristic appearance of the amorphous carbon film. In fact, the imaging of thin carbon films is essentially a weak phase-object problem and as such has been shown by Thon (27) to depend strongly on defocus, so that it may actually be used as an aid in focusing biological specimens. Such, however, could unfortunately not be the case in HVEM, since changing the conditions of defocus with an amorphous carbon film can hardly be appreciated by direct viewing.

The edge-width histograms corresponding to the micrographs of this sequence are assembled in Figure 16. Although these distributions consist of only 150–200 traces each, their peaks appear to define rather clearly the trend associated with increasing defocus. This focusing dependence of δ_e, hence of resolution, is shown in Figure 17, in which the vertical bars locate approximately the distribution peaks. The median value (m) for each distribution in Figure 17 gives an approximate idea of its sampling quality. It is interesting to note the match between visual focus in the focal series and the measured optimum resolution. A difference by a factor of about 2.5 in the edge-width change per step is evident between the underfocus and overfocus regions (about 0.22 and 0.55 nm/step, respectively). This difference is assignable to a rather complex interference mechanism by the waves generated at the thin Au-island edge (partially transparent rather than opaque at 1 MV as seen in

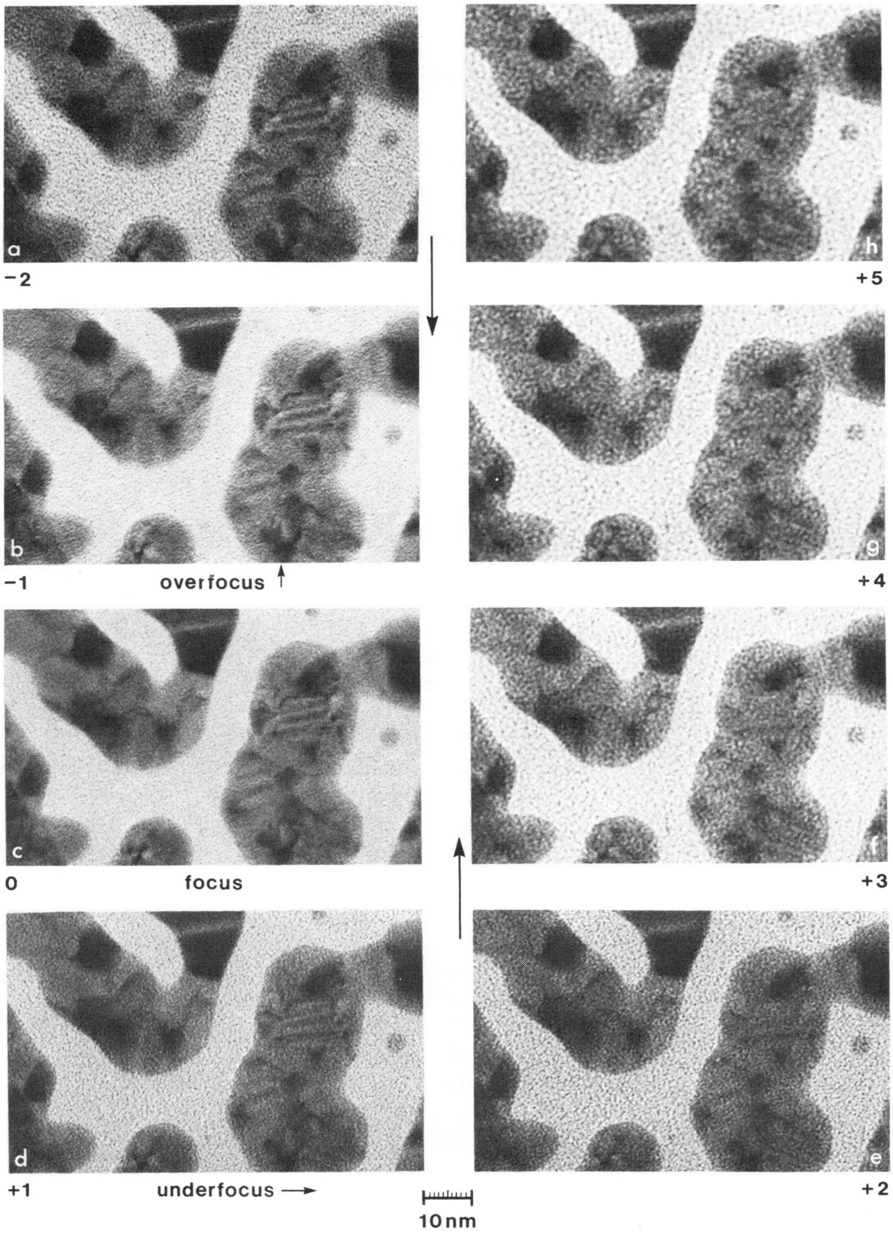

Figure 15 Fine focal series of Au-island substrate at 1 MV with fixed aperture (62 μm) in counterclockwise sequence in the direction of increasing defocus (long arrows), that is, from overfocus (*a*) to underfocus (*h*). This represents the finest routine focusing range ($\Delta f \sim 130$ nm/step) currently used with the JEM-1000 high-voltage electron microscope at the University of Colorado, Boulder. Number under each frame indicates defocus steps from optimum focus. Magnification: 620,000×.

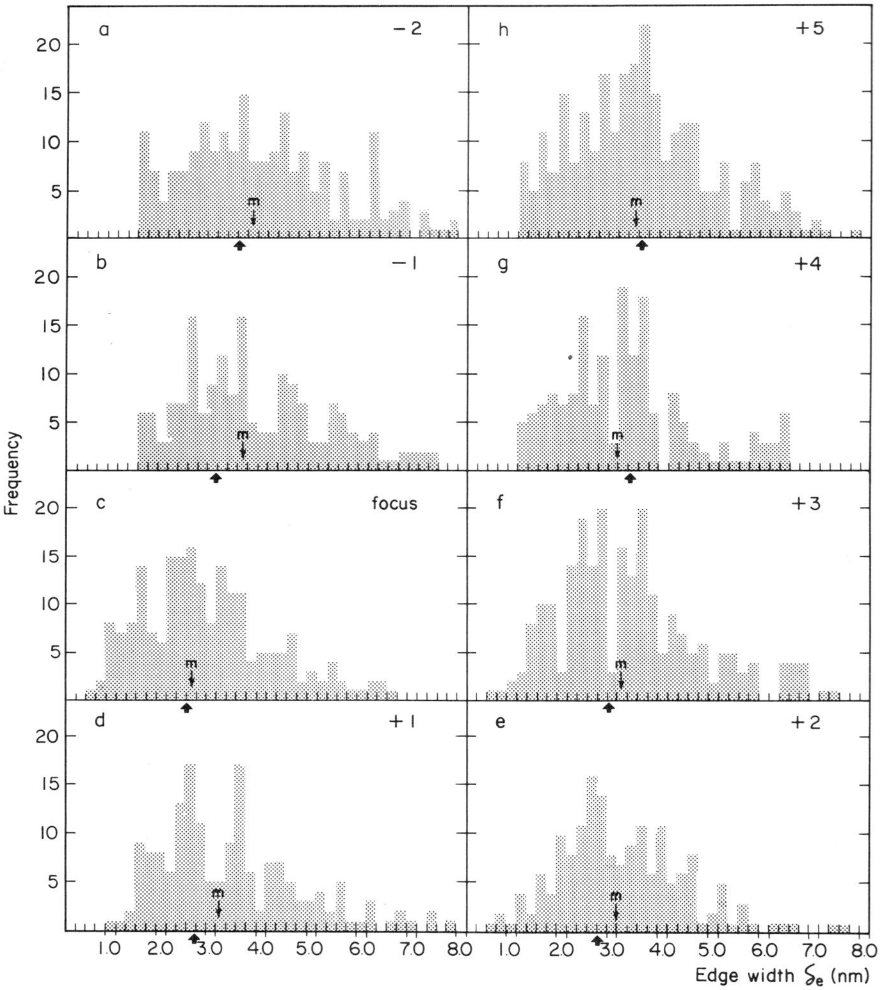

Figure 16 Histograms of the edge width δ_e measured on the micrographs of the focal series in Figure 15. The heavy arrow under each histogram locates the approximate peak, and the light arrow (m) indicates the median value for each distribution.

the micrographs) in which contributions from these waves inside the edge are unequal (larger and wider in overfocus, smaller and narrower in underfocus condition). In fact, since resolution along such a contour is determined by a mixture of absorption and phase contrasts, scanning with a narrow microdensitometer slit across the edge in essence represents a gradual transition from pure phase contrast (carbon film) to predominantly absorption contrast (Au island). This difference also corresponds to the empirical observation that the best focus reached from underfocus side has a wider and

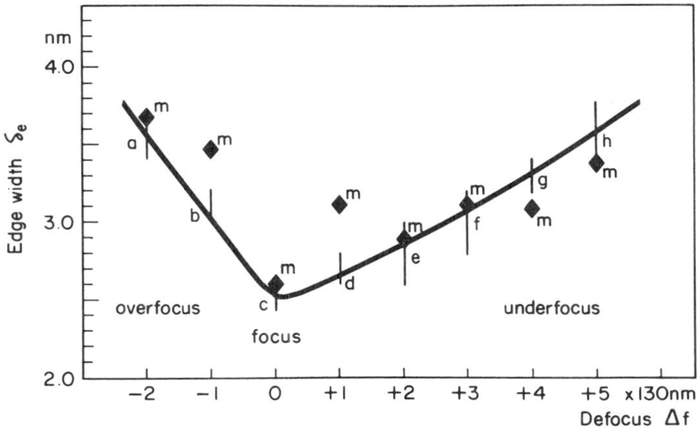

Figure 17 Focusing dependence of the edge width δ_e resulting from the approximate peaks of the corresponding histograms of Figure 16 and reflecting the variation of resolution with defocus. The median value (m) is also indicated for each aperture.

less sensitive margin of adjustment. Contour and phase contrast phenomena have been discussed in detail by Thon (27), Hillier and Ramberg (28), and Grinton and Cowley (29).

4.3 Astigmatism

As in any imaging instrument, astigmatism correction is an essential requirement for reaching best resolution with the high-voltage electron microscope. Stigmators are included below the pole pieces of the objective lens, and with two exceptions, the protocol for correction is very much as in the conventional instruments.

One is the relationship between defocus Δf and fringe width. Because of the reduction in electron wavelength at higher energies, the width of the Fresnel fringe (i.e., the distance between the centers of the main contiguous dark and light fringes closest to the edge) is smaller when the voltage is higher at fixed defocus. Since the visibility and separation of these fringes is often used as a criterion, if not for resolving power, at least for the performance of an instrument, including the HVEM, the defocus dependence of their width determined by the relation $x = \sqrt{\lambda \Delta f}$ given by Reisner (30) is illustrated in Figure 18 for accelerating voltages between 50 kV and 3 MV.

The second difference is in the direct visibility of Fresnel fringes on the microscope screen. As already mentioned, the separation of these fringes by direct viewing through binoculars even at high magnification is very ineffective mainly because of reduced contrast in the screen phosphor at 1 MV and also because of cross-talk between pixels due to beam broadening and diffusion of light. To reach satisfactory astigmatism correction, it is indispensable

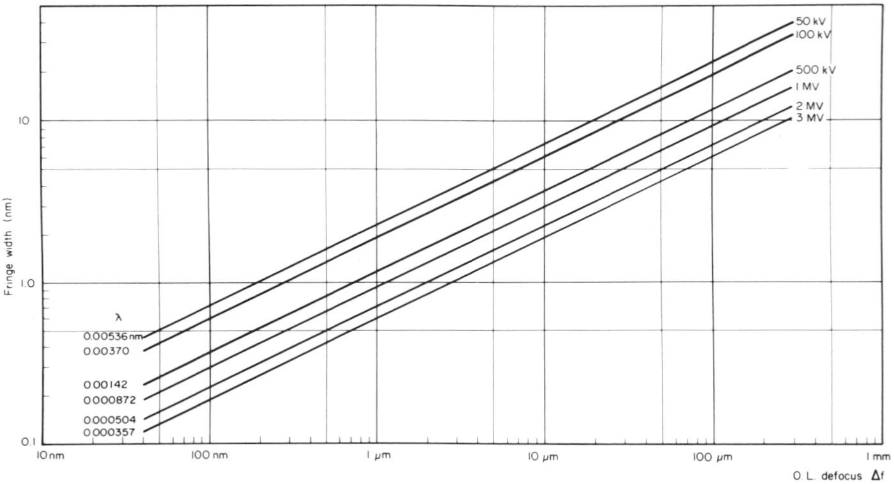

Figure 18 Width of Fresnel fringe versus defocus δf of the objective lens (O.L.) at several accelerating voltages.

to rely ultimately on a photographic exposure in which the fringe visibility remains clear. Correcting astigmatism in a high-voltage electron microscope is thus routinely a lengthy procedure by trial and error. It can be accelerated substantially by quantitatively determining the astigmatism parameters on the optical diffraction pattern of an amorphous carbon film as described by Krivanek et al. (31).

For the experimental circumstances that prevailed in this study, astigmatism was nevertheless corrected by the Fresnel fringe method. The degree of correction and the quality of instrumental performance thus realized with all apertures are illustrated by the set of hole micrographs in Figure 19, all of which were taken without anticontamination protection. These micrographs display a common level of astigmatism correction indicated by a 0.5 nm Fresnel fringe. Optimum resolution is obtained with an acceptance angle around 3 mrad as in the Au-island case shown in Figure 15. It appears very clearly from the images of Figure 19 that the contrast decreases as the aperture increases. The numbers on each frame represent the average optical densities of the carbon film and of the hole measured on the negative. The usual expression for contrast, $C = \Delta I/\bar{I}$, applied to these images leads to the aperture-dependent variation of contrast illustrated in Figure 20.

In connection with the requirement of optimum astigmatism correction, it is opportune to illustrate the sensitivity of the stigmator setting. Figure 21 shows that once a stigmator setting is obtained for good correction as in (*d*), very small current changes in the stigmator coils induce the appearance and substantial rotation of the two orthogonal directions (overfocus and underfocus) characteristic of astigmatism. Resolution and image quality as reflected in particular by grain isotropy in the carbon film are accordingly deteriorated.

The optimization of all critical parameters discussed so far is meaningful for practical applicability only to the extent that high-quality micrographs can be produced reliably and fairly routinely. An example of such performance is usually the type of test micrographs shown in Figure 22, in which the imaging of a standard lattice periodicity (here 0.31 nm in Sn/SnTe) is taken to reflect a corresponding resolution capability. It will be noticed that this periodicity appears at both 0.5 and 1.0 MV with comparable sharpness. No rigorous conclusions can be drawn from this comparison between performances at different voltages because these lattice portions came from different specimen areas and imaging conditions were not rigorously controlled. But since the resolving power δ_0 of a given instrument deteriorates as its operating voltage is reduced, as does the "resolution parameter" $C_s^{1/4}$ $\lambda^{3/4}$ of magnetic lenses described by Cosslett (32), the implication is that at the maximum operating voltage of this installation a lattice periodicity of 0.26–0.27 nm should be clearly visible.

More direct and significant evidence of high-resolution performance was obtained with an amorphous, thin carbon film shown in Figure 23. Point resolution as well as fringe width can be clearly observed and measured in the micrographs of this fine focal series taken with axial illumination and without anticontamination protection at instrumental magnification 100,000×. The areas shown are from different portions of a small hole. The micrograph contrast, as presented, is the same as in the original negatives. For easier identification and appreciation of Fresnel fringe profiles and point separations, several microdensitometer tracings are displayed with each micrograph. Although phase-contrast effects are reduced at higher voltages, the direct interpretation of micrographs still may not be sufficient at this resolution level, and a detailed analysis of the contrast transfer function may be necessary to identify the role, if any, of phase oscillations. Within such limitation in interpretation, nevertheless, the best evidence in this figure displays a point separation of 0.3–0.32 nm and a Fresnel fringe width of 0.27–0.3 nm.

The precautions and parameters analyzed above are of fundamental importance in obtaining optimum resolution with specimens of vanishing thickness. Although only a relatively small fraction of biological applications falls in this category, their relevance nevertheless extends to most of biological electron microscopy. To illustrate the merits and applicability of HVEM even in standard cases, Figure 24 compares the imaging of a thin section of guppy muscle at voltages in the conventional and megavolt ranges. The preparation of the 90 nm section was made routinely: fixed with glutaraldehyde, dehydrated, embedded in epon, stained en bloc with uranyl acetate at 60°C, and poststained with lead citrate for 20 minutes at room temperature. Micrographs were obtained at low voltage (100 kV) with JEM-100B and at high voltage (1 MV) with JEM-1000 under identical circum-

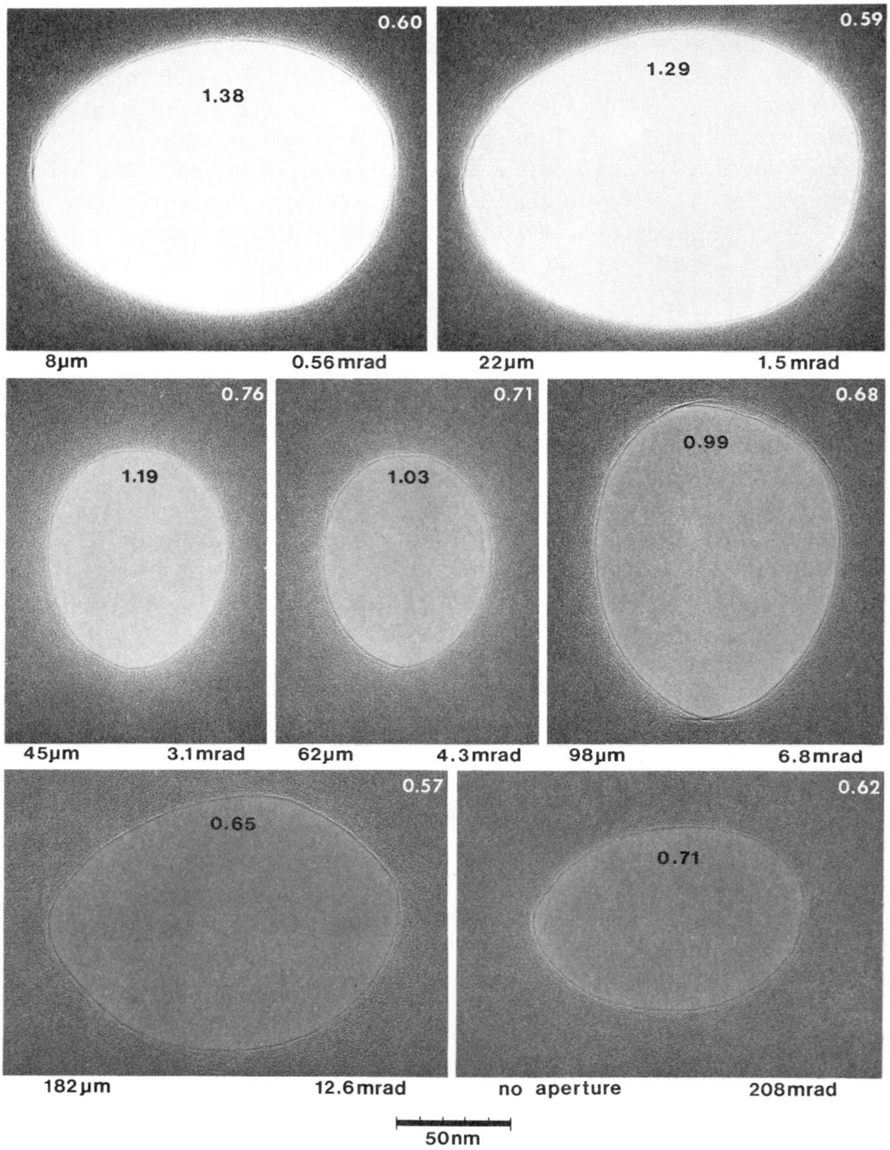

Figure 19 Test micrographs of holey carbon film illustrating the image characteristics and the quality of instrumental performance without anticontamination protection over the entire range of apertures used in this study with the JEM-1000 at 1 MV. The level of astigmatism correction with each aperture is indicated by an overfocus Fresnel fringe about 0.5 nm wide. The resolution discernible in these images displays an optimum in the vicinity of an acceptance angle $\alpha = 3$ mrad as predicted by theory, but deteriorates much less than expected at large angles. Decreasing contrast accom-

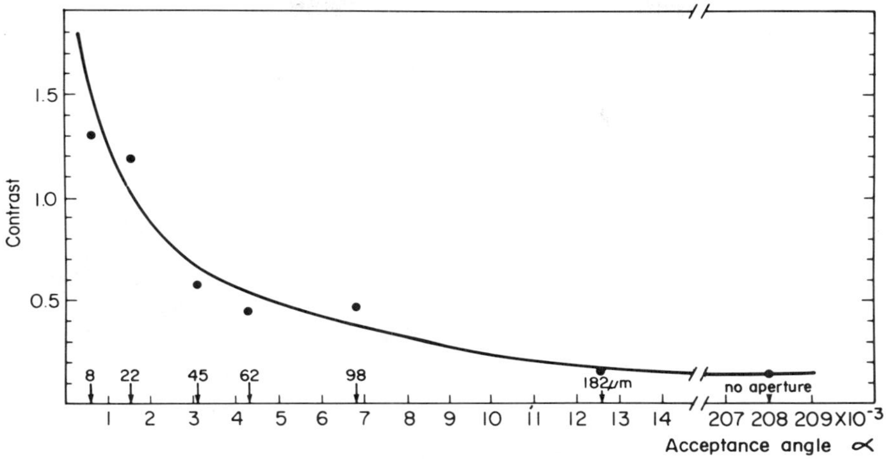

Figure 20 Variation of contrast with acceptance angle α (i.e., aperture) measured on the holey film test micrographs of Figure 19.

stances (focal series, optimum aperture, same film Kodak 4489, same overall optical density). The voltage-dependent difference in contrast is clearly visible (top row) since both micrographs were printed on the *same paper* (Polycontrast with filter 2.5). Because of this appearance, one may be misled and miss the information content in the high-voltage micrograph. However, when the printing is adjusted (bottom row) for *comparable contrast* (Polycontrast with filter 2.5 for low voltage and filter 4.0 for high voltage), the sharpness, that is, the resolution, of the high-voltage micrograph becomes clearly evident.

5 RESOLUTION IN THICK SPECIMENS

It was mentioned above that one of the main benefits of increased electron energy is the penetration, hence the inspection, of thicker specimens. The energy of imaging electrons determines the maximum specimen thickness to be used in the sense discussed above. In practice, however, it is the structure complexity in a three-dimensional specimen that imposes an upper limit to the usable thickness because of problems arising from unraveling the superposition of three-dimensional information projected onto two-

---

panies increasing aperture. Numbers represent the average optical density of the hole and of the surrounding carbon film away from the hole measured on each negative. Aperture diameter (*left*) and corresponding acceptance angle α (*right*) given under each frame. Negatives and prints processed under identical conditions. Magnification: 300,000×.

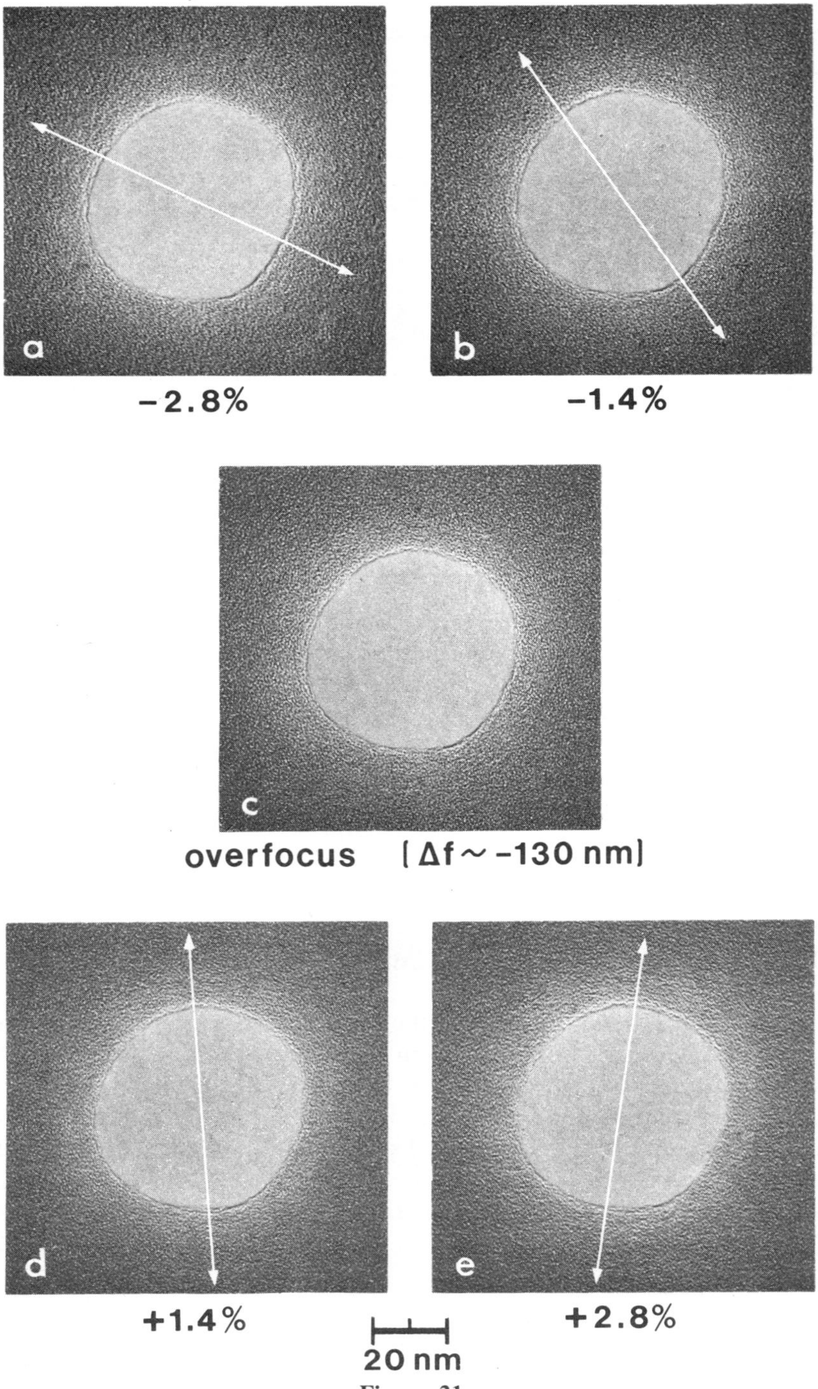

Figure 21

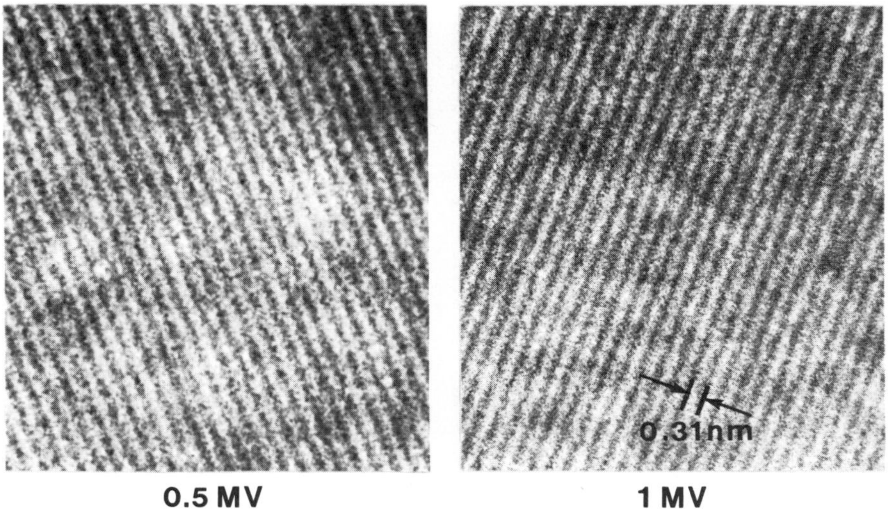

Figure 22 High-resolution performance test displaying line separation of 0.31 nm from Sn/SnTe lattice taken at 0.5 and 1.0 MV with the high-voltage electron microscope JEM-1000 of the University of Colorado, Boulder. Magnification: 176,000×.

dimensional micrographs. Some assistance in disentangling this complexity is provided by stereoscopic imaging and viewing, but such concern is outside the scope of this chapter. Furthermore, the determination of resolution as described here is restricted to the simplifying assumption of only one information source—the Au islands—within a specimen of a given thickness; the superposition of structures is not included. The requirements and imaging conditions with thick specimens are accordingly different in certain respects from those prevailing with thin specimens.

5.1 Contrast Mechanism

No phase contrast occurs with specimens that are considered thick in biological applications, ranging from about 0.2 μm to perhaps several micrometers. One is left only with amplitude contrast produced by absorption of the incident beam through scattering outside the objective aperture. Such absorption, if adequately monitored, can be used to determine the unknown thickness of a given specimen. To give an approximate idea of this absorp-

---

Figure 21 Sensitivity of stigmator setting at optimum astigmatism correction (*c*). Percentages indicate current change from optimum setting. The rotation of overfocus direction (marked) is about 72° over the entire range of current change shown (*a*)–(*e*). All micrographs taken at fixed defocus ($\Delta f \approx -130$ nm). Magnification: 500,000×.

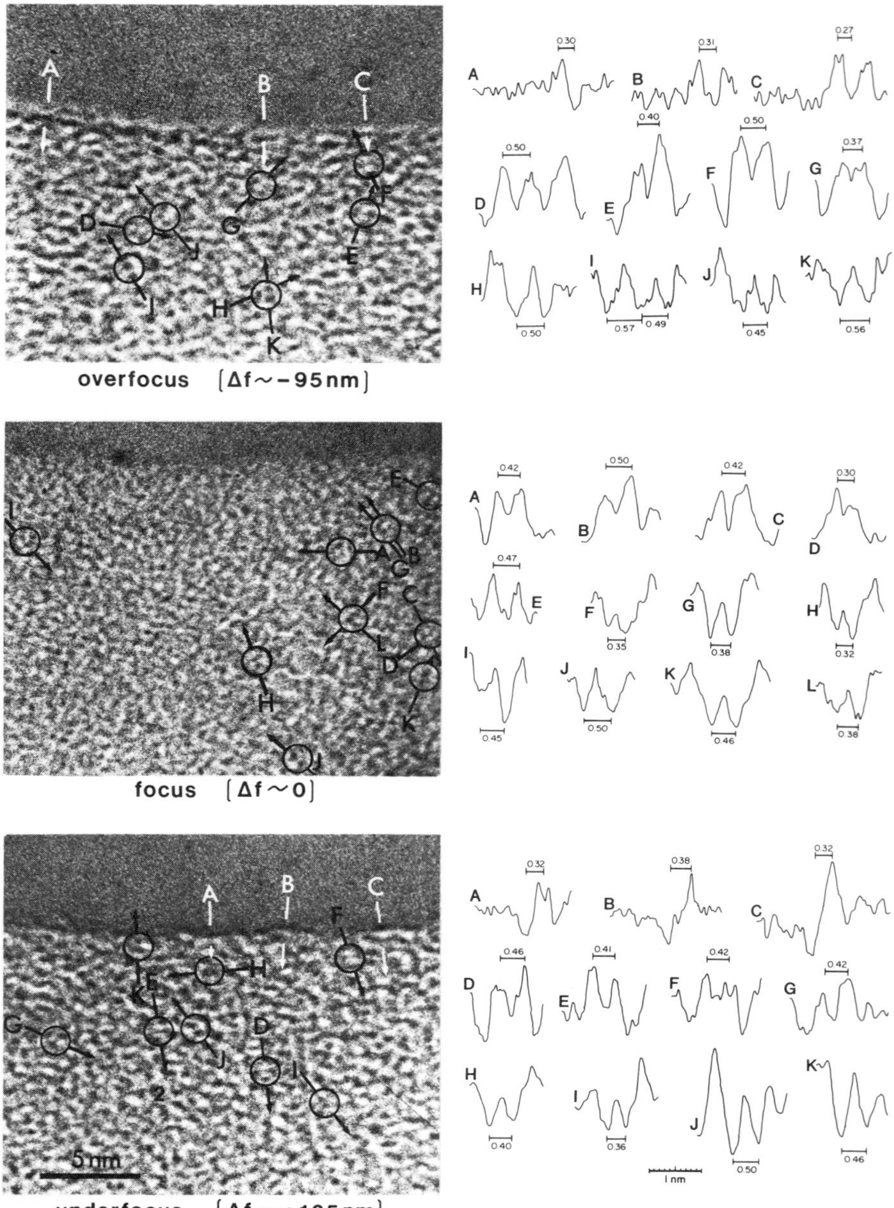

Figure 23 *Left:* portions of a small hole in an amorphous carbon film in fine focal series at very high magnification (2,600,000×) illustrating point resolution at 1 MV at the HVEM installation at the University of Colorado, Boulder. The print contrast has been reversed to appear as in the original micrographs. *Right:* typical microdensitometer tracings across Fresnel fringes and pairs of points identified by circles. Scanning is taken along the marked line in the direction of the arrow and away from the identifying letter. Numbers represent spacings in nanometers.

tion, Figure 25 plots the transmitted intensity of the electron beam reaching the image plane through epon slabs of increasing thickness and measured with the isolated screen as a Faraday cup as a function of objective aperture size. It should be pointed out that the uneven spacing of these curves, although corresponding to epon slabs cut nominally as 1, 2, 3, and 4 μm thick, is due to slightly different actual thicknesses, which have not been determined experimentally. Since beam absorption is exponential (i.e., of the form $I = I_0 e^{-\mu t}$), a very approximate value of the absorption coefficient μ for 1 MV electrons in epon derived from these curves is found to be in the range of 2.0–0.5μm^{-1} for acceptance angles α in the range 0.5–15 mrad, respectively. Transmission coefficients, and thus partial cross sections, for electrons in this energy range were discussed recently by Martinez et al. (33).

Since a given electron dose is needed to obtain a certain optical density in the photographic emulsion, these absorption curves also give a measure of the increase in exposure time or aperture size or both, within the requirements of other factors determining the image characteristics, that should be adopted with a specimen of a given thickness for optimum micrograph quality.

5.2 Aperture Dependence

In discussing the dependence of resolution on acceptance angle α in thin specimens, it is usually assumed that each scattered electron undergoes a single collision before passing through the aperture. The experimental evidence described above shows to what extent the resolution is deteriorated by the inelastic component, by diffraction, and by spherical aberration.

In thicker specimens the electron undergoes more than one collision, whether elastic or inelastic; thus the mixture of elastic and inelastic electrons accepted for imaging through the aperture varies in composition.

A reduction in resolution by chromatic effects takes place. From the familiar expression for the corresponding circle of least confusion, $r_c = C_c \alpha \Delta V/V$, where C_c is the chromatic aberration coefficient, it is readily apparent that the thicker the specimen, the higher the energy loss ΔE, or equivalently ΔV, through plural and multiple scattering; therefore, to improve resolution, it is necessary to reduce the acceptance angle α (aperture size) as much as possible. Such reduction, however, is accompanied by two unfavorable corollaries, namely, the increased predominance of the forward inelastic component and of the diffraction limitation, and the sharp reduction of beam intensities, leading to increasingly long exposures and thus possibly to increased radiation-induced specimen damage. At larger acceptance angles α, the energy loss of inelastic electrons increases because the trajectory inside the specimen is longer, hence the image displays less contrast and resolution. The net effect of aperture size on image quality with a thick specimen at 1 MV is shown in the sequence of micrographs assembled in Figure 26.

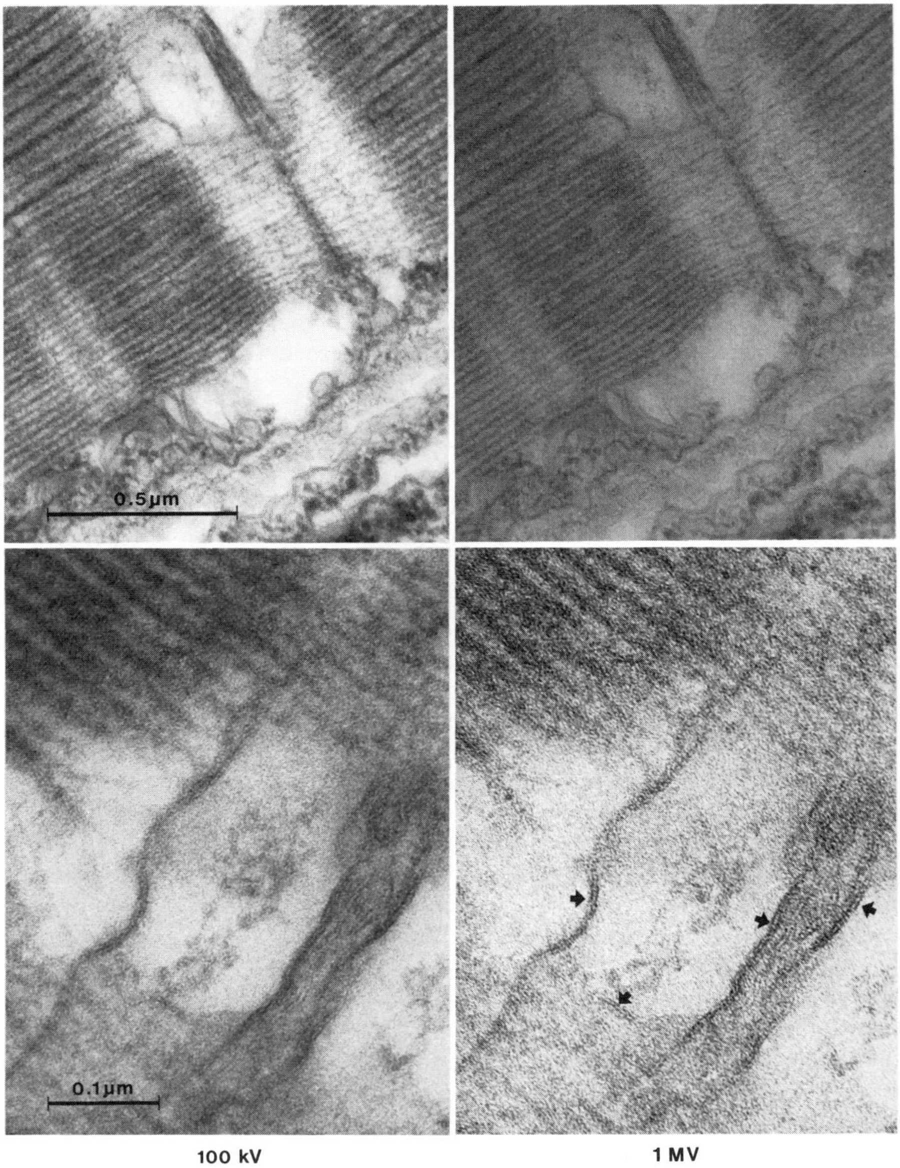

Figure 24 The change with accelerating voltage in micrograph characteristics of thin biological specimens. Best focus was obtained of the same guppy muscle section (~ 90 nm thick) by fine focal series with optimum aperture at each voltage. Exposures were adjusted for matching overall optical densities. When printed on the *same paper* (Polycontrast, filter 2.5) for comparable appearance (*top*), many details do not appear as clearly as in the negatives, and the reduced contrast at high voltage makes for a less striking impression than at lower voltage. However, when printing is

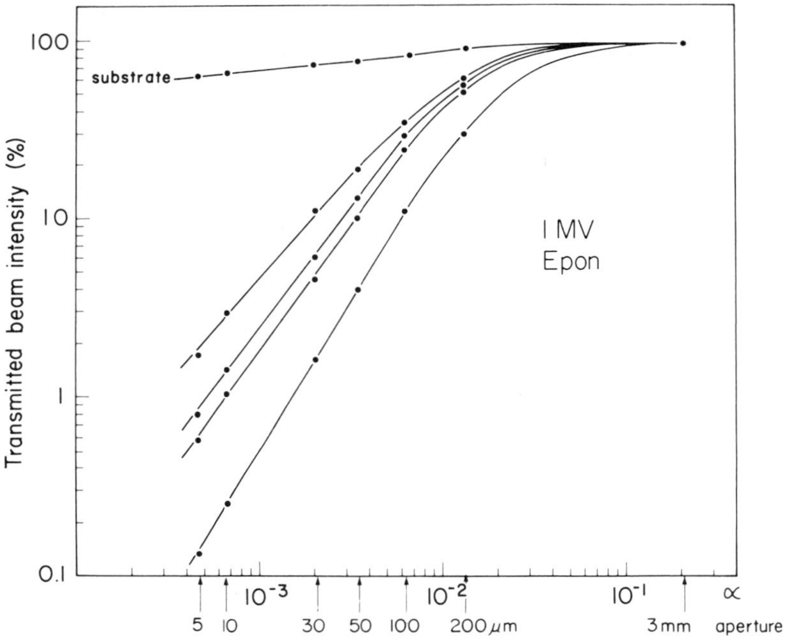

Figure 25 Approximate variation of transmitted beam intensity with collection half-angle α measured in epon slabs of nominal thickness 1, 2, 3, and 4 μm at 1 MV.

Several features in Figure 26 deserve special attention. It can be seen that the *same area* of the Au substrate superimposed on an epon slab of nominal thickness about 4 μm was imaged in both configurations (upstream and downstream) by focal series (6–8 exposures per aperture). Despite such extensive irradiation, the exposure times required to match the optical density of all negatives with an almost constant beam follow the aperture size in almost inverse proportionality in both orientations, indicating constant specimen integrity throughout the series. The contrast shows indeed an optimum around 50 μm aperture size (or 3.5 mrad for α) with Au substrate in either orientation, but it is more prominent in the downstream case. Furthermore, the order in which these micrographs were taken was such (with apertures in the sequence: downstream—5, 100, 200, 50, 15, 70 μm; upstream—70, 50, 15, 5, 100, 200 μm) that, if there had been any significant specimen alteration at any point during this series of exposures, an anomaly

---

adjusted (*bottom*) for *comparable contrast* (filter 2.5 for low voltage, filter 4.0 for high-voltage print), the overall image crispness at high voltage becomes clearly visible, and certain details such as actin and myosin filaments or membrane leaflets appear better separated (arrows). (Specimen courtesy of K. R. Porter, University of Colorado, Boulder.) Magnification: (*top*): 50,925×; (*bottom*): 145,000×.

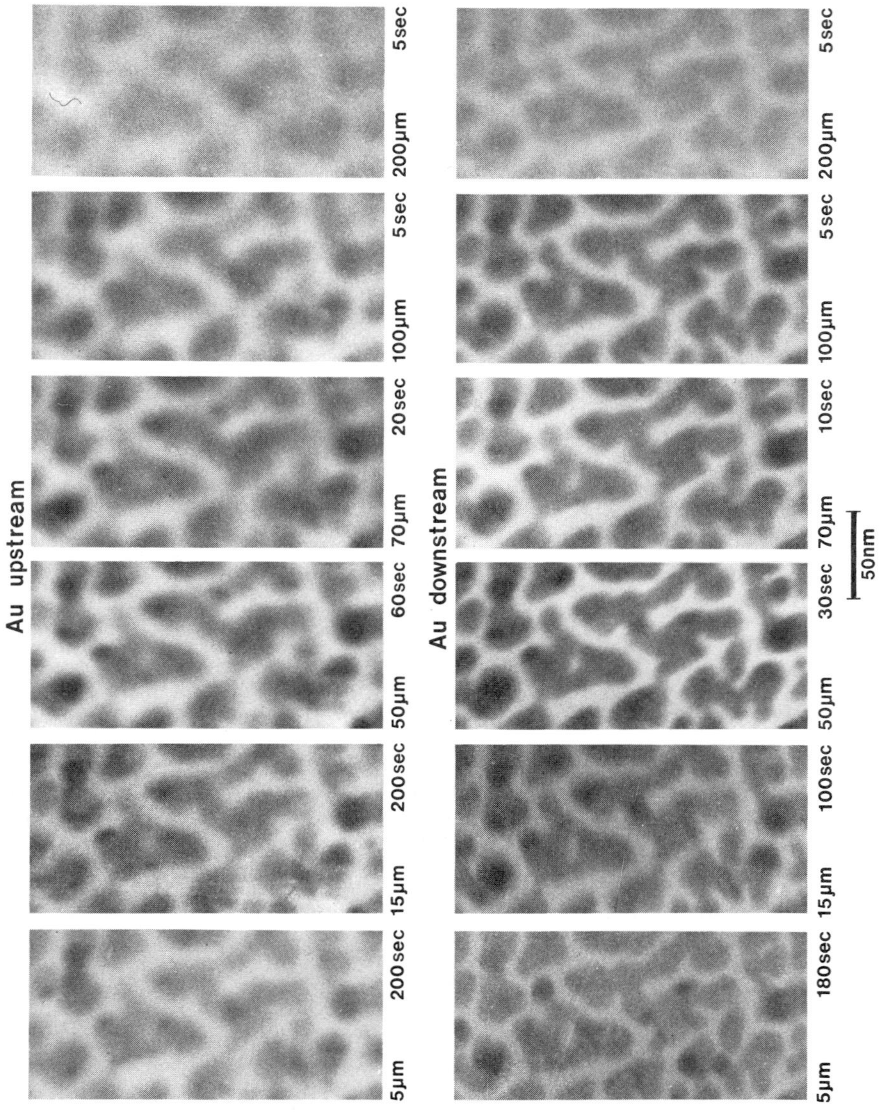
Au upstream
200μm 5sec
100μm 5sec
70μm 20sec
50μm 60sec
15μm 200sec
5μm 200sec
Au downstream
200μm 5sec
100μm 5sec
70μm 10sec
50μm 30sec
15μm 100sec
5μm 180sec
50nm

Figure 26

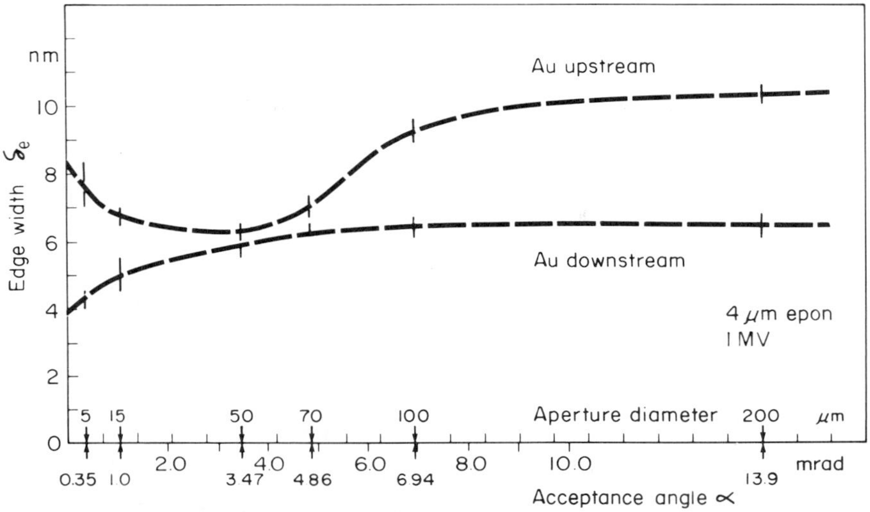

Figure 27 Variation of edge width δ_e with aperture and orientation in the beam at 1 MV measured on the sequence of thick-specimen micrographs shown in Figure 26.

would have appeared in some of the images. This, however, is not the case, and an overall internal consistency is apparent with respect to aperture size and orientation in the beam (all bottom images better than the top counterparts), as discussed below. It should also be noted that most of these exposure times, though commensurate with the posted aperture size and specimen thickness, are unusually long so that the corresponding micrographs thus reflect the high electrical and mechanical stability of the installation.

The qualitative illustration of the influence of aperture size on resolution in thick specimens presented above is confirmed by the corresponding edge-width measurements collected in Figure 27. Resolution in downstream

---

Figure 26 Sequence of micrographs of the same object area illustrating simultaneously the dependence of resolution on aperture (horizontally) and on orientation in the beam (vertically) in imaging thick specimens at 1 MV. Au-Island substrate superimposed on a thick epon slab (nominal thickness ~ 4 μm). All micrographs obtained with the JEM-1000 at 1 MV at optimum focus by focal series. Contrast increases with aperture diameter to a maximum near 50 μm and decreases again at large angles. The difference in resolution between upstream (*top row*) and downstream (*bottom row*) orientation of the Au substrate in the beam (top-bottom effect) is clearly visible with all apertures. Under each micrograph are indicated the nominal aperture diameter (*left*) and the exposure time (*right*) required for matching a common overall optical density with an approximately constant beam intensity. Note both the high instrumental stability (electrical and mechanical) and the absence of beam-induced distortions in spite of multiple and unusually long exposures. Magnifications: instrumental: 59,000×; illustration: 236,000×.

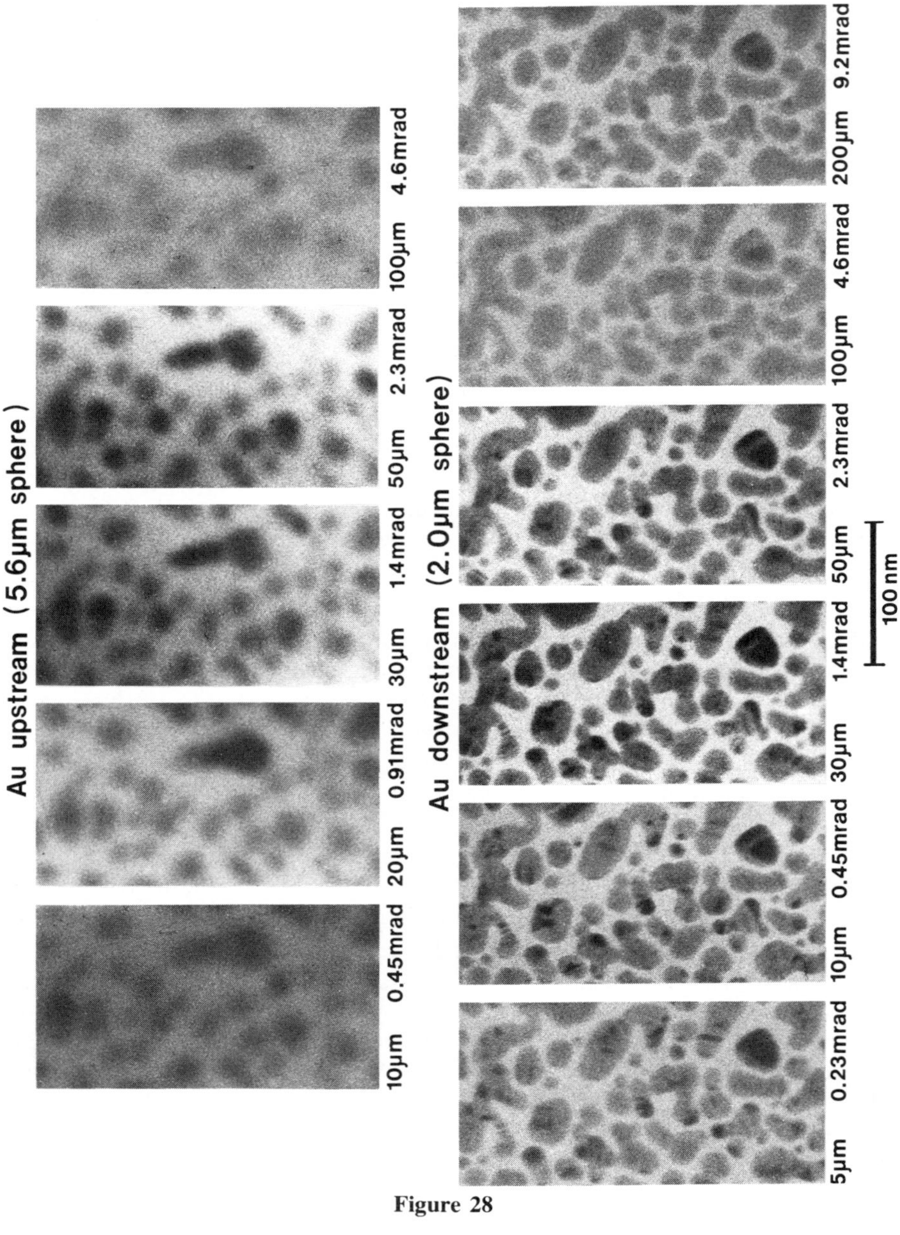
Au upstream (5.6µm sphere)
10µm 0.45mrad
20µm 0.91mrad
30µm 1.4mrad
50µm 2.3mrad
100µm 4.6mrad
Au downstream (2.0µm sphere)
5µm 0.23mrad
10µm 0.45mrad
30µm 1.4mrad
50µm 2.3mrad
100µm 4.6mrad
200µm 9.2mrad
100 nm

Figure 28

orientation is consistently better than in each upstream counterpart, although the difference appears less pronounced with acceptance angles of about 3.5 mrad, that is, with apertures about 50 μm in diameter. The constancy of δ_e beyond a certain angle (about 7 mrad in downstream orientation and 10 mrad in upstream orientation) would also indicate, as in the case of thin specimens illustrated in Figure 14, that the entire beam is accepted within this angle. The steady deterioration in resolution with increasingly larger apertures in downstream orientation clearly reflects the progressively longer trajectory of electrons through the specimen.

A similar pattern of contrast and resolution in thick specimens is also displayed at even higher voltages. Figure 28 presents a sequence of micrographs illustrating both aspects of this aperture dependence at 2.5 MV. Two different specimen areas—the centers of latex spheres of different size (top, 5.6 μm; bottom, 2.0 μm)—are imaged in this aperture-dependent sequence. Although a comparison as above is not quite possible because the two orientations apply to specimens differing in thickness, it is nevertheless clearly apparent that the contrast is highest with apertures around 30–50 μm (i.e., at acceptance angles α = 1.4–2.3 mrad) and decreases when small or large apertures are used. This reduction in acceptance angle for optimum contrast in thick specimens is the expected consequence of all scattering directed increasingly in the forward direction at higher energy. It cannot be concluded readily whether, as it appears, resolution is indeed almost independent of aperture. The radiation damage accumulated during long exposures is limited to only modest shrinkage, as briefly discussed below.

5.3 Depth of Field

The question of depth of field is usually of no consequence when viewing or imaging thin sections because it is implicitly fulfilled under normal operating circumstances. As specimens of increased thickness are routinely used in high-voltage electron microscopy, it is imperative to ascertain that the adapted aperture is compatible with both specimen thickness and desired resolution level. The depth of field D is related to the resolving power δ (or

---

Figure 28 Sequence of micrographs illustrating the aperture dependence of resolution and contrast in thick specimens with Au substrate in either orientation at 2.5 MV. Centers of latex spheres of different size were imaged in each configuration (*top*: 5.6 μm, *bottom:* 2.0 μm). As is also the case at lower voltage, contrast increases with aperture diameter to a maximum around 30–50 μm and decreases at larger values. All micrographs at optimum focus by focal series. Radiation damage accumulated during several dozen long exposures is limited to small shrinkage (7.5% in large sphere, 1.7% in small sphere). Dimensions under each frame indicate aperture diameter and corresponding acceptance angle α. HVEM at Laboratoire d'Optique Electronique, Toulouse, France, at 2.5 MV. Magnification: 198,000×.

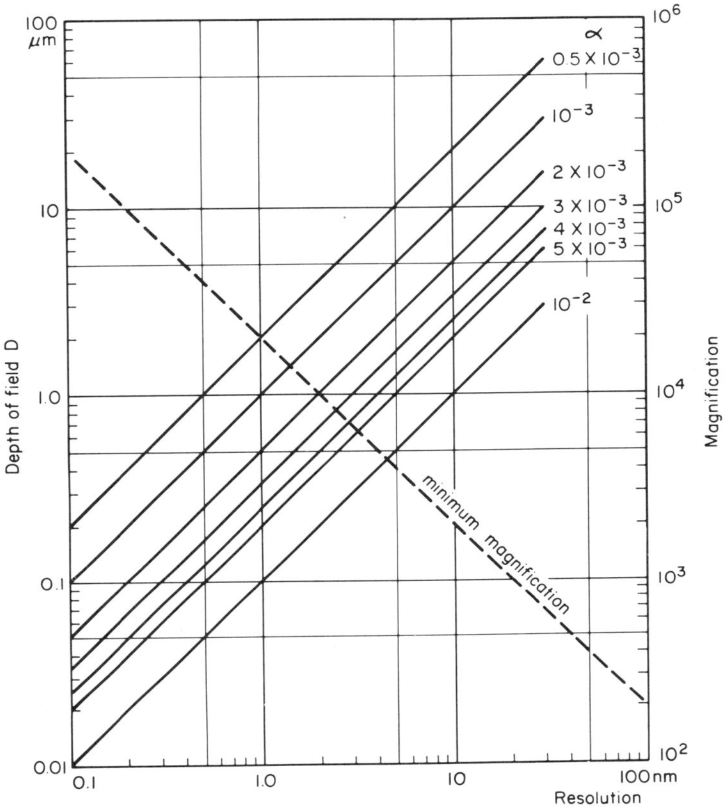

Figure 29 Relationship between depth of field D (*left scale*) and resolution for several values of the scattering angle α. The minimum magnification required by a resolving power of the photographic emulsion of about 50 lines/mm is also given (*right scale*).

resolution level) and to the acceptance angle α in a simple way: $D = \delta/\alpha$. This relationship is translated in graphic form in Figure 29. At high energies electron scattering occurs increasingly in the forward direction (i.e., within smaller α) so that the depth of field imaged at a given resolution increases. Since the thickness giving rise to a particular chromatic aberration increases with energy more rapidly than expected, the values obtained from this relation are readily modified by chromatic aberrations so that in practice a specimen of thickness higher than anticipated can yield a particular resolution. The experimental results presented in this section thus give an idea of the overall resolution that can be expected in practice with a particular aperture and from a particular specimen thickness.

5.4 Orientation in the Beam

As a specimen increases in thickness, scattering due to its various components introduces an overall lateral spread to the electron beam that leads to deterioration in resolution. The visibility of any given specimen component depends on the difference in scattering between it an the surrounding matrix. Furthermore, the lateral spread depends on the atomic number of each scattering center. It can thus be seen that, in a mixture of light and heavy components, a difference in resolution accompanies their relative orientation in the beam.

Such difference in resolution was advanced by Hashimoto (34) in connection with certain effects in the imaging of crystalline specimens by transmission electron microscopy and is known as the "top-bottom" effect. In amorphous specimens this difference simply comes from the order in which scattering occurs in each of the components, that is, the order in which chromatic effects are compounded. Best resolution occurs when the heaviest structure element scatters last. In the specimen configuration used in this study, this situation corresponds to the Au deposit oriented downstream in the beam (bottom). It is easy to see how the top-bottom effect is less pronounced in a given specimen at higher electron energies and more pronounced in thicker specimens at a given accelerating voltage.

Experimentally, this effect with the Au-island configuration was shown by Fotino (35) in a 4 μm epon slice by HVEM at 1 MV. A more complete illustration is clearly visible in Figure 26. Because it is restricted to the same specimen, the sequence of Figure 26 also unambiguously illustrates that the top-bottom effect occurs over the entire range of acceptance angles. To judge from the micrographs of Figure 26 and from the corresponding curves of Figure 27, the top-bottom effect appears to be smallest with an acceptance angle with which contrast is highest.

The simultaneous impact of both accelerating voltage and orientation in the beam on resolution in thick specimens is illustrated in Figure 30. Under the prevailing time limitation and other delicate experimental conditions, this is the closest that an assemblage of this sort could come to illustrating the same area of the same specimen in both orientations and at two different voltages. At 1 MV it was possible to image the same latex sphere area, as can be easily recognized, in both top and bottom orientations. As matching counterparts at 2.5 MV, other comparable spheres had to be selected, but with the discrepancy in size in a direction that, if normalized, would only make the comparison more striking and convincing. It can be readily concluded from Figure 30 that "top" resolution is improving with accelerating voltage faster than "bottom" resolution, so that with a given specimen thickness the top-bottom effect appears less strikingly at higher voltage. More systematic illustrations of the thickness dependence at 1 and 2.5 MV of the top-bottom effect in latex spheres at fixed aperture can be seen in Figure 32.

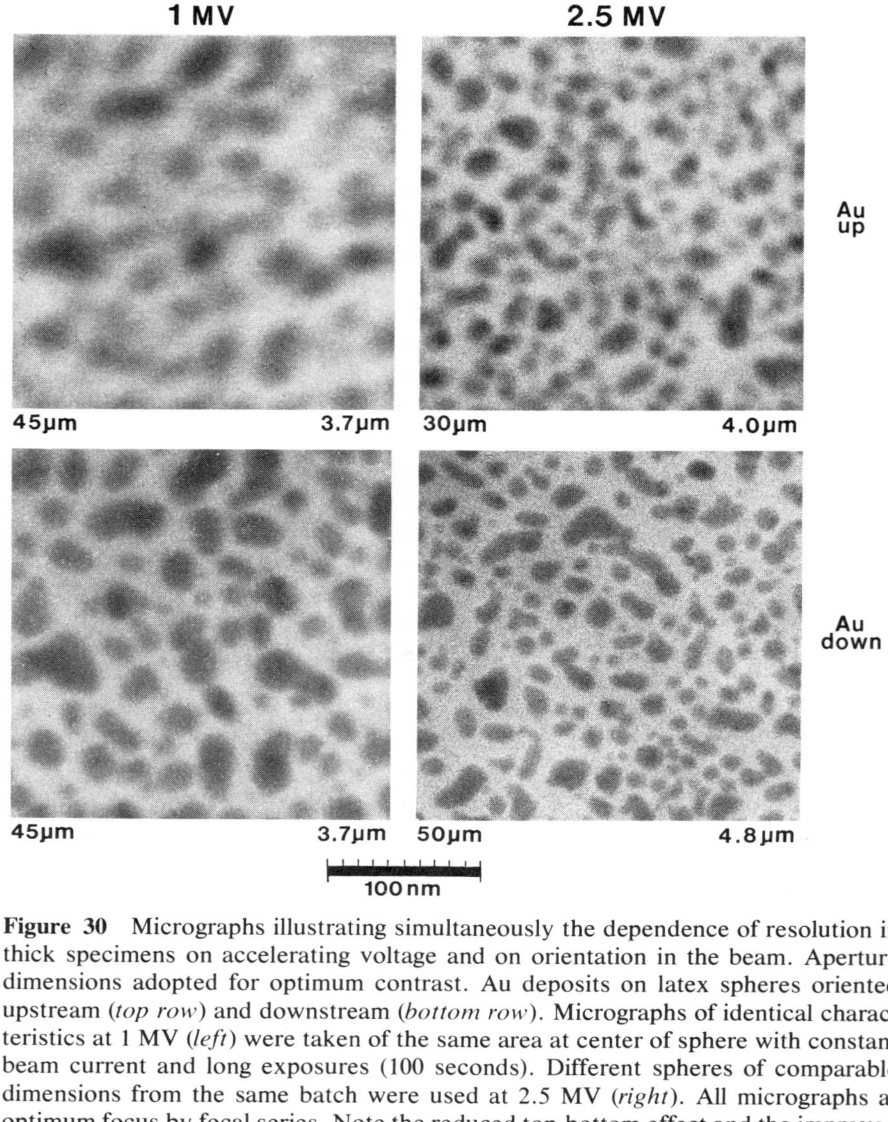

Figure 30 Micrographs illustrating simultaneously the dependence of resolution in thick specimens on accelerating voltage and on orientation in the beam. Aperture dimensions adopted for optimum contrast. Au deposits on latex spheres oriented upstream (*top row*) and downstream (*bottom row*). Micrographs of identical characteristics at 1 MV (*left*) were taken of the same area at center of sphere with constant beam current and long exposures (100 seconds). Different spheres of comparable dimensions from the same batch were used at 2.5 MV (*right*). All micrographs at optimum focus by focal series. Note the reduced top-bottom effect and the improved resolution in either orientation at the higher voltage in spite of larger object thickness. Dimensions under each frame indicate aperture diameter (*left*) and sphere diameter (*right*) assumed to represent object thickness. HVEM installations at University of Colorado, Boulder (1 MV), and at Laboratoire d'Optique Electronique, Toulouse, France (2.5 MV). Magnification: 200,000×.

The top-bottom effect occurs also in STEM. In this mode resolution is determined by the dimension of the electron probe and by multiple scattering in the specimen, but it appears reversed, as analyzed theoretically by Reimer et al. (36) and tested experimentally by Gentsch et al. (37), because the structure of interest is scanned by the smallest beam when oriented upstream and by a broadened beam when oriented downstream. Since there is no lens downstream from the specimen, chromatic effects are limited to those produced by scattering in the specimen.

The practical conclusion from this discussion is that for best resolution with a thick specimen in HVEM, the structure of interest should be oriented downstream in the beam.

5.5 Thickness Dependence

It is of utmost interest to identify the extent of the gain in resolution associated with higher voltages when imaging thick specimens. Such gain may be formulated alternatively as (1) the maximum thickness usable to reach a prescribed resolution level or (2) the resolution that can be reached with a specimen of given thickness.

Earlier measurements were done by Fotino (35) in epon and embedded rat liver, as plotted in Figure 31. It is found that resolution undergoes approximately a fourfold improvement that accompanies a tenfold increase in energy (i.e., from about 4.6 nm/μm at 100 kV to about 1.1 nm/μm at 1 MV). The measurements were made under optimum imaging conditions with JEM-100B and JEM-1000 instruments, respectively.

At accelerating voltages even higher than 1 MV, a further improvement in resolution takes place, as could be judged from the micrograph sequences with latex spheres in both orientations shown in Figure 32 at 1 and 2.5 MV.

The improvement with accelerating voltage displayed in this sequence of

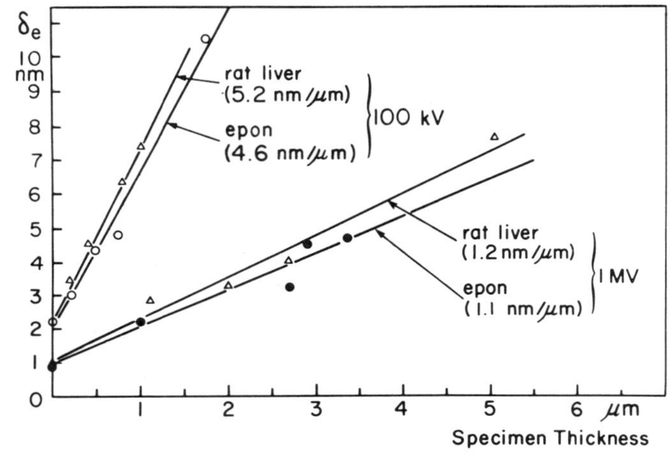

Figure 31 Dependence of edge width δ_e on specimen thickness at 100 kV and 1 MV.

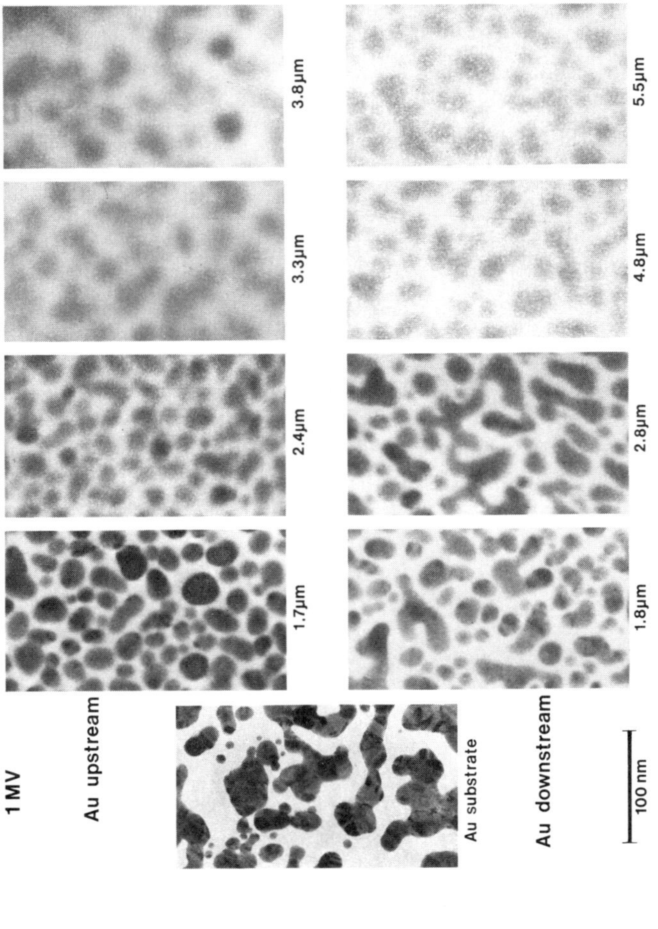
1 MV
Au upstream
1.7μm
2.4μm
3.3μm
3.8μm
Au substrate
Au downstream
1.8μm
2.8μm
4.8μm
5.5μm
100 nm

(*a*)

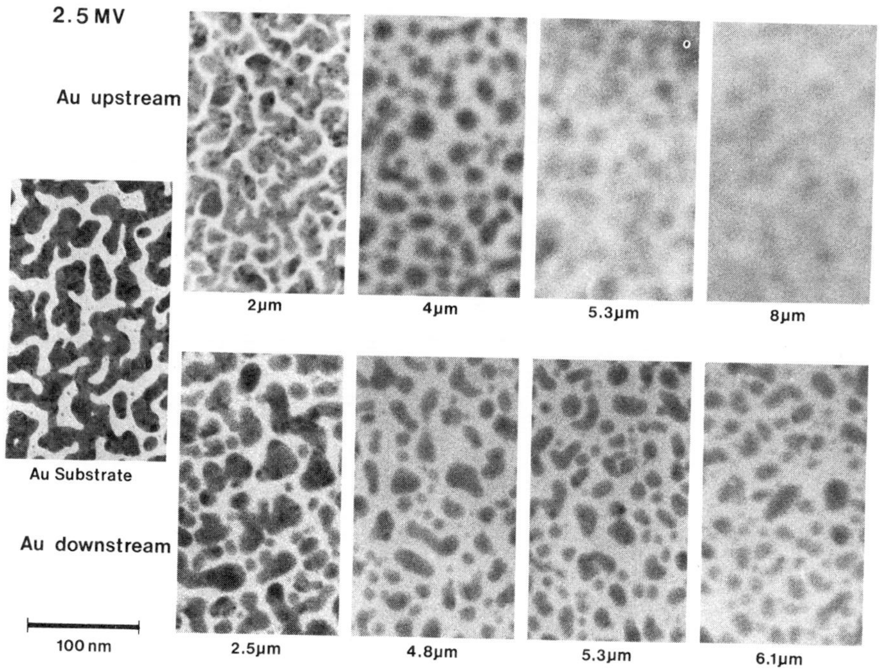

(b)

Figure 32 Sequences of micrographs illustrating the thickness dependence of resolution attainable in latex spheres of increasing diameter at 1 MV (*a*) and at 2.5 MV (*b*) with fixed aperture (~50 μm) and with Au islands oriented both upstream and downstream. Au substrate only is also illustrated for comparison. Micrographs taken with HVEM installations at the University of Colorado, Boulder (1 MV), and at the Laboratoire d'Optique Electronique, Toulouse, France (2.5 MV). Magnification: 150,000×.

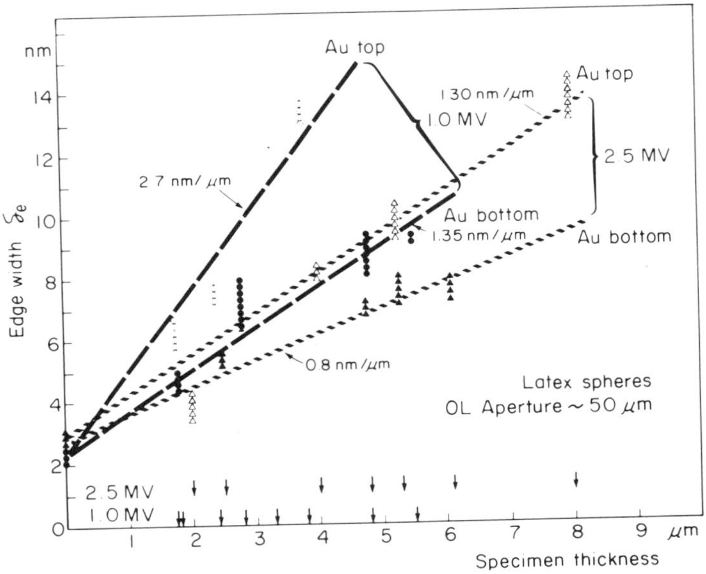

Figure 33 Variation of edge width δ_e with specimen thickness and accelerating voltage in latex spheres in either orientation resulting from measurements on the micrographs of Figure 32. Arrows at abscissa mark each specimen thickness used at 1.0 and 2.5 MV.

micrographs is confirmed by the measurements of δ_e shown in Figure 33. Although these data cover a somewhat limited sampling and display large uncertainties, the dependence of resolution on thickness appears rather clearly. It can thus be seen that an increase in accelerating voltage from 1 to 2.5 MV leads to an improvement in resolution rate in latex spheres that is higher in upstream orientation (factor ~ 2.1) than in downstream orientation (factor ~ 1.7). The spread in resolution between both orientations is approximately inversely proportional to the accelerating voltage over the entire thickness range. Since the micrographs illustrating this voltage dependence were taken at two different installations, a high-magnification check was made with the same object structure imaged with both instruments and showed them to agree within less than 2%.

The Au-island configuration used throughout this study has served the purpose of illustrating and studying, often quantitatively, many aspects concerning resolution in a consistent, reproducible, and normalizable way, while at the same time incorporating practical advantages such as ease of preparation, stability, good resistance to radiation damage, and good contrast in difficult experimental circumstances.

It is illuminating and fitting, nevertheless, to illustrate the characteristics of resolution in high-voltage electron microscopy with two examples of biological specimens. The first example may be considered to be representative

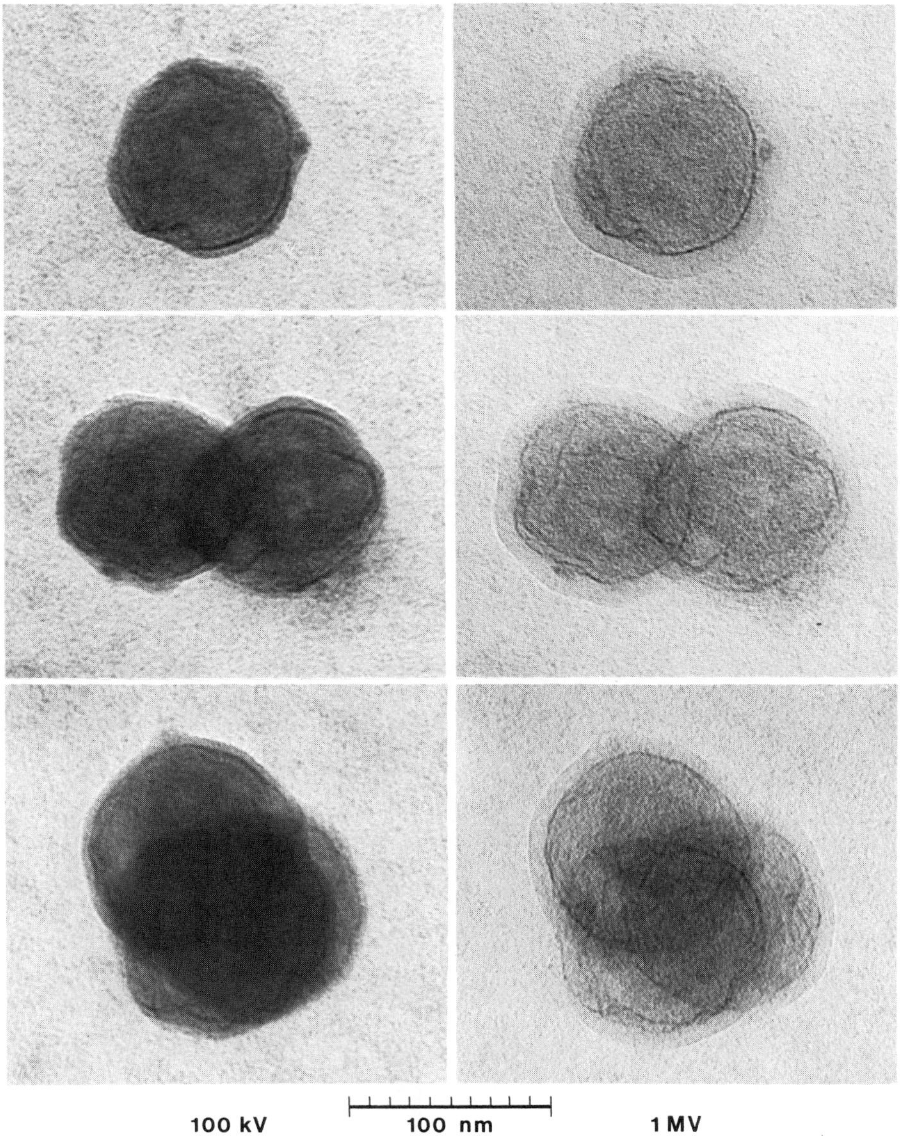

Figure 34 Example of simultaneous improvement in penetration and resolution attained by high-voltage transmission electron microscopy in a fairly thick and routine biological specimen under normal circumstances. Best focus images of whole-mount RNA murine tumor virus were taken at 100 kV and 1 MV by focal series with optimum aperture and matching exposures. Both sets of micrographs were printed on the same photographic paper (Polycontrast, filter 2.5) with matched overall densities. Instrumental magnification: 33,000×. (Specimen courtesy of E. de Harven, Sloan-Kettering Institute for Cancer Research, New York.)

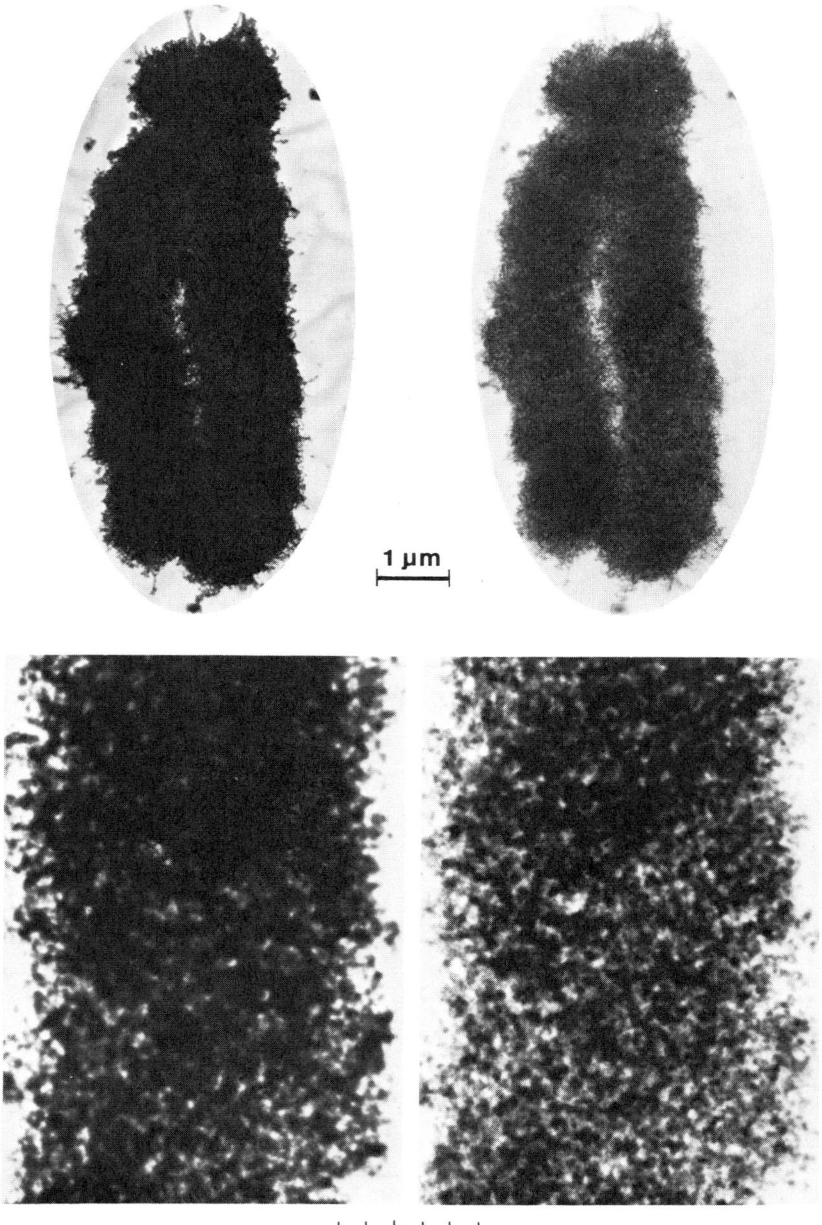

Figure 35 Comparison of the detail visibility accessible in a thick specimen by conventional and by high-voltage transmission electron microscopy. *Top:* normal exposures at 100 kV and 1 MV of a whole mount of *Vicia faba* metaphase chromosome expanded about 1.5 times by use of low-concentration Mg^{2+} and EDTA solutions and critical point dried. Magnification: 9,750×. *Bottom:* magnified portion

of a majority of biological applications of HVEM. Figure 34 shows qualitatively the benefits of improved penetration and resolution obtained at a high accelerating voltage in a fairly thick and delicate preparation in comparison with imaging by conventional electron microscopy. The same whole mounts of RNA murine tumor virus were imaged on the same film to matching exposures at low and high voltages by focal series with optimum aperture and printed under identical conditions to identical overall optical densities. The clarity and crispness at 1 MV immediately stand out. It can be unambiguously noted that rich detail can be seen in the three-virus cluster at 1 MV, whereas at 100 kV not much structural information is distinguishable even in a single virus.

The second example is meant to illustrate the difference in detail visibility in packed structures obtained by conventional and by high-voltage electron microscopy when seeking to circumvent the insufficient beam penetration in thick specimens. The *Vicia faba* metaphase chromosome shown in Figure 35 has been expanded by about 50% by means of low-concentration Mg^{2+} and EDTA solutions, followed by critical point drying, to render the fiber packing slightly more visible. The upper images at medium magnification were taken of the same preparation under identical conditions in all respects (in particular, the overall micrograph exposure) except for electron energy. At low voltage not much more than the chromosome outline is visible, whereas at high voltage an internal structure is clearly detectable in broad characteristics, although individual fibers remain undistinguishable at this magnification. The lack of penetration may be circumvented by longer exposures to make the dense areas more noticeable. This is what was done in the lower images at higher magnification with a chromatid segment. To reach a comparable improvement in overall optical density with optimum visibility, the exposure had to be increased about 23 times at 100 kV, but only four times at 1 MV. The visibility thus reached nevertheless remains different, as could be anticipated from the difference in electron scattering at both voltages, and more structural detail can be seen at 1 MV.

6 SPECIMEN DAMAGE

Whenever an object is imaged by electron microscopy, the energy transferred from the beam is likely to induce macroscopic alterations or structure modifications that are difficult to define and assess. This effect is particularly marked in amorphous and biological specimens; in crystalline materials the

---

of the right chromatid, overexposed to optimum visibility at its maximum thickness. At low voltage (*left*) the exposure has been increased 22.8 times; at high voltage (*right*) a comparable gain has required an increase of only four times, yet with finer detail and structure thus rendered visible. Magnification: 37,500×. (Specimen courtesy of P. S. Woods, Queens College, Flushing, NY.)

damage may be monitored by the changes and disappearance in time of diffraction spots, reflecting alterations in lattice integrity, as used by Kobayashi and Sakaoku (38) to determine its energy dependence. The many aspects of the mechanism of radiation damage include heating, mass loss, ionization, atomic or molecular excitation, breakup of chemical bonds, cross-linking, charging, and atomic displacement. They are all complicated, poorly known, and difficult to evaluate. But since the purpose of this study is not to analyze the effects of damage mechanisms in general or even in high-voltage electron microscopy, the reader is referred to thorough discussions of the subject in several fairly recent review papers, such as those by Stenn and Bahr (39), Glaeser (40), and Reimer (41).

What may be of interest here, however, are some facts and observations on specimen preservation or alteration that occurred during the experiments described in this chapter.

The most common observation was that the pattern and circumstances of occurrence of any beam-induced alteration varied over a wide range of specimen and beam characteristics. Larger or thicker specimens seemed to be more vulnerable, although there were many examples of just the opposite. In most cases the specimen could be stabilized or rendered more resistant by conditioning by low-intensity irradiation at low magnification for periods of tens of seconds to several minutes, and especially by slow changes in beam characteristics. As mentioned above, once a radiation-induced modification has been established and "frozen" in a rearranged form, the specimen often does not deteriorate further. Such, for instance, was the case of the 2 μm thick epon slice shown in Figure 36. The bubbles occurred in several seconds; but thereafter the unaltered, flat areas remained unchanged (arrow) and were actually used without further deterioration for long serial exposures.

The same occurred during exposures with latex spheres, the only difference being that because of isotropy of shape, some quantitative measure of alteration was readily identifiable in the microscope. As in the case of thick epon slices, alterations occur over a wide range of exposures and deformation. An example of massive specimen damage at 1 MV in a large latex sphere (5.2 μm diameter) is shown in Figure 37. Here the beam ($\sim$ 1 μA) was quickly condensed through crossover on the sphere, which instantly collapsed or disintegrated to the final shape with no change thereafter. At the other end of the spectrum, an example of insignificant specimen damage at 2.5 MV is shown in Figure 38. A comparison between the aspect of a relatively large sphere (2 μm diameter) before and after the irradiation reveals a shrinkage of only 1.7% in dimensions or 5% in volume in spite of the high dose (at least 100 C/cm^2) accumulated during more than 15 minutes of uninterrupted adjustments and production of the more than 50 high-magnification micrographs, which were needed for the sequence illustrating the aperture dependence in bottom orientation shown in Figure 28. A similar finding applies to the 4 μm thick epon slice with which all 91 micrographs were

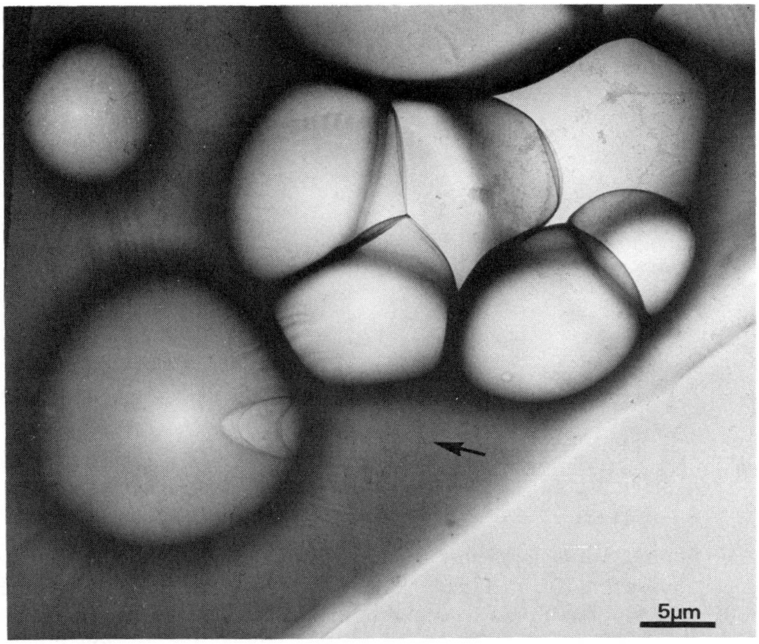

Figure 36 Segment of a 2 μm thick epon slice at low magnification (2000×) after moderate exposure in a fairly spread-out beam at 2.5 MV. Bubbles and craters developed after a few seconds, but many uniform areas were left unaltered (arrow), where high-magnification exposures were thereafter taken without further deterioration.

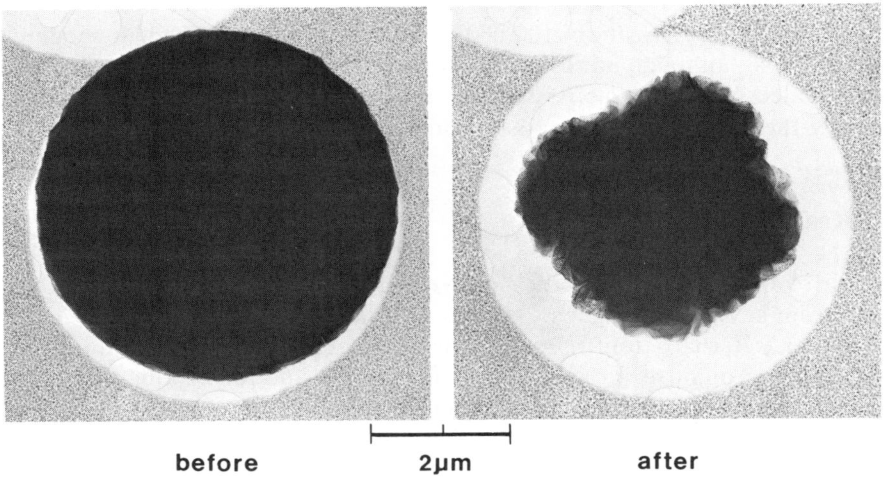

Figure 37 Example of massive specimen damage induced instantaneously in a large latex sphere (5.2 μm diameter) by rapidly condensing the 1 μA beam through cross-over. Micrographs taken with HVEM installation at University of Colorado, Boulder, at 1 MV. Magnification: 9,175×.

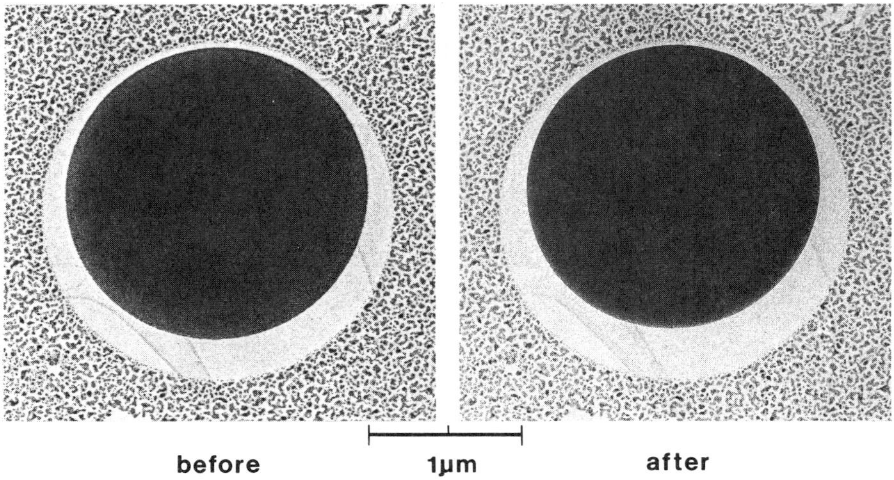

Figure 38 Example of insignificant specimen damage (1.7% shrinkage) in the relatively large latex sphere (2 μm diameter) that was used in the aperture dependence of resolution (bottom orientation) illustrated in Figure 28. The sphere remained almost unchanged in spite of prolonged irradiation after stabilization by slow conditioning and through minimization by gentle beam adjustments. Total dose accumulated during at least 15 minutes of continuous inspection, adjustments, and production of more than 50 high-magnification micrographs estimated to amount very approximately to at least 100 Cl/cm^2. Micrographs taken with HVEM installation at Laboratoire d'Optique Electronique, Toulouse, France, at 2.5 MV. Magnification: 20,250×.

taken at high magnification (100,000×) for the aperture-dependent sequence of Figure 26; although not illustrated here, the slice maintained its integrity, as revealed by the correctness of the top-bottom effect with each aperture and by the progression in exposure times.

ACKNOWLEDGMENTS

Much of the material presented above has been assembled with generous help from several colleagues.

I thank Bernard Jouffrey, director of the Laboratoire d'Optique Electronique in Toulouse, France, for his hospitality and for making time available on the 3 MV HVEM during a brief visit in 1978. My sincere appreciation goes to J. P. Martinez for interesting discussion and assistance, as well as to Christian Jouret and the entire crew of the 3 MV unit for their competent and efficient help in accumulating satisfactory data in the limited time available.

Special thanks are due George Wray for his cordial and dedicated participation in the collection and optimization of data and for consistently maintaining the Boulder HVEM in matchless condition. My appreciation goes to K. R. Porter, E. DeHarven, and P. S. Woods for use of their specimens. The technical assistance of Ross Lepera, Mary-Letitia Timiras, and James Newman is also gratefully acknowledged.

This work was supported in part by grant 5 P41 RR-00592 from the Division of Research Resources, National Institutes of Health, and by grant PCM-76-21853 from the National Science Foundation.

REFERENCES

1 Glaser, W. (1943) *Z. Phys.* **121:**647.

2 Berger, M. J. and Seltzer, S. M. (1964) *Tables of Energy Losses and Ranges of Electrons and Positrons,* NASA SP-3012, Washington, D.C.

3 Favard, P. and Carasso, N. (1973) *J. Microscopy* **97:**59–81.

4 Favard, P., Ovtracht, L., and Carasso, N. (1971) *J. Microscopie* **12:**301–316.

5 Glauert, A. M. (1974) *J. Cell Biol.* **63:**717–748.

6 Glauert, A. M. (1979) *J. Microscopy* **117:**93–101.

7 Peachey, L. D., Fotino, M., and Porter, K. R. (1974) "Biological Applications of High-Voltage Electron Microscopy," in *High-Voltage Electron Microscopy: Proceedings of the Third International Conference* (P. R. Swann, C. J. Humphreys, and M. J. Goringe, Eds.), Academic Press, New York, pp. 405–413.

8 Porter, K. R. and Wolosewick, J. J. (1977) *Proceedings of the Fifth International Conference on High-Voltage Electron Microscopy, Kyoto* (T. Imura and H. Hashimoto, Eds.), Japanese Society of Electron Microscopy, pp. 15–20.

9 Dupouy, G. (1968) "Electron Microscopy at Very High Voltages," in *Advances in Optical and Electron Microscopy,* Vol. 2 (R. Barer and V. E. Cosslett, Eds.), Academic Press, New York, pp. 167–250.

10 Soum, G., Arnal, F., Balladore, J. L., Jouffrey, B., and Verdier, P. (1979) *Ultramicroscopy* **4:**451–466.

11 Fotino, M. (1973) In *Proceedings of the 31st Annual Meeting of the Electron Microscopy Society of America* (C. J. Arceneaux, Ed.), Claitor's Publishing Division, Baton Rouge, p. 8.

12 Fotino, M. (1975) In *Proceedings of the 33rd Annual Meeting of the Electrical Microscopy Society of America* (G. W. Bailey, Ed.), Claitor's Publishing Division, Baton Rouge, p. 664.

13 Fotino, M. (1976) In *Microscopie Electronique à Haute Tension 1975* (Fourth International HVEM Conference, Toulouse) (B. Jouffrey and P. Favard, Eds.), Société Française de Microscopie Electronique, Paris, pp. 361–364.

14 Gentsch, P. and Reimer, L. (1973) *Optik* **37:**451.

15 Gentsch, P., Gilde, H., and Reimer, L. (1974) *J. Microscopy* **100:**81.

16 Knoll, M. and Ruska, E. (1932) *Z. Phys.* **78:**318.

17 von Ardenne, M. (1941) *Z. Phys.* **117:**657.

18 van Dorsten, A. C., Oosterkamp, M. J., and LePoole, J. B. (1947) *Philips Tech, Rev.* **9:**193–224.

19 Smith, K. C. A., Considine, K., and Cosslett, V. E. *Electron, Microscopy 1966*, Vol. 1, Maruzen, Tokyo, 1966, p. 99.

20 Dupouy, G. and Perrier, F. (1962) *J. Microscopie* **1**:167–192.

21 Dupouy, G., Perrier, F., and Durrieu, L. (1970) *J. Microscopie* **9**:575.

22 Sugata, E., Fukai, K., Fugita, H., Ura, K., Tadano, B., Kimura, H., Katagiri, S., and Ozasa, S. (1970) *Microscopie Electronique 1970*, Vol. 1, Société Française de Microscopie Electronique, Paris, p. 121.

23 Kobayashi, K., Suito, E., Uyeda, N., Watanabe, M., Yanaka, T., Etoh, T., Watanabe, H., and Moriguchi, M. (1974) In *Proceedings of the Eighth International EM Congress*, Vol. 1 (J. V. Sanders and D. J. Goodchild, Eds.), Australian Academy of Science, Canberra, p. 30.

24 Horiuchi, S., Matsui, Y., Bando, Y., Katsuta, T., and Matsui, I. (1978) *J. Electron Microsc.* **27**:39–48.

25 Dietrich, I., Fox, F., Knapek, E., Lefranc, G., Nachtrieb, K., Weyl, R., and Zerbst, H. (1977) *Ultramicroscopy* **2**:241–249.

26 Cosslett, V. E., Camps, R. A., Saxton, W. O., Smith, D. J., Nixon, W. C., Ahmed, H., Catto, C. J. D., Cleaver, J. R. A., Smith, K. C. A., Timbs, A. E., Turner, P. W., and Ross, P. M. (1979) *Nature (London)* **281**:49–51.

27 Thon, F. (1971) "Phase Contrast Electron Microscopy" in *Electron Microscopy in Materials Science* (U. Valdrè, Ed.), Academic Press, New York, p. 571.

28 Hillier, J. and Ramberg, E. G. (1947) *J. Appl. Phys.* **18**:48–71.

29 Grinton, G. R. and Cowley, J. M. (1971) *Optik* **34**:221.

30 Reisner, J. H. (1964) *RCA Sci. Instrum. News* **9**:1.

31 Krivanek, O. L., Isoda, S., and Kobayashi, K. (1977) *J. Microscopy* **111**:279

32 Cosslett, V. E. (1962) *J. R. Microsc. Soc.* **81**:1–10.

33 Martinez, J. P., Balladore, J. L., and Trinquier, J. (1979) *Ultramicroscopy* **4**:211–219.

34 Hashimoto, H. (1966) In *Proceedings of the AMU–ANL Workshop on HVEM*, p. 68.

35 Fotino, M. (1976) *Electron Microscopy 1976 (Proceedings of the Sixth European Congress on Electron Microscopy, Jerusalem)*, Tal International Publishing Company, p. 277.

36 Reimer, L., Gilde, H., and Sommer, K. H. (1970) *Optik* **30**:590.

37 Gentsch, P., Gilde, H., and Reimer, L. (1974) *J. Microscopy* **100**:81.

38 Kobayashi, E. and Sakaoku, K. (1965) *Lab. Invest.* **14** (G. F. Bahr and E. Zeitler, Eds., *Quantitative Electron Microscopy*), pp. 359–376.

39 Stenn, K. and Bahr, G. F. (1970) *J. Ultrastruct. Res.* **31**:526.

40 Glaeser, R. M. (1975) "Radiation Damage and Biological Electron Microscopy," in *Physical Aspects of Electron Microscopy and Microbeam Analysis* (B. M. Siegel and D. R. Beaman, Eds.), Wiley, New York, p. 205.

41 Reimer, L. (1975) "Review of the Radiation Damage Problem of Organic Specimens in Electron Microscopy," in *Physical Aspects of Electron Microscopy and Microbeam Analysis* (B. M. Siegel and D. R. Beaman, Eds.), Wiley, New York, p. 231.

5

MAPPING RNA:DNA HETERODUPLEXES BY ELECTRON MICROSCOPY

Louise T. Chow
Thomas R. Broker
Cold Spring Harbor Laboratory
Cold Spring Harbor, New York

CONTENTS

1 Introduction
2 Development of RNA:DNA heteroduplex techniques
3 Denaturation and reannealing of polynucleotides
  3.1 Parameters affecting the thermal stability of double helices
    The effects of base composition and formamide concentration on T_m.
    The effect of electrolytes on T_m
    The effect of duplex length on T_m
    The effect of partial homology on T_m
  3.2 Optimal conditions for hybridization
  3.3 R-loop mapping: Applications and limitations
  3.4 RNA: single-stranded DNA heteroduplex mapping: Applications and limitations
4 Hybridization materials
  4.1 Glassware and utensils
  4.2 Formamide
  4.3 Hybridization buffer
  4.4 DNA
  4.5 RNA
5 Hybridization methods
  5.1 R-loop formation
  5.2 RNA:single-stranded DNA heteroduplex formation
  5.3 Combination hybridization procedures
6 Electron microscopic procedures
  6.1 Mounting samples on grids
  6.2 Evaluation of specimens
  6.3 Photography and tracing
7 Data analysis
  7.1 Interpretation of R-loop structures
  7.2 Interpretation of RNA:single-stranded DNA structures
  7.3 Orientation
  7.4 Data presentation
  7.5 Standard deviation and statistical confidence intervals
  7.6 Systematic errors
  7.7 Quantitation
8 RNA:DNA heteroduplex mapping: Application to adenovirus-2
9 Ancillary procedures
  9.1 Enrichment of RNA after hybridization to DNA
    Density gradient centrifugation
    Gel filtration

9.2 Enrichment of DNA after hybridization to RNA
9.3 Direct visualization of single-stranded RNA
9.4 Labelling methods for EM visualization of RNA:DNA heteroduplexes
Ferritin labeling of RNA
Negative enhancement of short duplex regions with single-stranded DNA binding proteins

10 Conclusions
Acknowledgments
Abbreviations
References

1 INTRODUCTION

Electron microscopic observation of heteroduplexes formed between human adenoviral messenger RNA and viral DNA resulted in the discovery that eucaryotic mRNAs can be composed of sequences copied from separate segments of the genome (1, 2). Such RNAs are derived from continuous primary transcripts of the DNA by processes in which one or more internal deletions occur in the newly synthesized RNA and the conserved segments are religated or spliced together. The electron microscopic evidence for RNA splicing was dramatic and unequivocal, and was firmly correlated with several key biochemical observations (3–8). Taken together, the diverse evidence for the new concept of eucaryotic gene organization was compelling and, most importantly, provided the model experiments and expectations for analyses of RNAs from other systems, whether by electron microscopic or biochemical methods. Within months, evidence compatible with the phenomenon of RNA splicing was gathered from an impressive variety of eucaryotic RNAs (cf. reference 118). Electron microscopic analysis of RNA:DNA heteroduplexes has played a pivotal role in the elucidation of RNA splicing and what it achieves for the genetic versatility of eucaryotic organisms (for reviews, see references 9, 10).

This chapter reviews the basic techniques for forming, visualizing, and analyzing RNA:DNA heteroduplexes and their application with reference to the characterization of adenoviral transcripts, which are probably the most diverse, complex, and thoroughly studied set of eucaryotic messenger RNAs.

2 DEVELOPMENT OF RNA:DNA HETERODUPLEX TECHNIQUES

Kleinschmidt (11) and his colleagues established the basic procedures for mounting duplex DNA in positively charged protein films formed on the surface of aqueous solutions. The protein film serves to hold the DNA in a relaxed but extended configuration and allows the sample to be transferred without distortion to a hydrophobic EM grid touched to the solution surface. In a purely aqueous environment, single-stranded DNA and RNA collapse during spreading in the protein film because of adventitious intrastrand base pairing. Single-stranded DNA can be extended by the inclusion of formamide ($HCONH_2$) at greater than 30% v/v in the spreading solution (12, 13). (Methods for the visualization of single-stranded RNA are described in Section 9.3.) Formamide and other chaotropic agents decrease the thermal stability of duplex polynucleotides and are particularly effective in disrupting imperfect intrastrand secondary structures. The capability to extend single-stranded DNA in a nonreactive solvent allowed the immediate application of electron microscopy to the visualization and mapping of insertion, deletion, substitution, transposition, and inversion loops in heteroduplexes formed by hybridization of related DNA strands (12, 13).

The first R-loop structures created for the explicit purpose of accurately measuring the positions and lengths of RNA molecules hybridized to DNA were described by Hyman and Summers (14). To accentuate the single-strand/double-strand junctions at the ends of RNA hybridized to single-stranded DNA, they added the complementary DNA strand to the RNA:DNA heteroduplex reaction mixture and continued hybridization in 70% formamide. The DNA strands reannealed over their entire length except where the RNA had hybridized. This procedure allowed the RNA to remain associated with its complement, with little or no displacement by the homologous DNA strand. Meanwhile, Birnstiel et al. (15) discovered that RNA:DNA heteroduplexes were more stable than the comparable DNA:DNA duplexes in solvents containing high concentrations of formamide. This observation, in retrospect, explained why the RNA in the R-loop structures formed by Hyman and Summers was not displaced by the homologous DNA sequences. But the principle was not fully exploited for several years until Thomas et al. (16) and White and Hogness (17) took advantage of the exceptional stability of RNA:DNA heteroduplexes in the presence of formamide and introduced an important procedural modification. The DNA was held in 70% formamide near its melting temperature, partially (rather than completely) denatured, so that the RNA could pair with the complementary DNA sequences exposed in the denatured portions of the duplex (Figure 1*b*). Throughout this RNA assimilation process, the two DNA strands remain in register by virtue of pairing in the more stable, GC-rich domains. When the temperature is lowered, the complementary DNA strands reassociate rapidly, except at the positions of the bound RNA, where the DNA strands remain separated as a prominent loop which indicates the position and length of the RNA (Figures 1*c* and 2*a*). Thomas et al. (16) dubbed the structures "R-loops" by analogy to the DNA displacement loops found in replicative mitochondrial DNA (20). The stabilizing effect of formamide on RNA:DNA heteroduplexes has been analyzed systematically

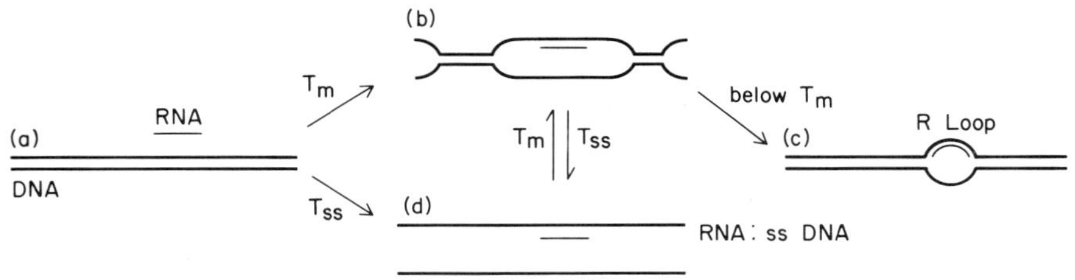

Figure 1 RNA:DNA heteroduplex formation. (*a*) RNA and double-stranded DNA. (*b*) A schematic structure of partially denatured DNA held near the average melting temperature (T_m^*). The AT-rich segments are largely denatured, but pairing persists in the GC-rich regions. In the presence of high concentrations of formamide, the T_m^* of RNA:DNA duplex is 10–15°C higher than that of the DNA:DNA analogue; therefore, RNA can pair with complementary DNA sequences exposed in denaturation loops. This structure is also an intermediate in the conversion of the RNA:single-stranded DNA heteroduplex (*d*) to the R-loop structures (*c*). (*c*) The R-loop structure formed from (*b*) upon cooling the hybridization mixture. (*d*) An RNA:single-stranded DNA heteroduplex formed upon incubating RNA with denatured DNA at or just above the strand separation temperature T_{ss} or upon heating structure (*b*) to T_{ss}. It can be converted to (*b*) and then to (*c*) by incubating the hybridization mixture at or slightly below T_m^*.

by Casey and Davidson (21). A number of laboratories, including our own, immediately applied the newly defined R-loop procedure to electron microscopic mapping of transcripts from many different systems and to the fractionation of RNA and DNA hybridized to each other.

A variation of RNA:DNA heteroduplex formation was also developed in which RNA was incubated in the presence of 70–80% formamide with fully denatured but unfractionated DNA at temperatures just above the melting temperature of the DNA (Figure 1*d*) (21–23). Because RNA:DNA heteroduplexes are more stable than comparable DNA:DNA duplexes under these conditions, the RNA can pair with its DNA complement. This method conveniently eliminates the need to fractionate the two DNA strands. The resulting RNA:single-stranded DNA heteroduplexes allow straightforward characterization of spliced RNAs (2) and other RNA molecules complementary to separate segments of the DNA (17, 22) because they constrain the DNA into diagnostic deletion loops (Figure 2*b*).

The two types of RNA:DNA heteroduplex are interconvertible (Figure 1). At the completion of R-loop formation, the temperature can be raised several degrees to dissociate the DNA:DNA duplex (24). Despite the separation of the DNA strands, the more stable RNA:DNA heteroduplexes will remain paired and DNA deletion loops indicative of RNA splicing can readily form. Conversely, RNA:single-stranded DNA heteroduplexes can be further incubated at slightly lower temperatures to allow the complementary DNA strands, present throughout the hybridization, to reassociate (25). This stepwise procedure allows the formation of R-loops in GC-rich as well as in

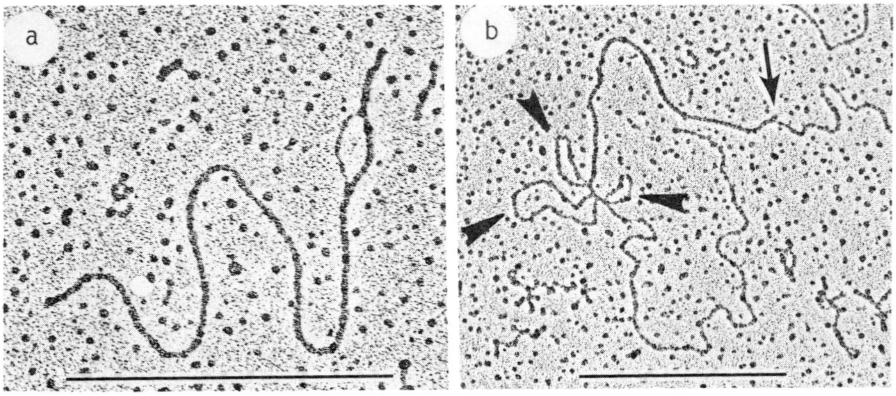

Figure 2 RNA:DNA heteroduplexes. (*a*) An R-loop formed between rabbit globin mRNA and a plasmid (pβG1) consisting of the vector pMB9 and a cDNA copy of the mRNA (cf. reference 18). (*b*) An RNA:single-stranded DNA heteroduplex formed between adenovirus RNA encoding the late 52-55K protein and adenovirus serotype 2 DNA (19). The RNA consists of three 5′-leader segments and a main coding body with its 3′-end indicated by the arrow. Upon hybridization, the spliced RNA constrained the DNA into three deletion loops (arrowheads) corresponding to the intervening sequences. Bar = ~ 2000 nucleotides or nucleotide pairs.

AT-rich regions. In all cases, the key factor is the extra stability conferred upon RNA:DNA duplexes by the high formamide concentration in the solution. Each method of preparing RNA:DNA heteroduplexes has important applications, and each has some disadvantages. Fortunately, the relative advantages and limitations of the RNA:single-stranded DNA and the RNA:double-stranded DNA hybridization and mapping procedures are nearly complementary and, together, they constitute a powerful tool for characterizing RNA structure and genome organization.

Our research interest in the electron microscopy of RNA:DNA heteroduplexes, which forms the basis of this chapter, has two aspects: the development of the technical potential of the methodology, and the exploitation of this capability to elucidate the organization and expression of a complex eucaryotic transcription system, that of the human adenoviruses (Figure 15, Section 8).

3 DENATURATION AND REANNEALING OF POLYNUCLEOTIDES

3.1 Parameters Affecting the Thermal Stability of Double Helices

The stability of duplex polynucleotides is a complex function of temperature, concentration of electrolytes, concentration of chaotropic agents such as formamide, urea, and sodium perchlorate, presence of di- or trivalent cations such as Mg^{2+}, Ca^{2+}, and Al^{3+}, AT/GC base composition of the polynucleotides, length of complementary strands or segments of strands, and degree of base complementarity. All these factors must be considered to establish optimal hybridization conditions for the RNA and DNA, and adjustments in one or more parameters usually must be made for every special application. Each element is considered.

DNA denatures gradually over a temperature range of 5–20°C, the breadth of which depends on the variation in AT and GC composition in different segments of the DNA. (During R-loop formation a segment can be considered to be approximately the size of an RNA transcript.) The melting temperature (T_m) is the midpoint of the helix-coil transition curve, and the strand separation temperature (T_{ss}) is the temperature at which the duplex completely denatures. Generally T_{ss} is about 7–10°C above T_m. As the temperature is raised, the most AT-rich regions are the first to denature, and the most GC-rich regions are the last. DNA held near its T_m is in a state of dynamic equilibrium of denaturation ⇄ renaturation, and the AT-rich regions spend a greater percentage of time denatured than do the GC-rich regions. Consequently, when RNA is mixed with DNA held near its T_m, the AT-rich RNAs will preferentially anneal in a shorter period of time.

The Effects of Base Composition and Formamide Concentration on T_m

The melting temperature of DNAs of various AT and GC base compositions dissolved in the standard solvents (0.18 *M* NaCl or 0.12 *M* Na phosphate or 0.15 *M* NaCl, 0.015 *M* Na citrate) can be calculated from the basic equation 1 (26):

$$T_m = 69.5°C + 0.41 \times (\text{percent GC}) \tag{1}$$

The chemical forces that stabilize duplexes include hydrogen bonds of base pairs between complementary strands (27) and the various hydrophobic van der Waals charge interactions associated with base stacking along the axis of the double helix (28–32). Water is a relatively poor solvent for bases exposed in single-stranded polynucleotides; conversely, it is a favorable solvent for duplex polynucleotides when sufficient electrolytes are present to neutralize the negative charge of the sugar-phosphate backbone of the strands. Formamide, urea, and certain other neutral organic solvents lower the T_m of duplex polynucleotides for several reasons:

1 They disorganize the supramolecular, hydrogen-bonded structure of water; the aromatic bases, which by themselves would have to disrupt water structure to dissolve, are more soluble when the structure is already broken down by other solutes (28).
2 Aqueous formamide has lower ion-solvating potential than do aqueous salt solutions, and native DNA, with the high charge of density of the

sugar-phosphate chains, is somewhat less soluble than single-stranded random coils, with lower charge densities (30).

3 Mixed organic-aqueous solvents provide a more polarizable medium than pure water; by solvating exposed, hydrophobic bases through transient dipole interactions (London dispersion forces), they can partially compensate for base-stacking interactions lost when a series of base pairs dissociate (32).

Replacement of interstrand hydrogen bonds by formamide or urea (which are analogues of the functional groups of the bases) does not seem to be a major factor in the dissociation of duplexes in the presence of these reagents (30, 33, 34), even though specific base-pair interactions are crucial to the registered association of polynucleotide strands in duplexes (27). The chaotropic anions perchlorate and trichloracetate, which lower the T_m of duplex polynucleotides, function by breaking down water structure, and they also "salt in" exposed nucleosides by establishment of ion-dipole interactions (34, 35).

The effective melting temperature, T_m^*, in formamide is related to T_m for DNA of 50% AT base composition by the relation summarized in equation 2 (36–38):

$$T_m^* = T_m - 0.62\ [\%\ \text{formamide}] \tag{2}$$

or by the more general equation 3 (21):

$$T_m^* = T_m - [0.5\ (\text{mole-fraction GC}) + 0.75\ (\text{mole-fraction AT})]\ [\%\ \text{formamide}] \tag{3}$$

Note that the base composition not only affects the overall T_m of the DNA (equation 1), but can impose a differential response to formamide in segments of the DNA that differ in AT and GC content (equation 3, Table 1).

The Effect of Electrolytes on T_m

NaCl and other monovalent cations affect hybridization in two ways. Electrolytes raise the T_m^* according to equations 4 (39) or 5 (40).

$$T_m^* = T_m - 18.5°\text{C} \times \log_{10} \frac{0.18}{[\text{Na}^+]} \quad \text{for } \text{Na}^+ \text{ or other monovalent cation expressed in molar concentration} \tag{4}$$

$$T_m^* = 16.6 \log_{10} [\text{Na}^+] + 0.41\ (\%\ \text{GC}) + 81.5°\text{C} \tag{5}$$

The relations are valid up to about 0.3 M Na^+.

The higher hybridization temperature due to an elevation in T_m^* is detrimental to the integrity of polynucleotide chains during annealing. But it is more than compensated by a substantial increase in the rate of reannealing (Table 2) (41–43).

Di- and trivalent cations stabilize duplex nucleic acids and significantly

TABLE 1 DNA Melting Temperature[a] in RNA Hybridization Buffer[b]: Effect of AT/GC Base Composition

	T_m (°C)		
GC:	40%	50%	60%
70% Formamide	47.5	53.4	59.2
80% Formamide	41.0	47.1	53.2

[a] Based on conversion equations 3 and 5.
[b] 0.4 *M* NaCl, 0.1 *M* HEPES, pH 7.9, 0.01 *M* Na EDTA (equivalent to 0.5 *M* Na^+).

raise the T_m^*. During heteroduplex formation, they should be chelated with EDTA.

The Effect of Duplex Length on T_m

The T_m^* is related to the length of duplex (existing or anticipated) by the approximate equation 6 (41):

$$T_m^* = T_m - \frac{500}{\text{nucleotides in duplex}} \tag{6}$$

This is a trivial effect for DNA or RNA segments the length of a collinear gene, but it becomes a significant adjustment for very short RNA segments such as those in some spliced RNAs. For instance, the length of the first

TABLE 2 DNA Melting Temperature[a] and Relative Reannealing Rates[b], in RNA Hybridization Buffer[c] with Different NaCl Concentration

		T_m^* (°C)	
[NaCl]	Approximate Relative Rates[b]	70% Formamide	80% Formamide
0.0	0.09	41.7	35.5
0.1	0.90	46.7	40.5
0.2	1.99	49.7	43.4
0.3	3.02	51.7	45.5
0.4	3.90	53.4	47.1
0.5	4.70	54.7	48.4
0.6	5.35	55.7	49.5

[a] T_m adjustment for salt and formamide is calculated for DNA of 50% GC content based on equations 3 and 5.
[b] Relative hybridization rates are based on the data of Wetmur and Davidson (41) and the table of Britten and Smith (42), but do not take into consideration possible nonlinear effects of formamide in the annealing mixture.
[c] 0.1 *M* HEPES, pH 7.9, plus 0.01 *M* Na^{3+} EDTA (equivalent to 0.1 *M* Na^+).

5′-leader segment of adenovirus late RNA is 41 nucleotides (44, 45), and the T_m of the RNA:DNA duplex is calculated to be about 12°C lower than long RNA:DNA duplex of the same base composition.

The Effect of Partial Homology on T_m

When imperfect heteroduplexes are expected between DNA strands with some sequence divergence, the T_m depression due to base-pair mismatch can be approximated by equation 7 (46–49):

$$T_m^* = T_m - 0.72°\text{C} \times (\% \text{ mismatch}) \tag{7}$$

Scattered (random) mismatches in very short heteroduplexes actually have a substantially bigger effect on lowering T_m than is estimated with equation 7 (50, 159). However, the assessment of T_m^* is complicated by competing intrastrand base pairing of the RNA as well as the DNA when the incubation temperature is lowered in attempts to promote pairing of partially mismatched RNA and DNA (cf. reference 15).

To a first approximation, all the equations for calculating T_m^* of DNA can be applied additively to estimate the stability of RNA:DNA heteroduplexes. But empirical observations of initial experiments should guide subsequent hybridization reactions.

3.2 Optimal Conditions for Hybridization

The T_m^* of RNA:DNA in aqueous solution containing 70% or more formamide has been measured to be about 15°C higher than the T_m^* of the DNA:DNA counterpart (21). DNA:DNA renaturation rates are maximal over a broad range of temperatures centered around −25°C (41). If this applies to RNA:DNA hybridization, the maximal rate might be expected to be about 10°C below the T_m^* of the DNA (15, 51), but for two reasons the hybridization is usually carried out between the T_m^* and T_{ss} of the DNA (16, 21, 52; and our own observations). First, RNA tends to have a rather stable intrastrand secondary structure, and this must be disrupted for the RNA sequences to be available for hybridization to DNA. Second, during R-loop formation, the two DNA strands must be partially dissociated (near or above the T_m^* of the DNA) to be available for pairing with RNA. Similarly, when hybridizing RNA to single-stranded DNA preparations (present at moderate to high concentrations) in which the complementary DNA strands are not physically fractionated, incubation usually should be above the T_m^* of the DNA or the strands will renature and become unavailable for hybridization with RNA. Only when separated DNA strands are used will RNA:DNA heteroduplex formation proceed with high efficiency at temperatures somewhat below the T_m^* of the DNA.

Another advantage of including formamide in the hybridization solution is that it lowers the T_m^*, and therefore the hybridization temperature, and reduces RNA and DNA degradation during prolonged incubations (53–55). Formamide itself is subject to nucleophile-catalyzed hydrolysis, which has three undesirable consequences: (*a*) the denaturing power of the solvent declines, and the T_m^* of the DNA correspondingly increases (equation 3); (*b*) the formamide dissociates to ammonium formate, an electrolyte, which also increases the T_m^* as it builds up (equations 4 and 5), and (*c*) the ammonium ion NH_4^+ is in equilibrium with NH_3 and H^+; the ammonia can evaporate, and the solution then becomes more acidic, with a possibility for polynucleotide depurination. Strongly nucleophilic amine buffers such as Tris cause significant formamide hydrolysis (16, 21, 56) and should be avoided. Sulfonate buffers, such as HEPES (pK_a 7.5) or PIPES (pK_a 6.8), are less likely to catalize hydrolysis. To provide good buffering capacity, we prefer 0.1 *M* HEPES adjusted to pH 7.9 or 8.0, where polynucleotide degradation is least likely.

The composition of the RNA:DNA hybridization solution is narrowly dictated by the high concentration of formamide required (and therefore the high percent volume taken up by the formamide). The concentration of NaCl should also be as high as practicable to promote the maximal rate of annealing. NaCl, HEPES or PIPES buffer, and EDTA must be made up as a nearly saturated stock solution to be added in a minimal volume. The RNA and DNA samples must be very concentrated, again to minimize the volume added. With all factors considered, the optimal RNA:DNA annealing mixture is 70–80% formamide, 0.4 *M* NaCl, 0.1 *M* HEPES or PIPES, 0.01 *M* EDTA, pH 7.9–8.0, 5–100 μg/ml DNA adjusted proportionally to its complexity), and RNA sufficient to have any given species present in, preferably, more than 0.1 times the concentration of the complementary DNA sequences (21).

As an alternative to the inclusion of formamide to promote RNA:DNA hybridization, Chien and Davidson (52) introduced the use of chaotropic perchlorate or trichloracetate anions, which stabilize RNA:DNA duplexes relative to DNA:DNA homologues by 5°C when present at 3–4*M* concentration.

3.3 R-Loop Mapping: Applications and Limitations

R-loops should be formed and analyzed in any of the following instances:

1 For a preliminary check of the quality of an RNA preparation: are the R-loops appropriately long and reproducible, or is there evidence of excessive RNA degradation?

2 During the preliminary mapping of a relative unknown transcription system; the loop distributions should give a good overall indication of the transcription pattern.

3 When convergent or divergent transcription units from opposite strands are suspected or known, and the common boundaries of those domains are to be mapped.

4 When rare RNAs are anticipated; R-loops are relatively easy to see and are unambiguous even when infrequent.

5 When the RNA is not spliced. Although spliced RNAs can form composite R-loops (cf. Figure 4*b*, V and Figure 6, Section 7.1) they are better studied by the RNA: single-stranded DNA method (see Figure 4*c*, VIII and IX).

6 When the DNA is somewhat nicked: upon snapback, it will be largely or completely restored to duplex.

7 When accurate measurements are desired from a rather large chromosome; all the relevant measurements are of duplex segments (DNA: DNA and RNA: DNA), and positioning is therefore more reliable than when RNA is annealed to single-stranded DNA.

Hybridizations performed with mixtures of different RNA samples such as those extracted at early and late times after virus infection can provide additional information when mapped relative to one another. For instance, in adenoviruses, the transcriptional strand switches occur between blocks of early and late genes (57, 58).

As with any technique, R-loop mapping has disadvantages; these are particularly evident in comparison with the analysis of RNA paired with single-stranded DNA:

(*a*) R-loops form poorly, if at all, in GC-rich regions, since these DNA: DNA segments remain paired during incubation near T_m^*. (*b*) Annealing requires many hours. (*c*) RNA, especially if it is shorter than about 500 nucleotides, can be partially or completely displaced when the sample is transferred to low percentage formamide solvents during spreading for EM analysis. Consequently, end point measurements can be in error. (*d*) Unless convergent (or divergent) R-loops are formed, R-loops do not provide information on the template strand for the RNA. (*e*) Spliced, multi-segmented RNAs do not hybridize readily to all their coding regions, particularly if the segments are short and/or are encoded by widely separated portions of the chromosome. (*f*) The complex topological constraints imposed when overlapping, spliced RNAs form multiple R-loops can result in uninterpretable tangles.

Some of these limitations can be minimized by various means and are discussed later.

3.4 RNA: Single-Stranded DNA Heteroduplex Mapping: Applications and Limitations

The study of RNA: single-stranded DNA heteroduplexes is, in most ways, more informative about the structures of the RNAs once a basic transcription map has been established by the R-loop technique. Its advantages are manyfold: (*a*) RNAs that are GC-rich as well as those that are AT-rich can be mapped, although the optimal temperature for the different transcripts may be many degrees apart; (*b*) all segments of spliced RNAs can, and usually will, hybridize to their complements, revealing the full complexity of their composition; (*c*) annealing times are much shorter than those for R-loop formation when RNA and DNA concentrations are comparable to those used in R-loop formation; (*d*) RNAs complementary to the same DNA strand can be identified and mapped simultaneously.

However, there are several disadvantages or technical limitations to the method: (*a*) any nicks in the DNA will result in the loss of the natural termini against which map position measurements are generally made (this problem is particularly severe with long DNA molecules); (*b*) the SS/DS distinction must be good to permit the visualization of the termini of the RNA, but poly(A) tails can help define 3′ ends; (*c*) unless the RNAs are long or are spliced and create DNA deletion loops, they are much less prominent than when they make R-loops (cf. Figure 4cX); (*d*) measurements along single-stranded DNA are inherently less reliable than those along duplex segments (as are used in the R-loop method) and may also be subject to differential contraction of GC-rich regions; (*e*) minor RNA species may not be driven into heteroduplexes during the short (30–60 minute) hybridization reactions possible; and (*f*) the instability of nucleic acids and formamide held at high temperature limits exhaustive hybridization. Fortunately, the relative advantages and limitations of RNA:DS DNA and RNA:SS DNA hybridizations are nearly complementary, and both can be employed to good advantage. Furthermore, the methods can be combined in multistage incubations, as discussed in Section 5, "Hybridization Methods."

4 HYBRIDIZATION MATERIALS

4.1 Glassware and Utensils

Glassware and utensils should be clean, free of grease and detergent, and baked at 350°C for several hours to inactivate RNAses. Siliconized glassware is not recommended, for transfer of small volumes of solutions would be difficult and inaccurate.

4.2 Formamide

Formamide should be spectral quality and may be used without deionization, distillation, or recrystallization as long as absorption of water is minimal and no ammonia smell is detectable. The commercial formamide we use has an absorbance at 270 nm of 1.7 optical density units and a conductivity of 120 μmho [which corresponds to an ammonium formate concentration of about 5 m*M* (21)]. After recrystallization, the conductivity can be reduced to 2 μmho and the A_{270} to less than 0.1 (21). These values set the working range of the formamide purity.

4.3 Hybridization buffer

Buffer stock of 7× contains 2.8 *M* NaCl, 0.7 *M* HEPES or PIPES, and 0.07 *M* NaEDTA. Dissolve the HEPES and EDTA in double-distilled water, and adjust pH to 7.9–8.0 with 4 *N* NaOH; heat to about 60°C until solution is achieved within 5/7 of the desired final volume. Add NaCl and water to bring the solution to full volume; heat, if necessary, to dissolve the salt. Filter the buffer through a prerinsed and sterile Millipore HAWP 0.45 μm filter in a Swinnex adaptor. Add several aliquots of diethylpyrocarbonate (DEPC) (caution: hazardous) totaling 1/2000 total volume, and mix vigorously after each addition. Finally, heat the solution to 50°C for a few minutes to decompose any residual DEPC. DEPC inactivates nucleases that might be present. Store the stock in the dark to decrease photooxidation of the buffer and at room temperature (rather than at 4°C) to prevent crystallization of the solutes.

4.4 DNA

The DNA should be a homogeneous population of molecules such as that of a purified restriction fragment, a cloned DNA segment, or an intact viral chromosome. Circular DNA should be linearized with a restriction endonuclease; for purposes of orientation, two samples of DNA should be prepared with enzymes that cleave at sites quite far apart, preferably not in the gene(s) of interest. It is sometimes necessary to orient linear molecules in a similar manner, as described in Section 7.3. The DNA should have few, if any, single-stranded nicks. Hence, when using circular DNAs, one should start with supercoiled material for endonuclease digestion. The quality of the DNA can be determined by denaturation followed by renaturation to 50% completion. The single-stranded DNA should be of homogeneous length by visual inspection. Renatured DNA should have few, if any, single-stranded ends or branches. Branched, renatured DNA is a result of multi-strand hybridization of random fragments due to single-stranded nicks in the starting material. Deletion, insertion, and substitution loops should also be rare or absent in the self-annealed DNA. A substantial percentage of DNA molecules with major structural mutations can severely complicate RNA:DNA heteroduplex mapping.

If the RNA to be mapped is highly purified and it is complementary to a tandem array of genes, the DNA can be of heterogeneous length but, preferably, enriched for the genes of interest. Measurements of the lengths and positions of RNA:DNA heteroduplexes in an array with respect to one another can reveal the reiterated organization of the gene cluster (22).

The DNA sample should be completely free of proteins such as restriction enzymes used in its preparation or covalently bound proteins such as those on the termini of native adenoviral DNA. The DNA stock should be

200 μg/ml or higher, dissolved in water or in low concentrations of buffer plus EDTA.

4.5 RNA

One of the great powers of RNA : DNA heteroduplex formation and analysis by electron microscopy is that the RNA usually need not be prepurified, although enrichment will reduce the background, the hybridization time, and the electron microscopic survey time. Any species that is 0.1–0.5% of the total RNA population can easily be mapped, and species that are 0.005% can be found with considerable effort. Two necessary precautions are to avoid using low molecular weight RNAs (tRNAs and 5S RNAs) as carrier during RNA preparation and to remove protein contaminants. Low molecular weight RNAs impair contrast in the EM sample, and protein contaminants cause aggregation during hybridization reactions. RNA preparations must be of high quality, with a low percentage of broken molecules, because analysis depends on statistical definition of reproducible populations. When a homogeneous RNA species can be prepared without carrier, its quality can be examined by EM after reaction of the RNA with glyoxal (59) or methylmercuric hydroxide (60, 61) (see Section 9.3). The RNA stock is generally prepared at concentrations up to 6 mg/ml (in water) when 90–95% of this RNA is heterologous carrier. When RNA or DNA is concentrated by ethanol precipitation, all ethanol should be removed by vacuum evaporation before attempting to redissolve the pellet.

5 HYBRIDIZATION METHODS

Hybridization is performed in sterile 6 × 50 mm culture tubes. If the RNA is not sufficiently concentrated to allow it to be added in a solution equivalent to several percent of the final volume, deliver it to the bottom of the culture tube and dry it down in a lyophilizer or, preferably, in a vacuum concentrator. Take it up in the formamide component (80% v/v) of the hybridization mixture. If the RNA is already sufficiently concentrated, transfer the formamide to the tube first, followed by the RNA, one-seventh volume of the 7× NaCl-buffer-EDTA solution, and finally, DNA to a concentration of 5–100 μg/ml. Thomas et al. (16) suggested overlaying the sample with mineral oil to prevent evaporation. We have not found this to be necessary. Tightly seal the tube with several layers of Parafilm. The hybridization mixture can also be sealed in capillaries. This minimizes evaporation, but prevents sampling the reaction at different times or performing multistep hybridizations. Incubation for up to 24 hours duration can be performed successfully in a 6 × 50 mm culture tube with 10 μl of reaction mixture, or with as little as 5 μl when the incubation time is less than a few hours. Clearly, measuring and transfer must be done with acute care. The samples should be incubated in covered water baths with good temperature control. During initial, exploratory experiments, several baths set about 2–3°C apart near and above the anticipated T_m^* (Equations 1–7; Tables 1 and 2) should be used to determine the optimal annealing temperature. Different RNAs complementary to the same genome may have substantially different optima, depending primarily on their relative AT/GC base compositions. Several water baths are also needed for multi-step hybridization procedures, as described in Section 5.3.

5.1 R-Loop Formation

To prepare R-loops, 8–20 hours incubation at T_m^* may be used. Samples can be taken periodically for electron microscopic examination to evaluate the extent of hybridization. In subsequent experiments make adjustments in the concentrations of DNA or RNA or in the duration or temperature of incubation to improve hybridization. Generally, it is not advisable to carry out the incubation longer than 24 hours because both the nucleic acids and the formamide begin to break down, and the effective T_m^* increases as the formamide is hydrolyzed, as discussed previously. Consequently, the DNA strands reassociate, and hybridization of RNA to DNA slows down or ceases. To terminate R-loop formation, the reaction tube can be chilled quickly in an ice-water bath. With some systems, it may be necessary to lower the annealing temperature several degrees for a few minutes to allow the partially denatured DNA strands to reassociate (16, 17). RNA displacement by renaturation of the homologous DNA strand is very slow at 4°C and can be prevented by the fixation of the displaced single-stranded DNA with glyoxal (62).

5.2 RNA: Single-Stranded DNA Heteroduplex Formation

In most respects, including the composition of the hybridization solution, the method of RNA: single-stranded DNA heteroduplex mapping is closely analogous to that with R-loop. The DNA may be denatured by brief incubation at alkaline pH greater than 12, followed by neutralization prior to adding the RNA. Or the DNA strands may be denatured and then fractionated by gel electrophoresis or by equilibrium density gradient centrifugation in the presence of either poly (U, G) (63) or heavy metals such as Ag^+ (64) or Hg^{2+} (65). But the easiest way to prepare the denatured DNA is to heat the complete hybridization mixture containing duplex DNA and the RNA to a temperature about 4–5°C above the DNA strand separation temperature (T_{ss}) for 5–10 minutes. The mixture can then be transferred to the desired RNA: DNA annealing temperature, which can range from T_{ss} to the T_m of the DNA (21, 23).

Even at T_m, DNA: DNA reannealing rates will be very slow during the short duration of the RNA-DNA hybridization reaction when the DNA concentration is moderate (41). Replicate samples should be annealed at several temperatures over this range if various RNA transcripts differ substantially in AT/GC base composition. Hybridization should not exceed 30–60 minutes; longer times at or above 60°C result in debilitating nicking of the RNA and of DNA the length of the adenovirus chromosome, for instance, or in significant renaturation of the DNA due to formamide degradation. With shorter DNA molecules and/or lower temperature, the incubation can be somewhat longer. The RNA:DNA annealing reaction can be quenched by 5–10× dilution with ice-cold water, and the sample may be stored for many days. This is in contrast to a continuing, slow DNA:DNA annealing (as with the pairing of inverted terminal duplications of adenoviral DNA) and suggests that intrastrand secondary structure of RNA effectively eliminates pairing with DNA at temperatures well below T_m. Therefore, to allow pairing of short or AT-rich RNA segments in spliced RNAs that may fail to anneal at T_{ss}, the incubation temperature can be lowered by a few degrees for 5–10 minutes prior to quenching. If DNA:DNA renaturation occurs during this secondary incubation, first dilute the sample severalfold into the same hybridization buffer-formamide solution, preheated to the same temperature. Since renaturation is a second-order reaction, dilution will decrease DNA:DNA renaturation considerably.

5.3 Combination Hybridization Procedures

To study spliced RNA of very low relative abundance, a combination of the R-loop method and single-stranded DNA method can be used when separated DNA strands are not available. RNA is first driven into R-loops by prolonged hybridization with double-stranded DNA, then the DNA strands are denatured by a brief incubation (7–10 minutes) at T_{ss} (24). Before the secondary incubation, it is advisable to dilute the R-loop mixture into fresh buffer-formamide solution to compensate for any loss of formamide during the long incubation for R-loop formation. Again, the temperature can be lowered for a few minutes before terminating the reaction to allow short spliced RNA segments to hybridize. Once the DNA strands have been separated, each segment of the spliced RNA can quickly anneal to its complement.

To form R-loops in the most GC-rich domains, RNA:single-stranded heteroduplexes can be prepared; the temperature is then lowered slightly to allow the complementary DNA strands to reassociate and form R-loops (25). Alternatively, R-loops can be formed in DNA which has been cross-linked with a photoreactive intercalating dye, psoralen (62, 66). Cross-links at one per several hundred to several thousand base pairs do not particularly interfere with RNA assimilation. Because the two DNA strands are irreversibly held in register by the psoralen, incubation at T_{ss} is possible, where all regions of the DNA, irrespective of relative AT/GC base composition, are continuously denatured and present at equal concentrations during the annealing reaction. Therefore, RNA:DNA association proceeds more rapidly than during standard R-loop hybridizations. Furthermore, several hours of incubation at or above 60°C is feasible because infrequent nicks in the DNA will not result in release of the cross-linked fragments.

6 ELECTRON MICROSCOPIC PROCEDURES

6.1 Mounting Samples on Grids (11–13)

Spreading should normally be carried out with a hyperphase composed, by sequential addition, of 0.1 *M* Tris-Cl, pH 8.5, 0.01 *M* EDTA, 50 μg/ml cytochrome *c*, 40% (or 45%) formamide, single-stranded and double-stranded DNA length standards such as ϕX174 SS DNA (5386 nucleotides) (67), colicin E1 DNA (6650 nucleotide pairs) (68), or SV40 DNA (5243 nucleotide pairs) (69, 70) at 0.1 μg/ml, and about 0.1–0.4 μg/ml sample DNA. Use less sample if the background RNA exceeds 2 μg/ml or if the DNA is short (< 10,000 base pairs). The hypophase should contain 0.01 *M* Tris-Cl, pH 8.5, 0.001 *M* EDTA, and 10% (or 15%) formamide. Spreading should be done within a few seconds of adding the RNA:DNA or R-loop sample to the hyperphase solution; with the latter type of sample, this will minimize displacement of the RNA by the homologous DNA strand (57, 58). Philippsen et al. (71) showed that spreading with a hyperphase containing 70% formamide further decreased RNA displacement, but the distinction between single-stranded and double-stranded polynucleotides degenerates. It is advisable to balance the formamide and electrolyte concentrations of the hyperphase and hypophase solutions at isodenaturing conditions (72), such as those achieved with the recipe above. Spreading on water can cause anomalous stretching of single strands. More seriously, the abrupt and substantial change in the helical pitch of heteroduplexes transferred from a high-salt hyperphase to a salt-free hypophase can cause strand twisting that may disrupt short (< 100 base pairs) duplex segments, for instance, those formed with RNA leader sequences. The sample film is picked up on a Parlodion-covered 200-mesh EM grid, stained immediately for 30 seconds with 5×10^{-5} *M* uranyl acetate and 5×10^{-5} *N* HCl in 90% ethanol, rinsed for 5 seconds in 90% ethanol, and rotary-shadowed at a 7° angle with platinum-palladium (80:20) wire supported on a tungsten filament.

6.2 Evaluation of Specimens

The DNA and DNA:RNA strands should appear as sinuous curves without particular orientation. If the overall contrast of the DNA and the distinction between single-stranded and double-stranded polynucleotides is not good, a new grid should be prepared. The concentration of RNA in the background must not impose on the free layout of the DNA; if this occurs, the sample should be respread at higher dilution. It is far easier and takes less scanning time to see unencumbered DNA at low concentration than DNA at higher concentration crowded by nonhybridized RNAs, which reduce the contrast and confuse the identification of heteroduplexes. The length-nucleotide ratio of double-stranded and single-stranded DNA should be within 5–10% of one another, especially in the RNA: single-stranded DNA method. A bigger difference is symptomatic of anomalous spreading, and measurements may not be accurate. When R-loops are being examined, and more than one per DNA molecule can be expected, hybridization to an average of several per molecule assists in orientation and measurement relative to one another, but too many RNAs per DNA molecule can actually reduce the information obtainable (cf. the discussion of expanded R-loops in Section 7.1). If so, the hybridization should be performed at lower RNA concentration or for shorter annealing times. If R-loops are not reasonably reproducible, consider the possibility that the RNA is partially degraded; it may be necessary to prepare a new RNA sample. It is worth noting that even in a good R-loop sample that has been well prepared for electron microscopy, not every DNA molecule will necessarily have RNA hybridized, nor will each be well laid out, if the DNA contains multiple R-loops that have complicated, overlapping splicing patterns, as with the adenovirus 2 late RNAs. Considerable patience while examining a grid is mandatory. With exceedingly rare RNA species, several hours of scanning may be needed to locate each example.

6.3 Photography and Tracing

Our individual preference is for large format sheet film over 35 mm roll film, although the latter is popular in some other laboratories. Advantages can be presented for either format. Roll film in the 70 mm size has many of the technical advantages of large format sheet film, plus the economy of 35 mm roll film. Molecules should be traced with an electronic planimeter or digitizer directly from a projected image of the negative. Since there is substantial stretching of photographic paper during processing and drying, prints of the micrograph should not be traced.

With R-loop molecules, all relevent portions of the structure are duplex. Segments are most accurately represented as percentages of the entire molecule, which, in turn, is calibrated against other added length standards (e.g., circular molecules) in the same micrograph to be sure of its intactness. RNA: DNA and DNA: DNA segments have approximately the same length/base pairs, within ±4% (16, 17, 57, 58, 71, 73, 74). Except when the RNA: DNA duplex amounts to a substantial portion of the entire measured contour, it may be assumed to have the same unit length as DNA: DNA duplex without introducing a systematic error larger than other measuring and interpretive uncertainties. In those cases, the RNA: DNA length can be normalized against comparable DNA: DNA structures on the same grid.

With RNA: single-stranded DNA heteroduplexes, the SS and DS segments can be evaluated with respect to SS and DS length standards present on the same grid and in the same micrograph. Generally, six or more standard molecules of each type should appear in each micrograph, a concentration that can be adjusted in successive sample preparations. However, we have found that small, circular single-stranded length standards are not always reliable for long, linear sample molecules, particularly if they differ substantially in average base composition. For instance, adenoviral DNA (58% GC; 36,500 nucleotides) is several percent shorter than expected when evaluated relative to ϕX174 DNA (50% GC, 5386 nucleotides). It may therefore be desirable to double-check length estimates by independent means—for instance, by mounting DS and SS sample DNA on the same grid as DS and SS standards, to define working ratios.

7 DATA ANALYSIS

7.1 Interpretation of R-Loop Structures

RNA annealed to double-stranded DNA can create many types of R-loops, depending on the number of RNA molecules annealed side by side to form the loops and on the structure of the RNAs. RNA transcripts that are collinear with the DNA will create simple R-loops in which a single RNA is completely paired with the DNA, and the homologous DNA sequences are displaced as a single-stranded loop (Figures 2*a*, 3, and 4*a*, II). If branch migration occurs, sequences at one or both ends of the RNA may be displaced to variable extents by partial reannealing of the homologous DNA strand, and RNA branches at the forks of the loop may be visible. However, the branches at the ends of R-loops may have other origins: at the 3′ ends, poly(A) tracts have no DNA complement and will form branches about 100–200 nucleotides long. If the RNA has sequences that are derived from remote segments of the genome, and fail to hybridize, branches will also be seen at the 5′ end of the R-loop (Figures 3*b* and 4*b*, VII).

In addition to simple R-loops, other types can form.

1 Two identical RNAs, or pieces of identical RNAs, can coanneal, creating a full size R-loop for that species, with branches (bushes) of unpaired RNA where the two molecules meet and overlap (Figure 4*a*, III). A similar structure can result if two different transcripts share some common sequences.
2 Two different but adjacent transcripts from the same strand can coanneal and form an expanded R-loop (Figure 4*b*, VI); these can be recognized because the loops will be smaller and will have new, reproducible termini when the hybridization is repeated at a lower RNA concentration.

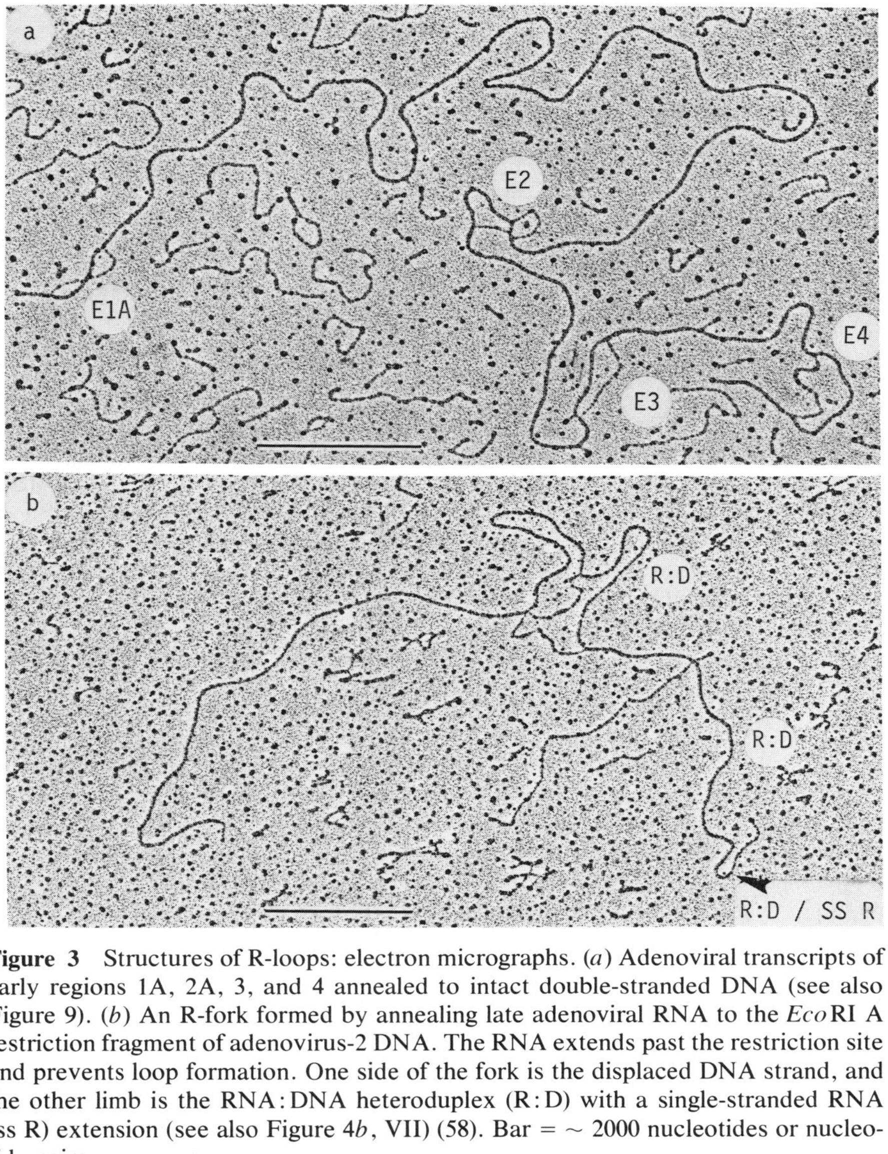

Figure 3 Structures of R-loops: electron micrographs. (*a*) Adenoviral transcripts of early regions 1A, 2A, 3, and 4 annealed to intact double-stranded DNA (see also Figure 9). (*b*) An R-fork formed by annealing late adenoviral RNA to the *Eco*RI A restriction fragment of adenovirus-2 DNA. The RNA extends past the restriction site and prevents loop formation. One side of the fork is the displaced DNA strand, and the other limb is the RNA:DNA heteroduplex (R:D) with a single-stranded RNA (ss R) extension (see also Figure 4*b*, VII) (58). Bar = ~ 2000 nucleotides or nucleotide pairs.

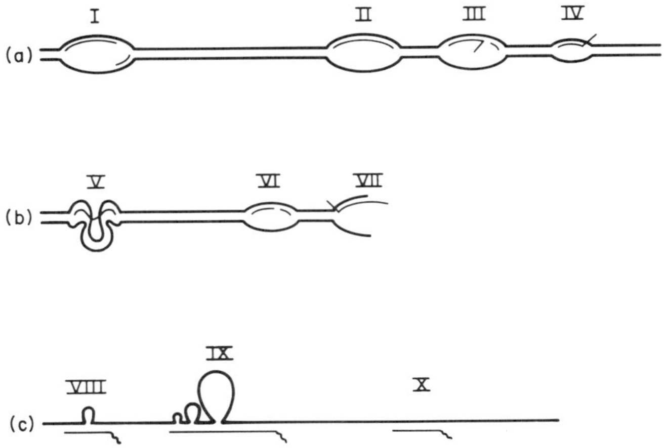

Figure 4 Structures of RNA:DNA heteroduplexes. (*a*) I, One type of convergent R-loop formed when RNAs complementary to opposite DNA strands meet at the transcriptional strand switch point (see Figure 5); II, a simple R-loop; III, a second type of convergent R-loop formed when complementing fragments of the same RNA species or of two distinct but overlapping transcripts coanneal (a short branch of redundant RNA can be observed somewhere on the RNA:DNA duplex side of the loop); IV, an R-loop with one end of the RNA displaced by partial reannealing of the DNA strands. (The sum of the length of the branch and that of the R-loop gives a more accurate value for the length of the transcript). (*b*) V, A composite R-loop formed by a spliced RNA annealing to its non-contiguous complements in the DNA (the intervening DNA sequence is constrained into a DNA:DNA loop closed by an RNA bridge; see Figure 6); VI, a third type of convergent R-loop formed by two discrete RNAs from adjacent regions of the same strand; VII, an R-fork formed by a DNA restriction fragment and a transcript which extends past the restriction site; displacement of the end of the RNA by branch migration is often observed because of the minimal constraint on axial rotation during strand interwinding at a fork compared to that at a closed loop: see Figure 3b). (*c*) RNA:single-stranded DNA heteroduplexes, the 3′-poly(A) tail on each RNA molecule remains unpaired; VIII, a single deletion loop created by hybridization of an RNA with a single splice (see Figure 12); IX, a triple deletion loop structure created by hybridization of an RNA with three splices (see Figure 2*b*); X, a simple heteroduplex with unspliced RNA. (Adapted from reference 58.)

3 Two convergent (or divergent) transcripts complementary to opposite strands can coanneal, forming an R-loop in which only part of one DNA strand of the loop is paired with an RNA and the remaining part of the other strand forming the loop is paired with the other RNA (Figure 4*a*, I, and Figure 5). The SS/DS junctions on the two arms of the loop indicate the strand switch point in the transcription map (57, 58).

4 An RNA annealed to a DNA molecule truncated by a restriction enzyme will extend past the restriction site and create an R-fork rather than a

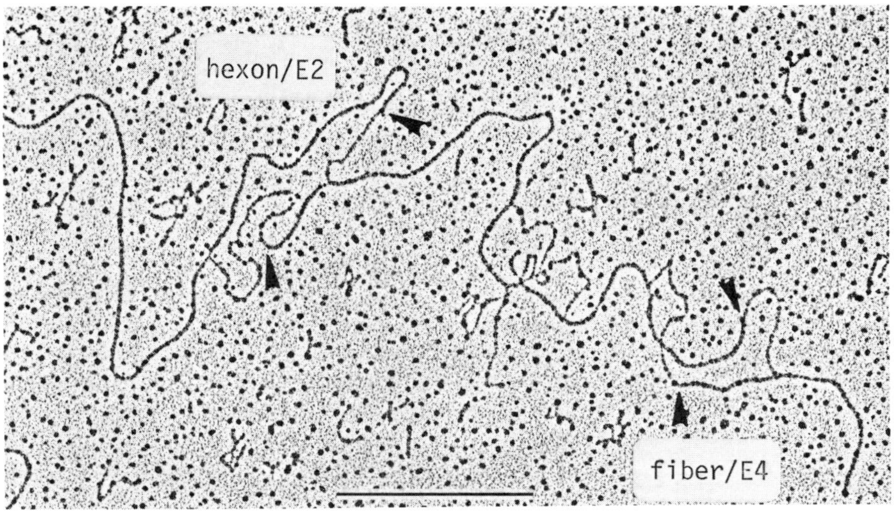

Figure 5 Electron micrograph of two transcriptional strand switches. The 3′ ends (arrowheads) of adenovirus hexon RNA (copied from the r-strand of DNA) and DNA-binding protein RNA (copied from the l-strand of early region 2A) converge at map coordinate 61.5. The 3′ ends (arrowheads) of fiber RNA (copied from the r-strand) and early region 4 RNA (copied from the l-strand) converge at coordinate 91.3 (see also Figure 4*a*, I). A portion of each end of the two R-loops is double-stranded RNA:DNA heteroduplex, and the remainder is displaced single-stranded DNA. The other R-loop present is complex because of coannealing of overlapping transcripts (Figure 4*a*, III) and the annealing of the intervening sequence of a spliced early region 3 transcript to its DNA complement in the opposing limb of the R-loop (Figure 4*b*, V) (58). Bar = ~ 2000 nucleotides or nucleotide pairs.

loop closed by DNA:DNA duplexes at both ends (Figure 3*b* and Figure 4*b*, VII) (25, 57, 58).

5 If the RNA is derived from noncontiguous segments of the DNA and both segments are reasonably long, a composite R-loop may form, with a segment of double-stranded DNA equivalent to the intervening sequence constrained into a duplex DNA loop between the R-loops (Figures 4*b*, V, and 6)(1, 17).

6 If one segment of a spliced RNA (such as a leader) is very short, it may pair without creating a visible R-loop.

7 If the intervening DNA sequence between R-loops of a spliced RNA is long, it may create a diagnostic supercoil because the segment is, topologically, a closed circle that may change helical pitch when the solvent is altered between hybridization and spreading (see Figure 2*b* in reference 1).

8 If the R-loop hybridization reaction is quenched quickly and the electron

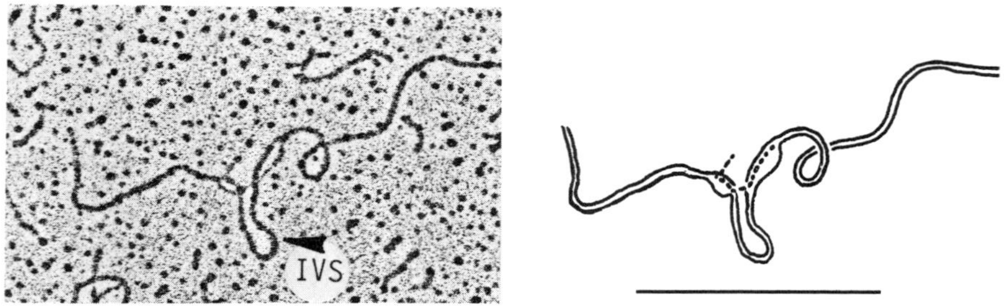

Figure 6 Composite R-loop formed by spliced RNA. The E15K message of adenovirus-2, composed of two coding segments separated by about 1350 bases of intervening sequence (IVS), causes the double-stranded DNA between the R-loop segments to loop out (arrowhead). A short "bridge" of RNA holds the two R-loops in proximity. The 5′ end of the RNA is partially displaced from the R-loop (1). An interpretative tracing is provided.

microscopic grids are prepared immediately, occasionally the reassociation of denatured, complementary segments that potentially should occur will not for kinetic reasons (Figures 3*e* and 12*b* in reference 58). This could happen in short internal segments flanked by RNA:DNA heteroduplex regions. Such expanded R-loops should be evident by careful examination for single-stranded/double-stranded junctions in the open portions of the heteroduplex. Failure of renaturation at the ends of the DNA result in single-stranded forks (Figure 3*e* in reference 58). These must not be confused with R-forks; in the R-fork, the RNA:DNA limb is longer than the displaced DNA, whereas in the DNA fork, both arms are of equal length even if one of the arms also contains some RNA:DNA duplex segment.

7.2 Interpretation of RNA:Single-Stranded DNA Structures

The sample preparation should be of sufficient quality that the double-stranded/single-stranded junctions are discernible. Most eucaryotic messenger RNAs have a 3′-polyadenylate tract 100–200 nucleotides long, which is added post-transcriptionally. This is often long enough to appear as a branch at the 3′ end of an RNA:DNA heteroduplex. A fully collinear transcript will make a simple RNA:DNA heteroduplex flanked by single-stranded DNA. Many eucaryotic mRNAs are composed of sequences derived from separate segments in the genome. Spliced RNA:DNA heteroduplexes generate diagnostic structures: each intervening sequence between coding regions for the RNA will appear as a DNA deletion loop in the heteroduplex (Figures 2*b* and 4*c*). Some adenovirus transcripts have been found with eight spliced segments and seven deletion loops (Figure 7). Other eucaryotic messages have been found to have even more splices. The lengths of internal

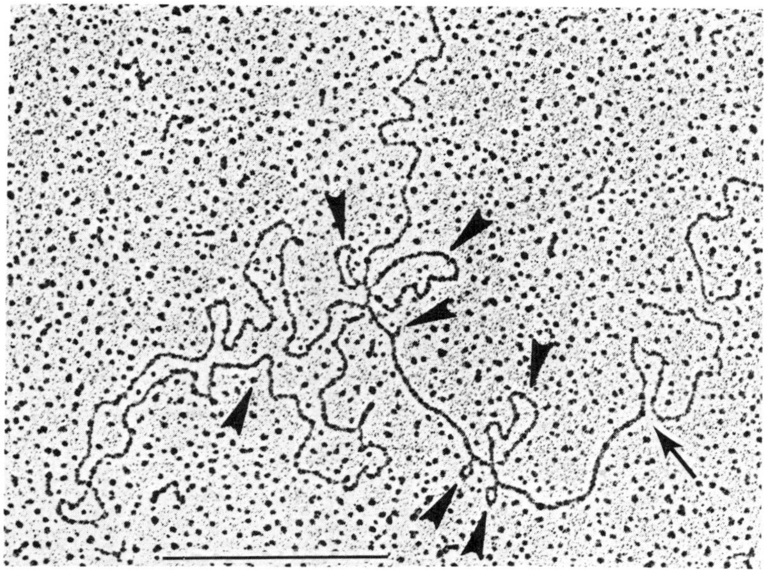

Figure 7 Multi-segmented human adenovirus fiber RNA. The incompletely spliced transcript retains eight segments joined by seven splices, as revealed by the seven deletion loops (arrowheads) constrained in the single-stranded DNA by the RNA (75). The 3′-end of the RNA is indicated by the arrow. Bar = ~ 2000 nucleotides or nucleotide pairs.

leader segments can be accurately estimated because they are clearly demarked by flanking deletion loops; the short 5′ terminal RNA:DNA duplex may not be measurable, although it will be evident from the DNA deletion loop caused (Figures 2*b*, 7, and Figure 13, Section 8). If, however, the 5′-terminal RNA components are very short, if shear forces or changes in helical pitch of duplex regions (e.g., when spreading on water) are extreme or if DNA used for hybridization does not extend as far as their coding regions, the terminal segments may remain unhybridized.

7.3 Orientation

Most aspects of data acquisition and analysis are identical for heteroduplexes between RNA and single-stranded or double-stranded DNA. The first basic problem is molecule orientation with respect to the genetic and physical maps, and the details depend, of course, on what is already known about the various maps. In our initial efforts to map adenoviral RNA, we found a distinct symmetry of the late RNA loops around the midpoint of the DNA, which hindered alignments. We solved the orientation problem by forming R-loops separately with two long restriction fragments containing sequences from map units 0.0 to 58.5 and from 45.9 to 100.0 units (Figure 8) (58). (This

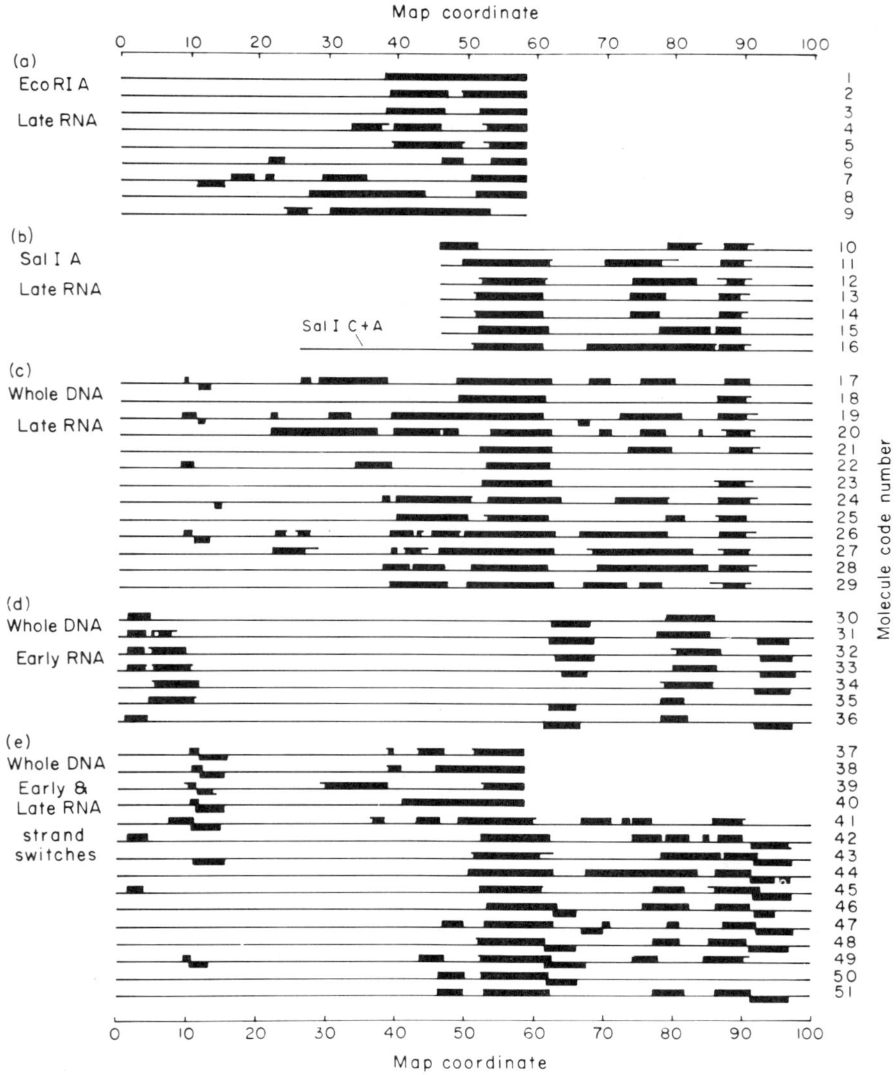

Figure 8 Array of adenovirus-2 DNA molcules with R-loops. Solid boxes designate R-loops or terminal R-forks. RNA branches extending from R-loops are indicated. The R-loop structures formed with late RNA in intact DNA (*c*) were oriented with respect to the genetic and restriction maps by establishing the R-loop patterns in (*a*) the *Eco*RI A restriction fragment (0–58.5 map units) and in (*b*) the *Sal*I A restriction fragment (45.9–100 map units). Early RNA formed an easily oriented, asymmetric pattern (*d*). Cohybridization using early and late RNA revealed many cases of convergent R-loops indicative of transcriptional strand switches (see Figures 4*a*, I, and 5) (58).

was done with the total digest because the desired fragments were substantially longer than all the others and were readily identified.) When accurate coordinates for the RNAs had been established from these overlapping left- and right-half fragments, individual transcripts could be recognized when hybridized to whole DNA by the slight differences in their distances from the ends of the DNA. Neuwald et al. (76) also prepared R-loops in restriction fragments using adenoviral nuclear RNA. Meyer et al. (57) took a different approach to the problem. They first prepared heteroduplexes between adenovirus DNA and adenovirus-SV40 hybrid virus Ad2$^+$ND4 DNA, in which the SV40 substitution sequences give rise to a distinct DNA:DNA substitution loop (77, 78). R-loops were then formed in such heteroduplexes and were readily oriented with respect to the asymmetrically disposed substitution loop. In another study, Ad2 RNA:single-stranded DNA heteroduplexes were first formed and subsequently annealed to Ad2$^+$ND4 single DNA strands (79).

Now that RNA splicing is known to occur in many eucaryotic RNAs, the pattern of RNA deletions can suggest polarity: short leader segments tend to be at the 5′ end of the RNA, and long, uninterrupted segments toward the 3′ end. Furthermore, the 3′ ends of RNAs hybridized to single-stranded DNA may be identified by their poly(A) tails 100–200 nucleotides long; the visibility of such poly(A) tracts can be enhanced by hybridization to poly(dT) or poly (bromouracil) linked to a label such as a small circular DNA (59, 80, 81).

With cloned DNAs, RNAs can be mapped accurately if the circular plasmid is linearized, in separate experiments, with two single-cut restriction enzymes that cleave the vector at two different distances from the gene(s) of interest.

7.4 Data Presentation

R-loop or RNA:single-stranded DNA heteroduplex data can be presented in a variety of ways, each of which is useful and gives an indication of the length and position of transcripts (58). The simplest description is an array or gallery of individual molecules with the position and length of the heteroduplex indicated (Figure 8). This display reveals the number of RNAs hybridized to each DNA and their relative spacing, allows a qualitative assessment of variability and reproducibility, and is particularly useful in the early stages of analysis to establish orientations, since multiple RNAs paired with the same DNA molecule reduce the uncertainty when all must be fit to a common, emerging pattern. Once oriented to a "best fit," arrays can be summarized by an aggregate positional histogram (Figure 9). Such displays have advantages and limitations. Unencumbered peaks reveal the level of reproducibility of a particular RNA species. But when different RNA transcripts are adjacent to one another, their boundaries become obscured by the aggregate representation. To resolve the individual RNA species, a histogram representing the 5′ and 3′ termini of transcripts can be constructed

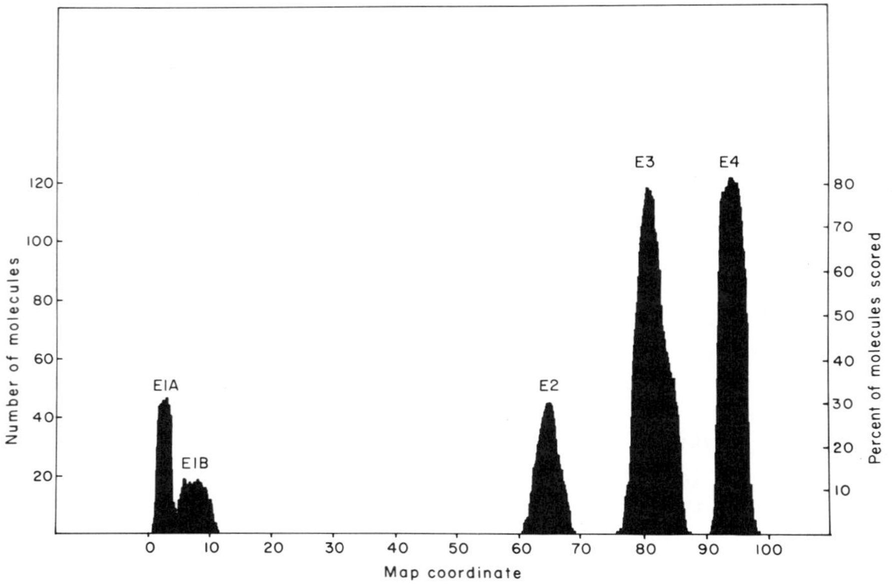

Figure 9 R-Loop distribution frequency of adenovirus-2 early RNA. The additive coverage of 148 DNA molecules is presented (see Figure 3*a*). Early regions 1A, 1B, 2a, 3, and 4 are clearly separate.* The biphasic slope on the 3′ side of region 3 is indicative of two alternative 3′ ends (see Figures 10 and 11) (58).

(Figure 10). In regions with clustered 5′ and 3′ termini, we represent 5′ ends as ticks above a line, and their 3′ ends as ticks below the line. The boundaries of adjacent transcripts complementary to the same strand then emerge. To represent data for convergent/divergent transcripts of opposite strands, the same approach can be used, or two separate panels can be prepared and aligned (Figure 10). Such end point histograms are particularly useful for evaluating adjacent transcripts and for detecting the occurrence of messages with one common end but alternative 5′ or 3′ ends. Histograms of map positions reflect, of course, the cumulative measuring errors between the site of interest and the index coordinate, and therefore may be subject to substantial standard deviations that can obscure innate RNA heterogeneity. Histograms of RNA length reduce such uncertainty because they eliminate the degree of freedom present when ends are evaluated independently, rather than pairwise (Figure 11). Evidence for overlapping adenoviral transcripts with a common 5′ end and variable 3′ ends (or with a common 3′ end and variable 5′ ends) was strengthened by such analyses. To reinforce such deductions, a correlated analysis can be performed in which the shorter individuals in the length histogram should also be the individuals contribut-

*Region 2 has recently been shown to be E2A (see Figure 15) (162).

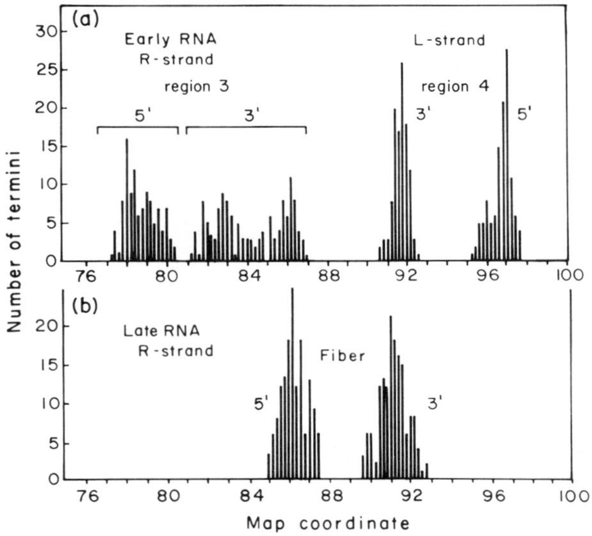

Figure 10 Distribution of R-loop 5′ and 3′ termini in the rightmost 23% of the adenovirus-2 chromosome. (*a*) The early RNAs from region 3 (r-strand transcripts) have alternative 3′ ends near coordinates 82.7 and 86.0 (see Figures 9 and 11). The main bodies of early RNAs from region 4 (l-strand transcripts) have two predominant 5′ ends near coordinates 97 and 96. (*b*) The main body of the late fiber RNA, transcribed from the r-strand, lies between coordinates 86.3 and 91.3. The 3′ end of early region 3 RNA abuts the 5′ end of fiber RNA, and the 3′ ends of early region 4 RNA and fiber RNA converge (58).

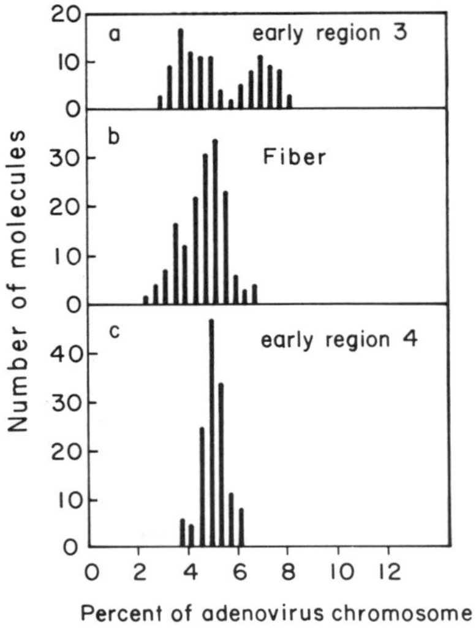

Figure 11 Lengths of adenovirus-2 R-loops formed with RNAs from early regions 3 and 4 and from the late fiber gene. The bimodal distribution of region 3 transcripts clearly indicates two major species that are shown in Figures 9 and 10 to have alternative 3′ ends (58).

ing most of the data to the proximal alternative end point peak, and the longer individuals should contribute most of the data to the more distal alternative end point. If this is not the case, a systematic error should be suspected.

Because of a multitude of potential anomalies in the molecule itself and errors in interpretation and measuring, statistical analysis based on many measurements is essential. For the number of measurements normally gathered in electron microscopic analysis (20–100 for a given coordinate), a proper choice of bin width in histograms is essential to approximate Gaussian curves. [Consult a reference on the application of statistical analysis to biological problems, such as Snedecor (82).] A reasonable choice of bin width is $0.4–0.5\sigma$, where σ is the standard deviation of the mean. By definition, about 68% of all measurements fall within 1σ of the mean, and 95% within 2σ of the mean. Therefore, a typical histogram with bin size 0.4σ would have virtually all measurements distributed into 10 bins, which would reveal a reasonable peak. Clearly if the bin size were 1.0σ, only four bins would have entries, and the data would suggest a nice, sharp peak, but resolution of alternative peaks within several σ of one another would be obscured. Similarly, choice of bin size 0.1σ (40 bins) would mean many bins having no entry or only one entry, and none having more than several; thus statistical fluctuation would obscure resolution of sites that are actually separate.

An important final point: the nominal accuracy of a mean cannot appreciably exceed the accuracy assigned to the input data. Rounding off measured numbers should be done to an accuracy realistically expected for the 95% confidence interval around the mean; in our analyses, numbers were rounded off to the nearest 0.1% of the adenovirus chromosome, or 36 nucleotides. To approach this accuracy, it is useful to make each individual measurement the average of several replicate measurements. Reproducibility in individual measurements can and should be within several tenths of a percent the length of adenovirus DNA, or ± 100 nucleotides. Expressed as raw data, if the projected image has a total contour length of one meter, replicate measurements of a given RNA:DNA heteroduplex junction should agree within several millimeters.

7.5 Standard Deviation and Statistical Confidence Intervals

Standard deviation is a measure of reproducibility, not accuracy. Precise measurements of about 20 individuals from a homogeneous sample will approximate a Gaussian distribution because of the natural scatter of measurements of individual specimens; the standard deviation is a measure of the peak width of the Gaussian curve. It does not appreciably improve with an increasing number of measurements, but the confidence interval does. This is related to the standard deviation by the following relation:

$$95\% \text{ confidence interval} = \text{mean} \pm \frac{T\sigma}{\sqrt{n-1}} \qquad (8)$$

where T is Student's test number. For measurements of about 17–101 individuals in the population, T is approximately 2. Therefore, the 95% confidence interval of a relatively few, somewhat scattered measurements (e.g., $\sigma = 0.5\%$) is equal to mean $\pm 2 \times 0.5\%/\sqrt{16} = 0.25\%$; that of a larger number of rather tightly distributed measurements (e.g., $\sigma = 0.2\%$) is equal to mean $\pm (2 \times 0.2\%)/\sqrt{100} = 0.04\%$. In practice, accuracies have been within these ranges, but the latter is approaching "hollow" resolution because systematic errors become significant, as described later. Adenovirus-2 DNA consists of about 36,500 base pairs. We have mapped the 5′ and 3′ ends and the splice junctions of most transcripts with a nominal standard deviation of ±0.2–0.5% (e.g., 73–182 nucleotides). A substantial portion of the adenovirus chromosome has since been sequenced, and RNA coordinates determined by electron microscopy have been correlated with the appropriate DNA sites. This has allowed an absolute test of the accuracy of electron microscopy that had not been possible previously. In most cases, it has indeed been within the 95% confidence interval calculated, as summarized in Table 3.

7.6 Systematic Errors

The apparent accuracy of any series of measurements can be compromised by systematic errors that affect each individual measurement in the same way (89). By correlating the EM measurements of adenoviral RNA transcript coordinates using nucleotide sequence analysis, certain systematic errors were recognized. Others were apparent from inconsistencies in the EM measurements themselves. Short gaps between RNA:DNA duplex segments tend to stretch slightly, possibly because of the concentration of shear forces in the intervening single-strand segment. Thus the gap between adenovirus early region 1A and region 1B RNAs measured 0.27 ± 0.13% and actually is 0.19% (a 29 nucleotide or 93 Å discrepancy) (Figure 12*a–d*) (24). Very short deletion loops tend to collapse or anomalously contract. Thus a deletion loop of 83 nucleotides (0.23%) (266 Å, which when doubled back on itself, is 133 Å) (86, 90) was not detected by EM (Figure 12*a–c*) (24). When the existence of such a loop is known from independent work, it can sometimes be recognized because it can cause a sharp bend in the DNA contour. Another deletion loop of 116 nucleotides (0.32%) (85, 90) appears as a knob (24) (Figure 12*b*). A third deletion loop of 254 nucleotides (0.70%) (85) is clearly visible but is measured to be 0.5% long (Figure 12*d*) (24). In this case, a correction to the loop length was made on the basis of measurements relative to coannealed RNA in an adjacent DNA segment. Longer deletion loops of 1.6% (600 bases) and 3.7% (1338 bases) are accurately measurable

TABLE 3 Test of Electron Microscopic Measuring Accuracy[a]

Event	EM Coordinate (Percentage Map Units)	Sequence Coordinate (Based on 365 Bases = 1%)
5′ End of E1A RNA	1.3	1.37
Upstream splice site, E1A RNA	2.6	2.67
Downstream splice site, E1A RNA	3.3	3.37
3′ End of E1A RNA	4.4	4.47
5′ End of E1B RNA	4.6	4.66
Upstream splice site, E1B RNA	6.1	6.18
Downstream splice site, E1B RNA	9.8–9.9	9.81, 9.85
5′ End of RNA IX	9.8	9.8
3′ End of convergent RNAs IX, IVa_2	11.2	11.16–11.2
Downstream splice site, IVa_2RNA	15.1	14.87
Upstream splice site, IVa_2RNA	15.7	15.64
5′ End of IVa_2RNA	16.1	16.00
Upstream splice site, first late leader	16.6	16.68
Downstream splice site, early region 4	95.7	95.92
Upstream splice site, early region 4	99.15	98.96
*Hin*dIII restriction site, coordinate 17.0	—	17.22

[a] The electron microscopic coordinates were reported by Chow, Gelinas, Broker, and Roberts (1), Chow and Broker (19), and Chow, Broker, and Lewis (24). DNA sequence was reported by van Ormondt et al. (83), Maat and van Ormondt (84), and Maat et al. (161). RNA splice junctions and sequences at the nucleotide level were determined by Perricaudet et al. (85, 86), Baker and Ziff (87), Aleström et al. (88), H. van Ormondt (personal communication), and D. Sciaky and N. Stow (personal communication).

(Figure 12*d*) (1, 19, 24, 86). In summary, small deletion loops seem to measure about 80 nucleotides shorter than they are and are not usually detectable below about 100 nucleotides long.

Measurements of R-loops are also subject to errors. Branch migration due to DNA:DNA renaturation can displace the ends of the RNA in the R-loop. This can be minimized by spreading immediately after dilution of the sample into the hyperphase. Heterogeneity in the RNA due to breakage can be masked by such branch migration and result in an average RNA length shorter than that obtained by RNA:single-stranded DNA heteroduplex methods.

Errors in map coordinates accumulate when multiple segments are measured to arrive at the desired coordinate (such as the situation posed by the series of deletion loops and duplexes formed in spliced RNA:DNA heteroduplexes). Errors also increase in proportion to the length of the measurement, especially with single-stranded DNA (13). Accumulated errors are

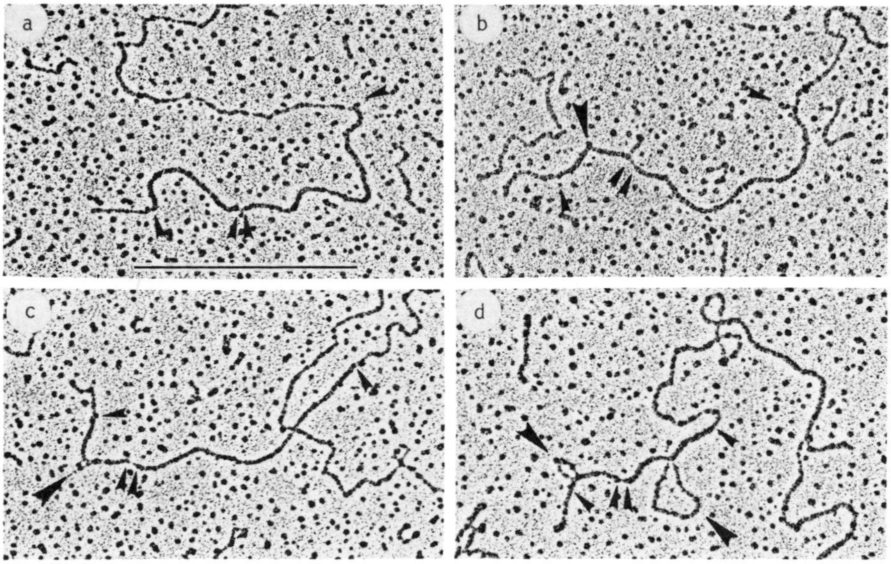

Figure 12 Alternatively spliced transcripts of the adenovirus-2 transformation genes. All species from early region 1A or from early region 1B share a common 5′- and a common 3′-terminus. The termini are indicated by small arrowheads. Large arrowheads point to the intervening DNA sequences. (*a*) 1A RNA and 1B RNA with no apparent splices; (*b*) 1A RNA with the smallest deletion loop of 116 nucleotides (85) and 1B RNA with no apparent splice; (*c*) 1A RNA with an intermediate size deletion loop of 254 nucleotides (85) and 1B RNA with no apparent splice; (*d*) 1A RNA with a deletion loop of about 600 nucleotides (24) and 1B RNA with a deletion loop of 1338 nucleotides (86). The apparently unspliced region 1B RNAs actually have a deletion loop of 83 nucleotides (86, 90), usually unresolvable by electron microscopy. Bar = ~ 2000 nucleotides or nucleotide pairs.

reflected in larger standard deviations of measurements of RNA coordinates near the center of the adenovirus-2 chromosome, for instance, than those near ends (19, 24, 58). To some extent, both types of error can be reduced by making and expressing measurements to a closer reference point, for instance, to a second RNA:DNA heteroduplex, to a secondary structure in single-stranded DNA, or to a restriction site in a truncated DNA molecule (the latter, of course, only being as accurate as the restriction map).

7.7 Quantitation

The complex sensitivity of hybridization rates to temperature, solvent, AT/GC base composition of the different segments of the DNA, and excluded volume considerations (91) that affect pairing of RNAs to DNA segments located internally in a molecule more than RNAs encoded by terminal

DNA segments all make it difficult to obtain absolute quantitation of RNA abundances by scoring RNA : DNA heteroduplexes. Nonetheless, two levels of relative quantitation are still possible:

1 Comparisons of the same RNA species present in different preparations of RNA (represented as a percentage of the total RNA) can be estimated. For instance, RNAs may be prepared at different times after infection or in the presence of different metabolic inhibitors (24, 76, 79). Because factors such as base composition cancel out, the only variables beside actual differences are the reproducibility of temperature and solvent.
2 The relative abundances of different RNAs in the same preparation can also be estimated, although with somewhat less accuracy. But changes in such relative abundances in different RNA preparations can indicate meaningful trends (24, 79). The values obtained are in good agreement with those achieved by filter hybridization studies. These latter values, of course, are subject to some of the same factors and variabilities as those scored by electron microscopy.

When attempting quantitation of relative abundances, the experiments should be carried out in the presence of DNA excess; hybridization of RNA to single-stranded DNA or formation of R-loops in psoralen cross-linked DNA at temperatures above T_{ss} are far superior to the standard R-loop method, in which hybridization of GC-rich transcripts are at a considerable disadvantage because of the stability of GC-rich duplex DNA under the annealing conditions.

8 RNA:DNA HETERODUPLEX MAPPING: APPLICATION TO ADENOVIRUS-2

When we began to focus attention on the human adenoviruses, the basic transcription map had been blocked out by hybridization of RNA to the separated strands of defined DNA restriction fragments (92–94). The lengths of various RNA molecules had also been estimated by sucrose gradient centrifugation (95) or by polyacrylamide gel electrophoresis (96, 97). Partial correlation of the physical maps of the DNA with the genetic maps (98, 99) and with RNAs and their encoded proteins (95, 100, 101) had generated the following basic organization of the genome: segments of both strands of the DNA are transcribed and represented by cytoplasmic RNA. Several widely separated regions of the DNA are transcribed at early times (1–8 hours) after infection (Figures 3*a* and 9). Thereafter most of the remaining portions of the genome are transcribed during the late phase from about 8 hours to cell death some 48–72 hours postinfection (Figure 8). From the general reproducibility of R-loops, we learned that a significant percentage of adenoviral transcripts are sufficiently intact to calculate the 5′- and 3′-end coordinates and the lengths of many species, with 95% confidence intervals generally about ±0.5% (±180 base pairs) (Figures 10 and 11) (58). This, in turn, allowed resolution of the individual mature RNA species that were located side by side within the blocks of early or late transcription regions. Early region 1 emerged as two adjacent regions 1A and 1B (Figure 9). A small late transcript for peptide IX was found to be coincident with the 3′ portion of early region 1B. Two late regions, in particular, were each found to consist of a family of 3′ coterminal RNAs with 3′ ends at coordinates 62 and alternative 5′ ends at 50 and 52 or common 3′ ends at coordinate 78 and alternative 5′ ends at 66, 68, and 74. Based on the EM mapping of the staggered 5′ ends, the coding capacities of the several RNAs, the lengths of RNAs specifying different proteins (95), and their approximate locations in the DNA (100), we successfully deduced the order of adjacent genes for pVI and hexon and for 100K and pVIII, respectively.

This conclusion coupled with the translation studies lent support to the emerging concept that in eucaryotic polycistronic RNAs, only the 5′ proximal message was translatable (102). We also correlated other, monocistronic messages and the proteins encoded by their locations, lengths, and minimal coding regions; these included genes for peptide IX, fiber (IV), DNA-binding protein (72K), and IVa_2 (58). These assignments have since been confirmed by the powerful hybridization-arrested translation (HART) method (103–105). The GC-rich interval from map coordinates 30–50 also gave evidence of preferred 3′ and 5′ ends and of families of related RNAs, but the difficulty in making R-loops there prevented detailed mapping.

At that time, a number of anomalous biochemical observations emerged that suggested that some adenoviral late RNA transcripts might have unusual structures. Briefly, many different late RNAs had the same nucleotide sequences present at their 5′ end (3, 4). Zain found that these sequences were not encoded immediately adjacent to the fiber gene, although they were present in the mature fiber message (4, 7, 106). And Klessig (4), Dunn and Hassell (5) and Lewis et al. (6) all found that many different late messages hybridized with low efficiency to adenoviral DNA sequences quite remote from their coding region, near coordinates 17–31 map units. Several alternative explanations for these unexpected findings were entertained. To test one exotic possibility—that the RNA sequences constituting mature messages were derived from separate segments of the genome—R. E. Gelinas, R. J. Roberts, and we (1) conceived and carried out the following electron microscopic experiment: R-loops were formed in whole adenovirus-2 DNA using late RNA. If short sequences at the 5′ ends of the RNAs were not encoded next to the main bodies of the various messages, they would appear as branches at the 5′ ends of the R-loops. [In fact, we had seen such short branches in our previous R-loop study (58), but had attributed them to branch migration displacement during spreading because their length could be minimized, down to a limit of about 200 bases, by rapid spreading after annealing.] After formation of the R-loops, the separated strands of defined

DNA restriction fragments collectively representing the entire adenovirus-2 chromosome were added in 16 individual annealing reactions, and hybridization was continued while the temperature was gradually decreased to promote the formation of heteroduplexes of unknown lengths and AT/GC base composition. Indeed, certain r-strand restriction fragments did anneal to the 5′ ends of late R-loops (Figure 13).

Two unanticipated conclusions were immediately drawn. First, not one, but three remote sites gave rise to sequences present at the 5′ ends of the late RNAs; all messages derived from the rightward-transcribed strand to the right of coordinate 36 (and now known to include all those to the right of coordinate 29) had the same three leader segments, and these were all transcribed from the r-strand at coordinates 16.5–16.6, 19.5–19.7, and 26.5–26.8, for a total length of about 200 nucleotides. The sequence of the short tripartite leader has since been determined and it is not translated (44, 45); but it does include a potential complement to the 3′ end of the 18s ribosomal RNA (107). Second, another message (for a 15K peptide) encoded near the left end of the genome had a unique, composite organization in which its 5′ and 3′ halves were of approximately equal length but derived from genomic coordinates several map units apart (Figure 6). This E15K message has subsequently been shown to contain coding sequences on both sides of the recombinant junction (86) and was the first example of what has become a common phenomenon of RNA splicing within protein coding regions.

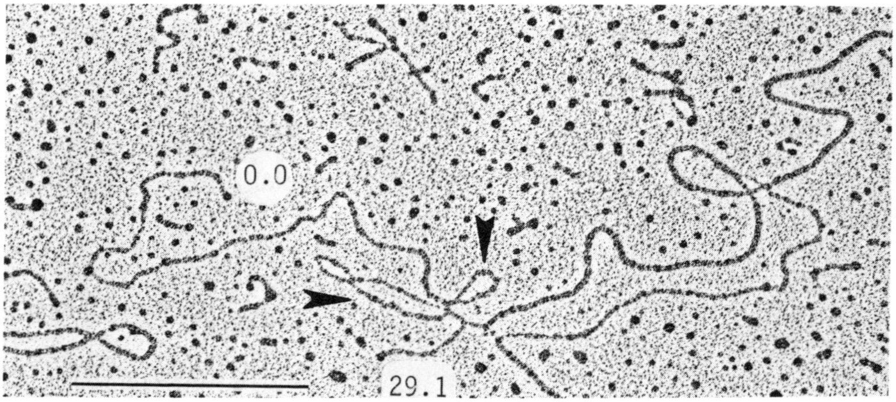

Figure 13 Multi-site hybridization of the 5′-leader sequences on adenovirus hexon RNA. The hexon RNA, an r-strand transcript, was annealed with double-stranded adenovirus-2 DNA to form an R-loop (coordinates 51.5–61.5). The purified r-strand of the *Bam*HI B restriction fragment (coordinates 0.0–29.1) was then added, and incubation was continued. Three segments at the 5′ end of the hexon RNA annealed with three separate sites in the restriction fragment, at coordinates 16.5–16.6/19.5–19.7 and 26.5–26.8, and constrained the fragment into two loops (1). Bar = ~ 2000 nucleotides or nucleotide pairs.

Using a slightly different EM technique—RNA: single-stranded DNA heteroduplex formation—Berget et al. (2) reached the same conclusion about the structure of the adenoviral hexon mRNA, one of the eight late RNAs that we had found to have a tripartite leader. It annealed to single-stranded adenovirus-2 DNA and constrained it into three deletion loops, indicative of the four-segmented nature of the RNA (cf. Figures 2*b* and 4*c*, IX). At about the same time, Darnell and colleagues (8) showed that the late r-strand adenoviral messages all originated from a common promoter near map coordinate 16. Combining these observations, the model in which long primarily nuclear RNA transcripts undergo selective deletions and re-ligations (splicing) to form segmented cytoplasmic messages emerged as the most likely of several alternative explanations (Figure 14) (4, 7, 108).

Shortly thereafter, Kitchingman et al. (117) reported that RNAs from each of the early regions also could exhibit a spliced structure, and RNA splicing was also found in a broad spectrum of other eucaryotic systems ranging from tRNA, rRNA, and other viral messages to single-copy eucaryotic genes for both the specialized products of differentiated cells and the more basic "housekeeping" proteins (118).

Our next efforts were to produce an exhaustive description of all the adenoviral transcripts made throughout the infection cycle, in the presence as well as absence of various metabolic inhibitors, and in different host cell types (19, 24, 75). Our aims were severalfold: (*a*) to reveal the full complexity of the organization and expression of the virus (cf. Figure 15), (*b*) to perturb the normal transcriptional and post-transcriptional processing pro-

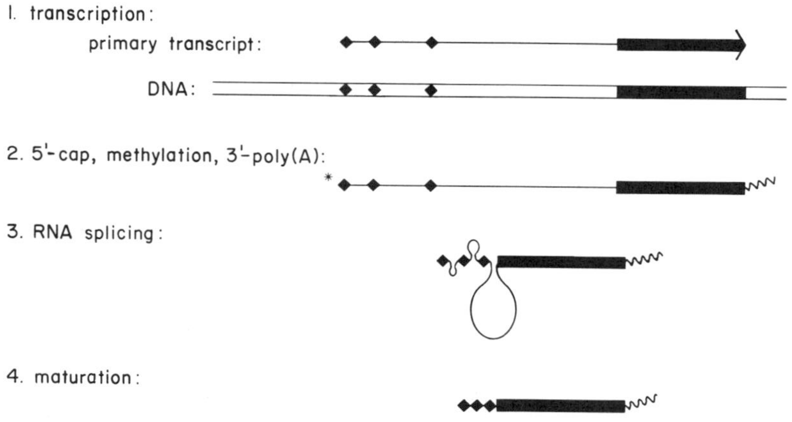

Figure 14 Model of adenovirus late RNA transcription and splicing: (1), nuclear transcription of double-stranded DNA; (2), post-transcriptional modification with 5′ caps (3, 4, 109–112), methylation of caps and internal sites (3, 4, 109–113), and 3′ polyadenylation (114, 115); (3), RNA splicing to delete the intervening sequences and conserve the leaders and main body of the message (4, 7, 108); (4), maturation and transport to the cytoplasm for translation on ribosomes. (Adapted from reference 10.)

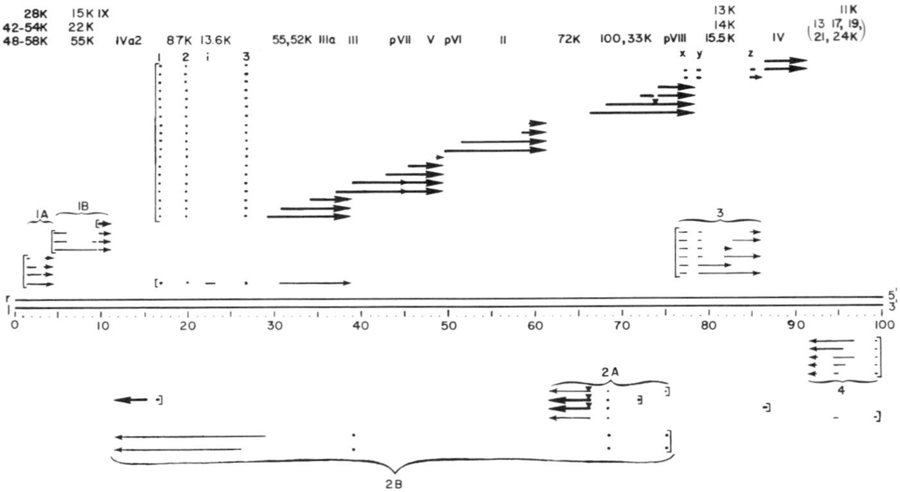

Figure 15 The map of human adenovirus serotype 2 cytoplasmic RNA transcripts, determined by electron microscopy of RNA:DNA heteroduplexes. The 36,500 base pair chromosome is divided into 100 map units. Arrows indicate the direction of transcription along the r-strand or l-strand of DNA. The 5′ ends of the cytoplasmic RNAs correspond to those of the primary nuclear transcripts and therefore refine the locations of transcription promoters (indicated by vertical brackets). The conserved segments constituting early RNAs (24; 162; are depicted by thin arrows and those in late RNAs (1, 19 and L. T. Chow and J. B. Lewis, unpublished results) by thick arrows. The promoter for the late RNAs at coordinate 16.5 is also active at early times, but transcription terminates at coordinate 39 (24, 131). Gaps in arrows represent intervening sequences removed from the RNAs by splicing. Early regions 1A, 1B, 2A, 2B, 3, and 4 are labeled. At intermediate and late times, region 2 is expressed from several additional promoters, and region 3 RNA can be made under the direction of the major r-strand late promoter (24). All derivatives of the late r-strand transcript have the same tripartite leader, the segments of which are labeled 1, 2, 3. They form a number of families of 3′-coterminal transcripts. All late r-strand messages from the major promoter can have the i-leader segment, but it is most commonly associated with the 55, 52K message, as shown, at early and intermediate times after infection, or in cells infected in the presence of cytosine arabinoside (24, 131). Some of the RNAs for protein IV (fiber) can also contain some combination of ancillary leader segments *x*, *y*, *z* (19). The correlations of mRNAs with encoded proteins were based on cell-free translations of RNA selected by hybridization to DNA restriction fragments (6, 100, 101, 119–124, 162). Proteins are designated by K (1000 daltons molecular weight) or by Roman numerals (virion components) (125). Alternatively spliced RNAs complementary to early regions 1A, 1B, 2B, 3, and 4 give rise to multiple proteins, some of which share common peptides (J. E. Smart and J. B. Lewis, personal communication).

grams to learn how they work and to what they respond, and (*c*) to refine the RNA map coordinates so that detailed correlations with DNA sequences could be drawn (116). For most of these studies, hybridization of RNA to single-stranded DNA has been the most productive technique because each RNA splice creates a DNA deletion loop in the heteroduplex, and this is usually easily scored and measured. Once all the RNA species were known and the investigator was very familiar with their heteroduplex positions and structures, quantitative estimation of the relative abundances of RNAs from different transcription units and of alternatively spliced products from the same transcription unit became possible by extensive visual (rather than photographic) survey of grids (24).

Numerous biological conclusions about the organization and function of adenoviral RNA synthesis and splicing have been drawn by correlation of the EM mapping data with information on DNA and RNA sequences, protein sizes and composition, and the effects of drugs, viral mutants, or alternative host types. For instance, broad regions have been identified to contain promoters for RNA transcription by estimating the coding regions for the 5′ ends of nascent, nuclear RNA (126–128). In all cases, the 5′ end of the mature cytoplasmic RNAs from each transcription unit (1, 2, 19, 24, 90, 117) falls within the map intervals estimated for the locations of the promoters. Thus, the two conclusions are drawn: (*a*) 5′ ends of nuclear RNA are conserved during post-transcriptional processing, in agreement with biochemical data on 5′-cap structures (3, 4, 109–112), and (*b*) the EM-derived coordinates for the 5′ ends of cytoplasmic RNAs provide a refined location ($\pm$ 70 bases) for promoters. Many of these have now been confirmed by direct nucleotide sequence analysis (87, 107).

The early adenoviral RNAs originate from five promoters. Each primary transcript is spliced in alternative ways (Figures 12 and 15) (24, 90), and each of these related species gives rise to alternative protein products (101, 119–123). One-to-one correlations of early RNA species and their protein products have been achieved in several cases. The relative abundances of the various RNAs change substantially during the course of infection and respond in different ways to various inhibitors of DNA replication and protein synthesis (24, 129, 130). Furthermore, the temporal program of viral development has emerged as much more intricate than the simple early-late pattern; immediate early, early, intermediate, and late phases are now evident (119, 122, 131–133). The late transcripts derived from the major r-strand promoter at coordinate 16.5 all have a tripartite leader (19, 24). They fall into five main families of 3′-coterminal transcripts, and several more minor families (Figure 15) (19, 115, 134–136). Some 15 or more RNAs constitute the sets of 5′- and 3′-coterminal families. Practically all of them have been correlated with the proteins encoded (Figure 15). Three events determine the messages that are produced: (*a*) positive and negative control of RNA promotion, (*b*) RNA termination and polyadenylation at one of several, alternative 3′ sites, and (*c*) alternative internal splicing (with the conservation of 5′ and 3′ ends) of the primary transcript. Thus gene regulation in the adenovirus system, at least, has several sites of action to allow very sensitive control. RNA splicing, as complex as it is, seems to follow a number of reasonably well understood principles, and it contributes to not one, but many different ways of extracting genetic information from a limited genome; it confers biological flexibility. We have reviewed these features elsewhere (9, 10).

9 ANCILLARY PROCEDURES

9.1 Enrichment of RNA After Hybridization to DNA

Following the formation of heteroduplexes with single-stranded or double-stranded DNA, RNA assumes the hydrodynamic properties of the much more massive DNA component. This property has been exploited to enrich the hybridized RNA by using equilibrium density gradient centrifugation or agarose gel filtration chromatography. Of immediate practical importance to electron microscopy, it is often necessary to start with large amounts of RNA when forming DNA heteroduplexes with a rare RNA species. Direct preparation of the sample for electron microscopy could then result in a dense RNA background, which makes scanning for the desired structures tedious and time-consuming. Fractionation of the heteroduplexes from the bulk of the unhybridized, input RNA can minimize this problem.

Density Gradient Centrifugation

Westphal and Lai (79) prepared RNA:SS DNA heteroduplexes with adenoviral RNA and DNA, then rapidly banded the entire mixture to pseudo-equilibrium in a Cs_2SO_4 step gradient to remove unhybridized RNA. The heteroduplexes can be examined by EM directly after recovery or can be converted to R-loops by reannealing the complementary DNA strand to the RNA:single-stranded DNA hybrids (14, 79). Similarly, Wellauer and Dawid (22) separated *Drosophila* ribosomal DNA R-loop structures from free RNA using either CsCl or Cs_2SO_4 gradients prepared after brief dialysis to remove the formamide used during hybridization. The molecules containing R-loops were slightly denser than free double-stranded DNA, and a sufficiently high fraction of each DNA molecule was paired with RNA that reasonable separation was obtained.

Gel Filtration

Several groups have devised methods for fractionation of R-loop structures by passage through agarose gel filtration columns (24, 25, 62, 137, 160). We prepared a 5 ml Sepharose 2B or 2B-CL (cross-linked) column and equilibrated it with the R-loop hybridization buffer: 70 or 80% formamide, 0.4 *M*

NaCl, 0.1 *M* HEPES, 0.01 *M* EDTA. In this solvent at 4°C, R-loops are stable, R-loop structures and duplex DNA are extended and move through the column in the void volume, whereas free RNA is largely collapsed and is included by one or more fractions of 0.25 ml. The molecules recovered in the void volume by elution with the R-loop buffer-formamide solution can be examined directly by electron microscopy, or the complementary DNA strands can be denatured before preparation for electron microscopy. It is important to note that the columns were equilibrated and eluted with 70 or 80% formamide to permit the second, strand-separating incubation.

Holmes et al. (25), Woolford and Rosbash (137), and Kaback et al. (62) also removed heterologous RNA from R-loop structures by passage through agarose bead (Sepharose 2B or Bio-Rad A150m) filtration columns. The columns in each of these studies were eluted with aqueous buffers lacking formamide. In this case, the single-stranded DNA limb of the R-loops can be reacted with glyoxal prior to gel filtration to prevent it from displacing the RNA (62).

RNAs annealed to very short single-stranded DNA molecules do not appear in the void column but are included in Sepharose 2B and are heavily contaminated with free RNA. Holmes et al. (25) resolved such heteroduplexes from free RNA and single-stranded DNA by retaining them on very small (20 μl) columns of hydroxylapatite. The desired heteroduplexes were then eluted with 0.5 *M* Na phosphate buffer.

9.2 Enrichment of DNA After Hybridization to RNA

Just as it can be important when starting with pure DNA and a crude mixture of RNAs to remove excess RNA from RNA:SS DNA or RNA:DS DNA heteroduplexes, it may also be desirable to separate the heteroduplexes from DNA that has not paired with RNA. Such gene enrichment can be done when starting with pure RNA and bulk genomic DNA or DNA partially enriched by other methods. The DNA may be single- or double-stranded. In the latter case, the DNA can subsequently be cleaved with restriction endonucleases and cloned, if desired. When the genes of interest are in tandem arrays, such as those for ribosomal RNAs, the heteroduplexes may have sufficiently altered buoyant density to allow direct separation by CsCl or Cs_2SO_4 equilibrium density gradient centrifugation, as just described (22). But with RNA probes for single copy genes or very short genes such as those for transfer RNAs, the potential density shifts are insignificant and the heteroduplexes would band coincidently with the bulk DNA. Nonetheless, the target genes and their flanking sequences can be enriched if the RNA is modified with appropriate reactive groups or affinity agents.

Thus Dale and Ward (138) used mercurated RNA and retained RNA:DNA heteroduplexes on sulfhydryl–Sepharose columns. Manning et al. (139) annealed *Drosophila* ribosomal RNA that had been covalently linked to the vitamin biotin to total single-stranded cellular DNA. The DNA segments containing clusters of rRNA genes were fractionated from the bulk DNA by either of two methods: (*a*) passage through avidin–glass bead columns, which retain the biotin-RNA: DNA complexes by virtue of the strong affinity association of avidin and biotin, or (*b*) association of the complexes with avidin linked to polymethacrylate microspheres (140), followed by buoyant density gradient centrifugation. Both the yield and the purity of the rDNA fraction were very high. To prepare double-stranded DNA for cloning, Pellegrini et al. (141) extended the same affinity selection methodology to enrich R-loops containing RNA-biotin.

9.3 Direct Visualization of Single-Stranded RNA

Examination of pure RNA preparations (as opposed to a mixture of many different RNAs) to determine their quality may be desirable prior to hybridization to DNA. Length distribution and secondary structure can be analyzed. The basic strategy in the preparation of free RNA for electron microscopy is to weaken or eliminate adventitious intrastrand base pairing that causes most single-stranded RNA to collapse into a bush, even under conditions in which single-stranded DNA is extended (142). Such base pairing can be disrupted with chaotropic solvents containing 3–6 *M* urea (143) and/or formamide (144, 145).

A simpler, and perhaps more reproducible, procedure is to react the amino groups present on guanine, adenine, and cytosine with formaldehyde (145, 146) or with glyoxal or kethoxal (59, 147). Alternatively (or, in addition), the ring imino (NH) groups of guanine and uracil can be reacted with methylmercuric hydroxide (CH_3HgOH) to denature fully the RNA in a reversible reaction (60, 61). (Note: methylmercuric hydroxide is very toxic and somewhat volatile and should be used with extreme caution in a well-ventilated fume hood.) All reactions and EM sample preparations that use aldehydes or methylmercuric hydroxide should, of course, be carried out in the absence of primary amine buffers such as Tris-Cl or ammonium acetate; the RNA stock solutions themselves should not carry such buffers into the reaction mixtures. Phosphate, sulfonate, carbonate, and tertiary amine buffers are satisfactory.

Reproducible RNA secondary structures can be achieved by controlled reaction with glyoxal before spreading (60), by careful adjustment of the urea-formamide solvent concentration (145), or by inclusion of 0.1–2.0 mm $MgCl_2$ during spreading (148).

Finally, single-stranded RNA, like single-stranded DNA, can be coated with bacteriophage T4 gene 32 protein and the complex stabilized by reaction with glutaraldehyde to cross-link the protein shell (149). Such samples can be spread in a hyperphase of 30% (or 40%) formamide onto a hypophase of 5% (or 10%) formamide. The RNA lengths will be reasonably reproducible, and incubation with strong denaturants can be avoided; this is useful if transcriptional complexes, for instance, are to be studied.

9.4 Labeling Methods for EM Visualization of RNA:DNA Heteroduplexes

Ferritin Labeling of RNA

Very small RNA molecules such as tRNAs and 5s ribosomal RNAs cannot be visualized directly when annealed to single-stranded DNA, nor do they make identifiable R-loops. They can be recognized indirectly if specifically coupled to an electron-dense or morphologically distinct label. For most work, ferritin, a multi-subunit iron-storage protein from spleen, has been ideal. Holoferritin has a molecular weight of about 900,000 daltons and a diameter of about 110 Å. Three schemes have been devised to couple ferritin to tRNAs. All involve sodium periodate oxidation of the 3′ ribose of the RNA, and the linkages are achieved by chemical reactions or by a combination of chemical reactions and biological affinity associations (150–153). The labeling usually does not go to completion because of slight instabilities of the linkages between ferritin and the RNA, because of the large size of the labels that decrease diffusion rates, and because of steric hindrance and electrostatic repulsions that can decrease or prevent free diffusion to sufficient proximity to permit reaction or affinity association. Furthermore, the molar concentrations of massive DNA molecules and labels are so low that second-order reactions are exceedingly slow. Each method can, with effort, result in the ferritin labeling of about 50% of the hybridized tRNAs. The methodology is fairly intricate, however, and should not be attempted casually.

Negative Enhancement of Short Duplex Regions with Single-Stranded DNA Binding Proteins

Small as well as large RNAs (or DNAs) annealed to single-stranded DNA can be demarked by binding the T4 gene 32 protein (154) or the *E. coli* single-stranded DNA binding protein (155) to the heteroduplex. The RNA:DNA hybrid is thin relative to the flanking SS DNA coated with the protein. The DNA binding protein can be further overcoated with specific antibody (155). One disadvantage of the *E. coli* DBP method is the substantial contraction of the SS DNA. As a result, accurate mapping of the RNA with respect to the ends of the DNA or other markers is somewhat compromised.

10 CONCLUSIONS

Electron microscopy of RNA:DNA heteroduplexes has been a major element of structural analysis of eucaryotic RNA transcripts. The methodology has been advanced profoundly by application of the observation that RNA:DNA heteroduplex stability exceeds that of DNA:DNA duplexes in the presence of high concentrations of formamide. We and others have developed coordinated variations in sample preparation (and complementary forms of data acquisition and analysis) to the point where the combination of sensitivity and accuracy exceeds all other RNA mapping methods except that of direct polynucleotide sequencing. The total characterization of a transcription system is most efficient when both the R-loop method and the RNA:single-stranded DNA method are used whenever practicable. Mapping resolution can reach ±50 nucleotides or better, which is sufficient to guide correlations with genomic DNA sequences. In contrast to methods of mass analysis, complex spliced RNAs can be deciphered readily with EM heteroduplex analysis, and minor RNA species can be mapped, with effort, in a background of 20,000 other RNAs. Very small amounts of RNA and DNA are needed for a complete project; only a few micrograms of early adenoviral RNA (consisting of a mixture of alternatively spliced RNAs from five major transcription units) were used to establish the transcription map (24, 58). Furthermore the RNA need not be pure; heterologous RNA will not hybridize to the DNA and simply appears in the background. Neither the RNA nor the DNA need be radioactively labeled, and mapping can be done without knowing restriction endonuclease cleavage maps of the DNA.

There are, of course, limitations to the EM methods. They can be extremely labor intensive, particularly for rare transcripts. Hybridization must be done with DNA excess in some experiments to quantitate the relative abundances of individual RNA species and with RNA excess in other experiments to reveal minor species. The RNA and DNA preparations must be of higher quality than is necessary for methods of mass analysis, since broken molecules enter into the initial data pool and must be processed until rcognized as spurious. RNAs shorter than several hundred nucleotides are not easily mapped, but spliced segments one-tenth that length can be recognized, since they cause diagnostic DNA deletion loops. The major technical limitation is in identifying and measuring small DNA deletion loops (intervening sequences) of 100 nucleotides or less; S1 nuclease methods (94, 156–158) are more sensitive in this respect. Radioactive labeling can be important in following the kinetics of appearance, processing, and turnover of transcripts; these cannot be followed by electron microscopy. Alternative methods of RNA analysis: nuclease–gel electrophoretic analysis, pulse-chase labeling, chemical analysis of modified nucleotides, cloning of cDNA, and nucleotide sequence analysis all contribute important and unique information to the total description of an RNA transcript. Together with these methods, RNA:DNA heteroduplex mapping using electron microscopy has revealed much of what we know about gene organization in many systems and about the remarkable discontinuous structures of many eucaryotic genes.

ACKNOWLEDGMENTS

We thank James B. Lewis for providing most of the adenoviral RNA used in the transcript mapping studies, and also our many colleagues who contributed to other aspects of the project. Our research has been supported by a National Cancer Institute Cancer Center Grant to Cold Spring Harbor Laboratory.

ABBREVIATIONS

RNA	Ribonucleic acid
mRNA	Messenger RNA
tRNA	Transfer RNA
rRNA	Ribosomal RNA
DNA	Deoxyribonucleic acid
SS	Single-stranded (nucleic acid)
DS	Double-stranded (duplex) (nucleic acid)
R:D	RNA:DNA heteroduplex
AT	Adenosine:thymidine bases
GC	Guanine:cytosine bases
EDTA	Ethylenediamine tetraacetic acid
Tris	Tris(hydroxymethyl)aminomethane buffer
HEPES	*N*-2-Hydroxyethylpiperazine-*N*′-2-ethanesulfonic acid
PIPES	Piperazine-*N*,*N*′-bis(2-ethanesulfonic acid)
DEPC	Diethylpyrocarbonate
EM	Electron microscopy, or electron microscopic
T_m	Duplex nucleic acid melting temperature under standard solvent conditions (e.g., 0.18 *M* NaCl or 0.15 *M* NaCl, 0.015 *M* Na citrate or 0.12 *M* sodium phosphate, pH 6.8).
T_m^*	Duplex nucleic acid melting temperature under nonstandard solvent conditions
T_{ss}	Strand separation temperature

REFERENCES

1 Chow, L. T., Gelinas, R. E., Broker, T. R., and Roberts, R. J. (1977) *Cell* **12:**1.

2 Berget, S. M., Moore, C., and Sharp, P. A. (1977) *Proc. Nat. Acad. Sci. U.S.A.* **74:**3171.

3 Gelinas, R. E. and Roberts, R. J. (1977) *Cell* **11:**533.

4 Klessig, D. F. (1977) *Cell* **12:**9.

5 Dunn, A. R. and Hassell, J. A. (1977) *Cell* **12**:23.

6 Lewis, J. B., Anderson, C. W., and Atkins, J. F. (1977) *Cell* **12**:37.

7 Broker, T. R., Chow, L. T., Dunn, A. R., Gelinas, R. E., Hassell, J. A., Klessig, D. F., Lewis, J. B., Roberts, R. J., and Zain, B. S. (1977) *Cold Spring Harbor Symp. Quant. Biol.* **42**:531.

8 Darnell, J. E., Evans, R., Fraser, N., Goldberg, S., Nevins, J., Salditt-Georgieff, M., Schwartz, H., Weber, J., and Ziff, E. (1977) *Cold Spring Harbor Symp. Quant. Biol.* **42**:515.

9 Chow, L. T. and Broker, T. R. (1980) In *Gene Structure and Expression,* (D. H. Dean, L. F. Johnson, P. C. Kimball, and P. S. Perlman, Eds.), Ohio State University Press, Columbus, OH., p. 175.

10 Broker, T. R. and Chow, L. T. (1980) *Trends Biochem. Sci.* **5**:174.

11 Kleinschmidt, A. K. (1968) In *Methods in Enzymology*, Vol. 12*B* (L. Grossman and K. Moldave, Eds.), Academic Press, New York, p. 361.

12 Westmoreland, B. C., Szybalski, W., and Ris, H. (1969) *Science* **163**:1343.

13 Davis, R. W., Simon, M., and Davidson, N. (1971) In *Methods in Enzymology*, Vol. 21 (L. Grossman and K. Moldave, Eds.), Academic Press, New York, p. 413.

14 Hyman, R. D. and Summers, W. C. (1972) *J. Mol. Biol.* **71**:573.

15 Birnstiel, M. L., Sells, B. H., and Purdom, I. F. (1972) *J. Mol. Biol.* **63**:21.

16 Thomas, M., White, R. L., and Davis, R. W. (1976) *Proc. Natl. Acad. Sci. U.S.A.* **73**:2294.

17 White, R. L. and Hogness, D. S. (1977) *Cell* **10**:177.

18 Maniatis, T., Sim, G. K., Efstratiadis, A., and Kafatos, F. C. (1976) *Cell* **8**:163.

19 Chow, L. T. and Broker, T. R. (1978) *Cell* **15**:497.

20 Robberson, D. L., Kasamatsu, H., and Vinograd. J. (1972) *Proc. Natl. Acad. Sci. U.S.A.* **69**:737.

21 Casey, J. and Davidson, N. (1977) *Nucleic Acids Res.* **4**:1539.

22 Wellauer, P. K. and Dawid, I. B. (1977) *Cell* **10**:193.

23 Vogelstein, B. and Gillespie, D. (1977) *Biochem. Biophys. Res. Commun.* **75**:1127.

24 Chow, L. T., Broker, T. R., and Lewis, J. B. (1979) *J. Mol. Biol.* **134**:265.

25 Holmes, D. S., Cohn, R. H., Kedes, L., and Davidson, N. (1977) *Biochemistry* **16**:1504.

26 Marmur, J. and Doty, P. (1962) *J. Mol. Biol.* **5**:109.

27 Watson, J. D., and Crick, F. H. C. (1953) *Nature (London)* **171**:964.

28 Herskovits, T. T., Singer, S. J., and Geiduschek, E. P. (1961) *Arch. Biochem. Biophys.* **94**:99.

29 T'so, P. O. P., Helmkamp, G. K., and Sander, C. (1962) *Biochim. Biophys. Acta* **55**:584.

30 Levine, L., Gordon, J. A., and Jencks, W. P. (1962) *Biochemistry* **2**:168.

31 DeVoe, H. and Tinoco, I., Jr. (1962) *J. Mol. Biol.* **4**:500.

32 Hanlon, S. (1966) *Biochem. Biophys. Res. Commun.* **23**:861.

33 Herskovits, T. T. (1962) *Arch. Biochem. Biophys.* **97**:474.

34 Hamaguchi, K. and Geiduschek, E. P. (1962) *J. Am. Chem. Soc.* **84**:1329.

35 Robinson, D. R. and Grant, M. E. (1966) *J. Biol. Chem.* **241**:4030.

36 McConaughy, B. L., Laird, C. D., and McCarthy, B. J. (1969) *Biochemistry* **8**:3289.

37 Blüthmann, H. B., Brück, D., Hübner, L., and Schöffski, A. (1973) *Biochem. Biophys. Res. Commun.* **50**:91.

38 Tibbetts, C., Johansson, K., and Philipson, L. (1973) *J. Virol.* **12**:218.

39 Dove, W. F. and Davidson, N. (1962) *J. Mol. Biol.* **5**:467.

40 Schildkraut, C. and Lifson, S. (1965) *Biopolymers* **3:**195.

41 Wetmur, J. and Davidson, N. (1968) *J. Mol. Biol.* **31:**349.

42 Britten, R. J. and Smith, J. (1968) *Carnegie Inst. Washington Year Book* **68:**384.

43 Hutton, J. R. (1977) *Nucleic Acids Res.* **4:**3537.

44 Akusjärvi, G. and Pettersson, U. (1979) *Cell* **16:**841.

45 Zain, S., Sambrook, J., Roberts, R. J., Keller, W., Fried, M., and Dunn, A. R. (1979) *Cell* **16:**851.

46 Smiley, B. L. and Warner, R. C. (1979) *Nucleic Acids Res.* **6:**1979.

47 Laird, C. D., McConaughy, B. L., and McCarthy, B. J. (1969) *Nature (London)* **224:**149.

48 Ullman, J. S. and McCarthy, B. J. (1973) *Biochim. Biophys. Acta* **294:**416.

49 Hutton, J. R. and Wetmur, J. G. (1973) *Biochemistry* **12:**558.

50 Engler, J. A., Chow, L. T., and Broker, T. R., (1981) *Gene*, in press.

51 Nygaard, A. P. and Hall, B. D. (1964) *J. Mol. Biol.* **9:**125.

52 Chien, Y.-H. and Davidson, N. (1978) *Nucleic Acids Res.*, **5:**1627.

53 Bonner, J., Kung, G., and Bekhor, I. (1967) *Biochemistry* **6:**3650.

54 Gillespie, S. and Gillespie, D. (1971) *Biochem. J.* **125:**16.

55 Friedrich, R. and Feix, G. (1972) *Anal. Biochem.* **50:**467.

56 Pinder, J. C., Staynov, D. Z., and Gratzer, W. B. (1974) *Biochemistry* **13:**5367.

57 Meyer, J., Neuwald, P. D., Lai, S.-P., Maizel, J. V., Jr., and Westphal, H. (1977) *J. Virol.* **21:**1010.

58 Chow, L. T., Roberts, J. M., Lewis, J. B., and Broker, T. R. (1977) *Cell* **11:**819.

59 Hsu, M.-T., Kung, H.-J., and Davidson, N. (1973) *Cold Spring Harbor Symp. Quant. Biol.* **38,** 943.

60 Kung, H.-J., Bailey, J. M., Davidson, N., Nicolson, M. O., and McAllister, R. M. (1975) *J. Virol.* **16:**397.

61 Bailey, J. M. and Davidson, N. (1976) *Anal. Biochem.* **70:**75.

62 Kaback, D. B., Angerer, L. M., and Davidson, N. (1979) *Nucleic Acids Res.* **6:**2499.

63 Szybalski, W., Kubinski, H., Hradecna, Z., and Summers, W. C. (1971) In *Methods in Enzymology*, Vol. 21 (L. Grossman and K. Moldave, Eds.), Academic Press, New York, p. 383.

64 Jensen, R. H. and Davidson, N. (1966) *Biopolymers* **4:**17.

65 Grunwedel, D. W. and Davidson, N. (1967) *Biopolymers* **5:**847.

66 Cech, T. B. and Pardue, M. L. (1976) *Proc. Natl. Acad. Sci. U.S.A.* **73:**2644.

67 Sanger, F., Coulson, A. R., Friedman, T., Air, G. M., Barrell, B. G., Brown, N. L., Fiddes, J. C., Hutchinson, C. A., III, Slocombe, P. M., and Smith, M. (1978) *J. Mol. Biol.* **125:**225.

68 Ohtsubo, E., Zenilman, M., and Ohtsubo, H. (1980) *Proc. Natl. Acad. Sci. U.S.A.* **77:**750.

69 Fiers, W., Contreras, R., Haegeman, G., Rogiers, R., van de Voorde, A., van Heuversyne, H., van Herreweghe, J., Volckaert, G., and Ysebaert, M. (1978) *Nature (London)* **273:**113; also, in *DNA Tumor Viruses* (J. Tooze, Ed.), Cold Spring Harbor Laboratory, Cold Spring Harbor, NY. 1980, Appendix A, p. 799.

70 Reddy, V. B., Thimmappaya, B., Dhar, R., Subramanian, K. N., Zain, B. S., Pan, J., Ghosh, P. K., Celma, M. L., and Weissman, S. M. (1978) *Science* **200:**494.

71 Philipssen, P., Thomas, M., Kramer, R. A., and Davis, R. W. (1978) *J. Mol. Biol.* **123:**387.

72 Davis, R. W. and Hyman, R. W. (1971) *J. Mol. Biol.* **62:**287.

73 Hyman, R. W. (1971) *J. Mol. Biol.* **61:**369.

74 Griffith, J. D. (1978) *Science* **201:**525.

75 Klessig, D. F. and Chow, L. T. (1980) *J. Mol. Biol.* **139**:221.

76 Neuwald, P. D., Meyer, J., Maizel, J. V., Jr., and Westphal, H. (1977) *J. Virol.* **21**:1019.

77 Kelly, T. J., Jr. and Lewis, A. M., Jr. (1973) *J. Virol.* **12**:643.

78 Mulder, C., Sharp, P. A., Delius, H., and Pettersson, U. (1974) *J. Virol.* **14**:68.

79 Westphal, H. and Lai, S.-P. (1977) *J. Mol. Biol.* **116**:525.

80 Bender, W. W. and Davidson, N. (1976) *Cell* **7**:595.

81 Engel, J. D. and Davidson, N. (1978) *Biochemistry* **17**:3883.

82 Snedecor, G. W. (1956) *Statistical Methods*, 5th ed., Iowa State College Press, Ames, Iowa.

83 van Ormondt, H., Maat, J., de Waard, A., and van der Eb, A. J. (1978) *Gene* **4**:309.

84 Maat, J. and van Ormondt, H. (1979) *Gene* **6**:75.

85 Perricaudet, M., Akusjärvi, G., Virtanen, A., and Pettersson, U. (1979) *Nature (London)* **281**:694.

86 Perricaudet, M., Le Moullec, J. M., and Pettersson, U. (1980) *Proc. Natl. Acad. Sci. U.S.A.* **77**:3778.

87 Baker, C. and Ziff, E. (1979) *Cold Spring Harbor Symp. Quant. Biol.* **44**:415.

88 Aleström, P., Akusjärvi, G., Perricaudet, M., Mathews, M. B., Klessig, D. F., and Pettersson, U. (1980) *Cell* **19**:671.

89 Lang, D., Bujard, H., Wolff, B., and Russell, D. (1967) *J. Mol. Biol.* **23**:163.

90 Berk, A. J. and Sharp, P. A. (1978) *Cell* **14**:695.

91 Wetmur, J. G. (1971) *Biopolymers* **10**:601.

92 Philipson, L., Pettersson, U., Lindley, U., Tibbetts, C., Vennström, B., and Persson, T. (1974) *Cold Spring Harbor Symp. Quant. Biol.* **39**:447.

93 Sharp, P. A., Gallimore, P. H., and Flint, S. J. (1974) *Cold Spring Harbor Symp. Quant. Biol.* **39**:457.

94 Pettersson, U., Tibbetts, C., and Philipson, L. (1976) *J. Mol. Biol.* **101**:479.

95 Anderson, C. W., Lewis, J. B., Atkins, J. F., and Gesteland, R. F. (1974) *Proc. Natl. Acad. Sci U.S.A.* **71**:2756.

96 Craig, E. A. and Raskas, H. J. (1976) *Cell* **8**:205.

97 Büttner, W., Veres-Molnár, Z., and Green, M. (1976) *J. Mol. Biol.* **107**:93.

98 Williams, J. F., Young, C. S. H., and Austin, P. E. (1974) *Cold Spring Harbor Symp. Quant. Biol.* **39**:427.

99 Grodzicker, T., Williams, J., Sharp, P., and Sambrook, J. (1974) *Cold Spring Harbor Symp. Quant. Biol.* **39**:439.

100 Lewis, J. B., Atkins, J. F., Anderson, C. W., Baum, P. R., and Gesteland, R. F. (1975) *Proc. Natl. Acad. Sci. U.S.A.* **72**:1344.

101 Lewis, J. B., Atkins, J. F., Baum, P. R., Solem, R., Gesteland, R. F., and Anderson, C. W. (1976) *Cell* **7**:141.

102 Kozak, M. (1978) *Cell* **15**:1109.

103 Paterson, B. M., Roberts, B. E., and Kuff, E. L. (1977) *Proc. Natl. Acad. Sci. U.S.A.* **74**:4370.

104 Hastie, N. D. and Held, W. A. (1978) *Proc. Natl. Acad. Sci. U.S.A.* **75**:1217.

105 Miller, J. S., Ricciardi, R. P., Roberts, B. E., Paterson, B. M., and Mathews, M. B. (1980) *J. Mol. Biol.* **142**:455.

106 Zain, B. S. and Roberts, R. J. (1979) *J. Mol. Biol.* **131**:341.

107 Ziff, E. B. and Evans, R. M. (1978) *Cell* **15**:1463.

108 Berget, S. M., Berk, A. J., Harrison, T., and Sharp, P. A. (1977) *Cold Spring Harbor Symp. Quant. Biol.* **42**:523.

109 McGuire, P. M., Piatak, M., and Hodge, L. D. (1976) *J. Mol. Biol.* **101**:379.

110 Wold, W. S. M., Green, M., and Munns, T. W. (1976) *Biochem. Biophys. Res. Commun.* **68**:643.

111 Moss, B. and Koczot, F. (1976) *J. Virol.* **17**:385.

112 Sommer, S., Salditt-Georgieff, M., Bachenheimer, S., Darnell, J. E., Furuichi, Y., Morgan, M., and Shatkin, A. J. (1976) *Nucleic Acids Res.* **3**:749.

113 Chen-Kiang, S., Nevins, J. R., and Darnell, J. E., Jr. (1979) *J. Mol. Biol.* **135**:733.

114 Philipson, L., Wall, R., Glickman, G., and Darnell, J. E. (1971) *Proc. Natl. Acad. Sci. U.S.A.* **68**:2806.

115 Nevins, J. R. and Darnell, J. E. (1978) *J. Virol.* **25**:811.

116 Broker, T. R. and Chow, L. T. (1979) In *Eucaryotic Gene Regulation, ICN-UCLA Symposia on Molecular and Cellular Biology*, Vol. 14 (R. Axel, T. Maniatis, and C. F. Fox, Eds.), Academic Press, New York, p. 611.

117 Kitchingman, G. R., Lai, S.-P., and Westphal, H. (1977) *Proc. Natl. Acad. Sci. U.S.A.* **74**:4392.

118 Abelson, J. (1979) *Annu. Rev. Biochem.* **48**:1035.

119 Lewis, J. B., Esche, H., Smart, J. E., Stillman, B., Harter, M. L., and Mathews, M. B. (1979) *Cold Spring Harbor Symp. Quant. Biol.* **44**:493.

120 Harter, M. L. and Lewis, J. B. (1978) *J. Virol.* **26**:736.

121 Green, M., Wold, W. S. M., Brackman, K. H., and Cartas, M. A. (1979) *Virology* **97**:275.

122 Spector, D. J., Halbert, D. N., Crossland, L. D., and Raskas, H. J. (1979) *Cold Spring Harbor Symp. Quant. Biol.* **44**:437.

123 van der Eb, A. J., van Ormondt, H., Schrier, P. I., Lupker, J. H., Jochemsen, H., van den Elsen, P. J., DeLeys, R. J., Maat, J., van Beveren, C. P., Dijkema, R., and deWaard, A. (1979) *Cold Spring Harbor Symp. Quant. Biol.* **44**:383.

124 Pettersson, U. and Mathews, M. B. (1977) *Cell* **12**:741.

125 Ishibashi, M. and Maizel, J. V., Jr. (1974) *Virology* **57**:409.

126 Berk, A. J. and Sharp, P. A. (1977) *Cell* **12**:45.

127 Evans, R. M., Fraser, N., Ziff, E., Weber, J., Wilson, M., and Darnell, J. E. (1977) *Cell* **12**:733.

128 Wilson, M. C., Fraser, N. W., and Darnell, J. E., Jr. (1979) *Virology* **94**:175.

129 Spector, D. J., McGrogan, M., and Raskas, H. J. (1978) *J. Mol. Biol.* **126**:395.

130 Wilson, M. C., Nevins, J. R., Blanchard, J.-M., and Darnell, J. E., Jr. (1979) *Cold Spring Harbor Symp. Quant. Biol.* **44**:447.

131 Chow, L. T., Lewis, J. B., and Broker, T. R. (1979) *Cold Spring Harbor Symp. Quant. Biol.* **44**:401.

132 Berk, A. J., Lee, F., Harrison, T., Williams, J., and Sharp, P. A. (1979) *Cell* **17**:935.

133 Shenk, T., Jones, N., Colby, W., and Fowlkes, D. (1979) *Cold Spring Harbor Symp. Quant. Biol.* **44**:367.

134 McGrogan, M. and Raskas, H. J. (1978) *Proc. Natl. Acad. Sci. U.S.A.* **75**:625.

135 Fraser, N. and Ziff, E. (1978) *J. Mol. Biol.* **124**:27.

136 Berget, S. M. and Sharp, P. A. (1979) *J. Mol. Biol.* **129**:547.

137 Woolford, J. L., Jr. and Rosbash, M. (1979) *Nucleic Acids Res.* **6**:2483.

138 Dale, R. M. K. and Ward, D. C. (1975) *Biochemistry* **14**:2458.

139 Manning, J., Pellegrini, M., and Davidson, N. (1977) *Biochemistry* **16**:1364.

140 Manning, J. E., Hershey, N. D., Broker, T. R., Pellegrini, M., and Davidson, N. (1975) **53**:197.

141 Pellegrini, M., Holmes, D. S., and Manning, J. (1977) *Nucleic Acids Res.* **4**:2961.

142 Davis, R. W. and Hyman, R. D. (1970) *Cold Spring Harbor Symp. Quant. Biol.* **35**:269.
143 Granboulan, N. and Scherer, K. (1969) *Eur. J. Biochem.* **9**:1.
144 Robberson, D., Aloni, Y., Attardi, G., and Davidson, N. (1971) *J. Mol. Biol.* **60**:473.
145 Wellauer, P. K. and Dawid, I. B. (1974) *J. Mol. Biol.* **89**:379.
146 Chi, Y. Y. and Bassel, A. R. (1974) *J. Virol.* **13**:1194.
147 Staehelin, M. (1959) *Biochim. Biophys. Acta* **31**:448.
148 Jacobson, A. (1976) *Proc. Natl. Acad. Sci. U.S.A.* **73**:307.
149 Delius, H., Westphal, H., and Axelrod, N. (1973) *J. Mol. Biol.* **74**:677.
150 Wu, M. and Davidson, N. (1973) *J. Mol. Biol.* **78**:1.
151 Broker, T. R., Angerer, L. M., Yen, P. H., Hershey, N. D., and Davidson, N. (1978) *Nucleic Acids Res.* **5**:363.
152 Angerer, L., Davidson, N., Murphy, W., Lynch, D., and Attardi, G. (1976) *Cell* **9**:81.
153 Sodja, A. and Davidson, N. (1978) *Nucleic Acids Res.* **5**:385.
154 Wu, M. and Davidson, N. (1975) *Proc. Natl. Acad. Sci. U.S.A.* **72**:4506.
155 Ferguson, J. and Davis, R. W. (1978) *J. Mol. Biol.* **123**:417.
156 Shenk, T. E., Rhodes, C., Rigby, P., and Berg, P. (1975) *Proc. Natl. Acad. Sci. U.S.A.* **72**:979.
157 Berk, A. J. and Sharp, P. A. (1978) *Proc. Natl. Acad. Sci. U.S.A.* **75**:1274.
158 Flaveloro, J., Treisman, R., and Kamen, R. (1980) In *Methods in Enzymology,* Vol. 65 (L. Grossman and K. Moldave, Eds.), Academic Press, New York, p. 718.
159 Howley, P. M., Israel, M. A., Law, M.-F., and Martin, M. A. (1979) *J. Biol. Chem.* **254**:4876.
160 Persson, H., Perricaudet, M., Tolun, A., Philipson, L., and Pettersson, U. (1979) *J. Biol. Chem.* **254**:7999.
161 Maat, J., van Beveren, C. P., and van Ormondt, H. (1980) *Gene* **10**:27.
162 Stillman, B. W., Lewis, J. B., Chow, L. T., Mathews, M. B., and Smart, J. E. (1981) *Cell*, in press.

IMAGE PROCESSING OF BIOLOGICAL SPECIMENS: A BIBLIOGRAPHY

6

Timothy S. Baker
The Rosenstiel Basic Medical Sciences Research Center
Brandeis University
Waltham, Massachusetts

CONTENTS

1 Introduction

2 Table 1: Theory and methods of electron micrograph image processing

  2.1 Optical diffraction

  2.2 Optical filtration

  2.3 Digital Fourier methods

  2.4 Real-space and other reconstruction methods

3 Table 2: Topics related to image processing

  3.1 Diffraction and symmetry

  3.2 Specimen preparation

  3.3 Imaging conditions

  3.4 Image interpretation

4 Table 3: Biological applications of image processing

5 Image processing of specimens of different types

  5.1 Helical

  5.2 Sheets

  5.3 Three-dimensional crystals

  5.4 Rotationally symmetric particles

  5.5 Asymmetric particles

6 Progress and prospects

  6.1 Radiation effects

  6.2 Dehydration damage

  6.3 Image distortions

  6.4 Aperiodic specimens

  6.5 Missing three-dimensional data

  6.6 Treatment of the constant term of diffraction data

  6.7 Correlation with other data

  6.8 Technology

  6.9 Interpretation of image reconstructions

Acknowledgments

References and bibliography

1 INTRODUCTION

Studies of biological structure by image processing of electron micrographs have rapidly expanded since the rotational superposition method was introduced in 1963 by Markham et al. (671) and the optical diffraction method in

1964 by Klug and Berger (589). More than 700 articles on the theory and applications of image processing have been published, and this chapter attempts to organize this sometimes bewildering volume of information. Citations are included through December 1979* for most English-language journals of molecular and cell biology, biochemistry, virology, bacteriology, microscopy, and crystallography. The major concepts of image-processing theory, methods, and applications are presented in outline form with appropriate references to the literature. The main intent is to provide, for experts and novices of image processing, a guided entry to the literature.

Image-processing methods are aids for examining and altering the information contained in images. Electron microscopists know that an electron micrograph cannot exactly represent the two-dimensional projection of the mass distribution in the specimen existing in the microscope environment. Astigmatism, defocusing, spherical aberration, speciman and focal drift, multiple scattering, and the nonlinear response of the photographic emulsion are among several factors contributing to differences between specimen and image. Furthermore, specimen structure is often radically altered from its initial state owing to damage caused by the effects of dehydration and the electron beam. Image-processing methods provide powerful and convenient tools to assess and recover structural information obscured by "noise" contained in all micrographs (see Table 2.III.E). These methods both allow detection of and correction for the imaging aberrations and provide a way to reconstruct and visualize the average features of a specimen.

Several excellent review articles introduce the theory, methods, and applications of image processing: Lake (622), Horne and Markham (500), Amos (21), and Crowther and Klug (189). Review articles on more specific aspects of image processing are cited in the tables.

The remaining sections in this chapter briefly highlight the table outlines and introduce the novice to the more commonly applied image-processing methods. Of course, this material is no substitute for practical experience. Obviously, not all possibilities have been covered here, nor can they be found in the literature. Even though most image-processing methods are well established and routine, the reader is cautioned against the temptation to process images blindly or to interpret results without a basic understanding of the theoretical and practical principles. In some instances (in Tables 1–3), where multiple citations are given, suggested reading is signified by boldface type if one or more articles seem to be more appropriate for initial inquiries. Some citations, included for completeness or historical interest, may not provide practical information. For example, catalase has been widely studied by image processing (Table 3), but mainly as a test specimen.

*Additional abridged lists, including 1980 references, appear at the end of the references and bibliography section.

2 TABLE 1: THEORY AND METHODS OF ELECTRON MICROGRAPH IMAGE PROCESSING

Table 1 outlines several of the central topics concerning the theory, methods and applications of image processing. Most entries cite several literature articles where additional information, clarification or examples may be found. The Fourier methods (I–III) are presented in greatest detail because they are the most powerful and widely used techniques in the study of periodic specimen structure. Fortunately, a large range of intensely studied biological specimens occur *in vivo,* or can be isolated *in vitro* as regular arrays and are amenable to this type of analysis. Viruses, muscle proteins, membranes, microtubules, ribosomes, flagella, and enzymes have all been successfully examined by Fourier image-processing techniques (Table 3). The outline and bibliography pertaining to alternate image-processing methods (IV) is less comprehensive than for the diffraction methods (I–III) but should provide adequate access to additional literature. The reader is urged to consider the advantages and disadvantages of various techniques before studying a particular specimen, since the physical nature of the specimen may dictate which procedure or procedures are preferred (see discussion on specimen types, Section 5).

The most common form of image processing involves one or more of the reciprocal space or Fourier methods (optical or digital diffraction and filtering, and three-dimensional reconstruction). The major advantage of these methods arises from the two-stage nature of the processing in which the selection and assessment of images and the production of averaged images are separate steps. In the first step (diffraction or forward transformation step), the diffraction pattern of the image is formed. The pattern conveniently separates the noise (random fluctuations) and signal (pertaining to average specimen details) components in the image so they can be identified. The diffraction pattern is useful because it provides an objective measure of the lattice dimensions and symmetry in the images of periodic specimens. It also reveals the specimen preservation (Table 1.I.B.1.d), and instrumentally introduced aberrations in the image. In the second stage (reconstruction or back-transformation step), noise is removed with a mask and a filtered image is formed by optical or digital rediffraction of the unobstructed portion of the pattern. Multiple views of the structure can be combined in the computer to reconstruct the three-dimensional specimen structure.

2.1 Optical Diffraction

Optical diffraction is the simplest and most widely practiced image-processing technique, and is usually the initial step of most image-processing studies. The main advantage of this technique is its ability to objectively assess and reveal periodic structural information. Additional applications are listed in Table 1.I.B.

Klug and Berger (589) were first to study biological structure in electron micrographs by optical diffraction. The technique is straightforward and easy to perform: the electron micrograph is illuminated, usually with a parallel beam of monochromatic laser light, and its diffraction pattern is formed by the interference of the light waves passing through the micrograph. The pattern may be viewed or recorded behind the micrograph. The experimental apparatus, an optical diffractometer, may be designed in several different ways depending on its intended use. Details of the design, operation, components, alignment, and calibration of optical diffractometers are given in the articles cited in Table 1.I.C.

A diffractometer of reasonable quality, suitable for simple experiments such as screening for best images or detecting and locating periodicities in images, can be built or purchased for less than $2000. A more expensive diffractometer ($5000–$10,000) is usually built for easier and more precise operation and alignment, and to produce higher quality diffraction patterns and reconstruction images than are possible with less expensive designs. This type of apparatus usually includes an image reconstruction system (with a high-quality, corrected, doublet lens), a pinhole spatial filtering system to remove noise in the illumination beam, a moderate- to high-power laser (1–50 mW), high-quality, high-reflectance mirrors (if the optical path is folded), fully adjustable, precision holders for all components, and an image and diffraction pattern recording system.

Many laboratories prefer to construct their own diffractometers with specifications dictated by the intended use of the instrument. For example, if most image processing is done by computer, a simple, inexpensive, linear diffractometer for surveying images will suffice. Liquid gates (Table 1.I.C.2.e) are now rarely used to iron out inhomogeneities in the micrograph emulsion (and glass or gelatin backing) to produce better diffraction patterns. They are inconvenient to use and, in any event, fast and reliable digital processing systems are now often preferred over high-quality optical processing systems.

Optical diffraction provides useful information about the geometrical arrangement of subunits in the specimen. These structural details often cannot be discerned from simple, visual inspection of the original micrograph. For example, the presence of rotational screw axes or pseudo-symmetries may go undetected without the information provided by the optical diffraction pattern. Such structural information is determined by correctly indexing the pattern, that is, by defining a lattice (or lattices for multilayered or helical particles) that accurately defines the location of all diffraction spots. Indexing is an essential step for correct application of optical or digital filtering, or three-dimensional reconstruction techniques (see below). Indexing patterns from most two-dimensional-type specimens, except for some helical and multilayered particles, is quite straightforward. The papers by Finch et al. (288), Moody (725, 726), Kiselev and Klug (570), Mikhailov and Belyaeva (707), DeRosier and Klug (206), Lake (622), Leonard et al. (637), and Unwin and Taddei (1004) give excellent, specific examples of pattern indexing (see additional examples cited in Table 1.I.D.8). Misell (720) devotes an entire section to a discussion of the theoretical and practical problems of indexing. It is always useful to examine the optical diffraction pattern for artifacts (Table 1.I.E) because they may lead to difficulties in indexing.

Other topics, related to the principles and practice of optical diffraction, are outlined in Table 1.I.F. Of particular interest is the extensive application of optical diffraction in small-molecule crystal structure determinations (Table 1.I.F.6), many of which predated the development of applications to biological macromolecules. This literature includes most of the original papers on the history and development of the optical diffraction apparatus.

2.2 Optical Filtration

Optical filtering, independently introduced by Klug and DeRosier (592) and Bancroft et al. (55), is suitable only for the study of periodic specimens with translational symmetry. The main advantage of this method is that it provides a simple way to remove the contributions from noise in micrographs and thereby reveal clearer images of specimen structure. In addition, it is a powerful method for separating moiré images of multilayered specimens. Table 1.II.B outlines several other applications.

The review by Erickson et al. (263) and papers by Klug and DeRosier (592) and Fraser and Millward (326) are excellent introductions to the theory and techniques of optical filtration. The basic principle of the technique is straightforward and easy to understand, but in practice, the method can easily lead to erroneous results, especially when used by uninformed novices who are unaware of the types of artifact that can occur.

The first, and most important step, is to correctly index the optical diffraction pattern of the specimen. A pattern is considered to be successfully indexed if it is possible to determine which spots arise from image noise (aperiodic details) and which spots are attributed to the periodic nature of the specimen. Although it is unnecessary to attempt to identify all the noise components in the unprocessed image, a correctly performed filtration experiment requires some understanding of how noise and signal components may be distinguished in the diffraction pattern. In most examples there is a crystalline specimen that gives rise to a clearly identified, periodic array of spots. Noise, or aperiodicities in the image, produce spots in all parts of the pattern. Note that noise near or at the identified lattice points cannot be removed by filtering (sometimes referred to as "periodic" noise). Systematic specimen flattening or staining artifacts are examples of situations that produce periodic-type noise. Other major sources of noise are listed in Table 2.III.E.

If the diffraction pattern is difficult to index, it may be that the chosen lattice is incorrect (e.g., because a superlattice has been missed). Some strong, but non-indexible spots may be attributed to multiple scattering

(Table 1.I.D.7), or they could arise from a strong aperiodicity in the specimen. The temptation at first may be to ignore the images with non-indexible patterns, but indexing difficulties may often be a clear indication of important but overlooked structural information. The image-processing novice is urged to study several of the indexing examples cited in Table 1.I.D.8.

Once the diffraction spots have been found to be consistent with a given lattice, a filter mask is designed with holes positioned to allow unobstructed passage of the diffraction spots at the lattice points (or lattice lines for helical particles). The mask is accurately located in the diffraction plane of the optical diffractometer so all spots at the lattice points are allowed through. Opaque regions of the mask block out most of the diffraction pattern noise arising from the nonperiodic image components. A reconstruction lens, placed behind the mask, refocuses the unobstructed rays and forms a filtered image. An unfiltered image is formed if the mask is absent. This illustrates the double-diffraction phenomena of image formation: the diffraction pattern of the micrograph is formed in the first stage (forward transformation), and in the second stage, the reconstruction lens acts to rediffract the diffracted rays (back- or reverse transformation) to form an image (filtered or unfiltered). Thus, an image (filtered or unfiltered) is the result of rediffraction of the diffraction pattern (masked or unmasked) of the object (micrograph).

Filtration experiments are performed with the aid of an optical diffractometer, to which has been added a reconstruction lens (or lenses). The reconstruction system must be of high optical quality to minimize image distortions (e.g., phase errors due to spherical aberration). Camera lenses are often suitable as reconstruction lenses, although they usually are an unnecessary expense and are not ideally designed for optical reconstruction experiments (i.e., camera lenses are generally designed for optimum transmission of light, but not for flatness of field). An inexpensive, but high-quality, corrected doublet, with a large usable aperture, is suitable for producing high-resolution, reconstruction images.

Most image-processing laboratories have adopted the folded design both for survey and reconstruction work, mainly because it is more convenient to operate than the linear-type designs. The major disadvantage of the folded design is the requirement for mirrors to bend the optical light path. Mirrors add extra optical surfaces, which may collect dust or become scratched and thus deteriorate the image or pattern quality. Expensive, high-quality (high-reflectance and optically flat) mirrors are recommended for optimum results. A useful way to assess the quality of the optical bench is to compare an unfiltered reconstruction with the original image.

Reduction in image noise by optical filtering is a consequence of averaging among neighboring, periodically repeated units in the array. As the size of holes in the filter mask is reduced to block out more of the noise in the diffraction pattern, the extent of local averaging increases. That is, the image of a given unit in the array is averaged with more of its neighbors. The holes should not be made smaller than the size of the diffraction spots or the signal-to-noise ratio will decrease (Table 1.II.F.1.d).

Filter mask design and fabrication (Table 1.II.E) usually constitute the rate-limiting steps in a filtration experiment, once a consistent indexing scheme has been established. Etched masks may be made with great precision, but they require more tedious and demanding manufacturing skills than are needed for punched or drilled masks. Erickson et al. (263) describe a simple apparatus and method that facilitate the production of suitable masks in a matter of minutes. This method has the additional advantage of using the original, recorded pattern as a template.

Filtered reconstructions often contain undetected, erroneous details as a result of several types of artifact. Three obvious sources of artifact include pattern misindexing (resulting in incorrect mask design), incorrect positioning of the mask in the diffraction plane (causing spots to be partially or totally blocked), and mispositioned or misshaped holes in the mask (making it impossible to align the mask so that all spots pass through simultaneously). Other, less obvious, but significant sources to consider, are indicated in Table 1.II.F. Some authors (83, 441, 500, 952) have suggested that all reconstructions are, at least in part, erroneous.

The translational photographic-superposition method (linear integration) produces results analogous with but not identical to those of optical filtering. The translational parameters (lattice repeat and geometry) are best determined by optical diffraction rather than by a subjective, trial-and-error method (Table 1.IV.B.1.a). Despite the procedural differences, optical and digital filtering methods produce remarkably similar results (11, 720). There are considerable advantages to digital processing (see below), even though structural details may be reliably represented by either method.

2.3 Digital Fourier Methods

There are several advantages to processing images by digital rather than optical Fourier methods. The main benefits derive from the quantitative nature and flexibility of the digital methods. In addition, three-dimensional reconstruction and rotational filtering are impractical or impossible using the optical Fourier techniques. Quantitative analysis or manipulation of data is also not practical by optical means. For example, it is easier to digitally correct for image aberrations such as astigmatism and defocusing, or specimen distortions such as crystal lattice imperfections and curved or bent filamentous structures (Table 1.III.B.6). Diffraction amplitudes and phases can be quantitatively measured and altered, if necessary (e.g., to correct for contrast transfer effects: see Table 1.III.C.3.g). Another advantage of digital processing is that individual image reconstructions can be averaged together and a quantitative measure of their agreement assessed (Table 1.III.B.3.b,c). Digital processing offers virtually infinite flexibility in the manipulation of data. For example, in the digital equivalent of optical filtering, filter masks can be designed with an infinite variety and combination of hole sizes, shapes, and transparencies.

Computer image processing is rapidly circumventing the need for high-quality and expensive optical systems. Nevertheless, there are certain disadvantages. The main drawback of digital procedures arises from the necessity for discrete sampling of the data. This produces aliasing artifacts (transform overlap) that can be reduced, but never totally removed, by judicious choice of scanning conditions. DeRosier and Moore (208) define and discuss the aliasing problem of digital image processing.

The initial costs in time and money of establishing a working digital system may be prohibitive unless a large fraction of one's research efforts are devoted to image-processing studies. It makes little sense to develop a digital system whose main purpose is to provide qualitative examination of specimen diffraction patterns. An optical diffractometer is not only inexpensive, but it operates at the speed of light! In addition, the qualitative results of careful optical filtering studies are comparable to those determined by digital methods (11, 720), even though digital aliasing errors will always produce small differences. Screening for best images is facilitated by optical diffraction because the method is quick and inexpensive and far superior to its digital analog. Aebi et al. (11), Misell (720), and Table 1.III.D identify several advantages and disadvantages of computer and optical image processing systems. Additional applications and selected examples of digital processing are outlined in Table 1.III.B.

A typical digital processing procedure includes the following steps. Several (usually 50 or more) micrographs are examined by optical diffraction (Table 1.III.C.2.a) to select a small subset (< 10) of images of the best preserved specimens for digital processing. (Optical diffraction is unsuitable for selecting particles for digital, rotational filtering. The rotational power spectrum is used instead.) The micrograph is then densitometered to convert optical densities on the photographic emulsion to a digital image (a numerical array corresponding to the relative optical densities in the image). Images are scanned at raster settings corresponding to one-third or less of the expected resolution in the image, to minimize aliasing artifacts (Table 1.III.C.2.c). Regions in the digital image outside the area of interest are set to zero (equivalent to masking out an area of the micrograph for optical diffraction) and the numerical image is "floated" (Table 1.III.C.3.b) by subtracting the average image intensity around the perimeter of the boxed area from all image intensities. (Floating helps suppress the strong diffraction spikes which arise from the edges of the masked area and are often the most prominent feature in optical patterns.)

The numerical array is then transformed, usually by fast Fourier methods (Table 1.III.C.3.c), and the diffraction amplitudes and phases are displayed. The pattern is indexed (if it had not previously been determined from optical patterns), and the transform is set to zero amplitude at all but the reciprocal lattice points (or finite regions (holes) surrounding the ideal lattice point locations). The filtered diffraction pattern is back-transformed in the computer, and the results are displayed in any of a number of ways (character overprinting on a line-printer, contour plotting, cathode ray density plotting or film writing: see Table 1.IV.B for examples). If a particle is studied by three-dimenstional reconstruction, the diffraction phases and amplitudes (structure factors) must be measured in all regions of the three-dimensional diffraction pattern by combining data from the two-dimensional patterns of several independent views of the specimen. The rationale for collecting and combining information from different views partially depends on which class of specimen is studied (see Section 5).

Two major disadvantages of digital processing are the expense and complexity of the required hardware (microdensitometer and computer) and software (programs for carrying out the image processing procedures). Most protein crystallography laboratories are equipped with the needed hardware, and probably have programs (e.g., fast Fourier transform, film scanning, and plotting routines) that are easily adapted for most basic image processing needs. Microdensitometers cost from \$50,000 to more than \$150,000, and a minicomputer suitable for image processing costs \$50,000 or more. Rotating-drum-type microdensitometers are less expensive and generally scan images more quickly than the flatbed designs, which have greater precision and flexibility (larger choice and range of scanning parameters). Rotating drum densitometers are obviously unsuitable for scanning glass plates (the image may be copied onto film, but this is generally avoided to prevent the introduction of contrast changes). The flatbed densitometer is useful for examining data from unstained specimens studied by low-dose imaging techniques (1002). Here, images recorded at medium-low magnification (10,000–40,000×), containing medium-high resolution details (0.5–2.0 nm), require a small scan raster and aperture ($< 20\ \mu m$). Also, the spots in electron diffraction patterns (for determining diffraction amplitudes from specimens studied by low dose: 1002) are usually smaller than 20 μm in diameter and must be scanned with high precision on a very fine raster ($< 5\ \mu m$).

Many laboratories engaged in digital processing studies prefer to develop their own computer software systems because they can be designed according to the specific needs of the user to efficiently study the particular specimens of interest. In this way, results are more easily understood and correctly interpreted by the user. If a uniquely designed system is not essential, it may be easier to incorporate an established, portable system (Table 1.III.C.3.a). This has the advantage of saving considerable effort (and frustration) in the development and testing of programs, but it has the drawback of possibly being incorrectly implemented by the "black-box" novice user.

2.4 Real-Space and Other Reconstruction Methods

Although the Fourier methods of image processing are those most widely practiced, several real-space and other methods are available and may be preferred for the study of certain specimens. The rotational (Table 1.IV.A) and translational (Table 1.IV.B) photographic superposition methods are quite popular because they are easy to understand and apply, and they lead to results comparable to those obtained by optical (digital) filtering and digital rotational filtering. In addition, the experimental apparatus is not expensive or difficult to construct.

The main purpose of the superposition methods (Table 1.IV.A,B) is to produce an averaged image by forming on a single photographic plate a composite image of the symmetry equivalent parts of the specimen image. Thus the micrograph (or final photographic plate) is rotated or translated to superimpose the images of individual asymmetric units in the specimen. The results of such manipulations will be erroneous if the rotational symmetry center or number, or the translational period and direction, is incorrectly deduced. The sometimes subjective nature of these methods is one of their major drawbacks. Of course, results will be erroneous with any incorrectly applied processing method. The Fourier methods are generally preferred because of their objective, two-step nature. The superposition methods are reliable and powerful when the specimen symmetry is obvious or well established by other methods, or when applied in conjunction with the results of optical or digital diffraction (Table 1.IV.B.1.a). In other words, the diffraction pattern (or power spectrum) directly and objectively indicates how the superposition should be performed.

Three-dimensional image reconstructions may be determined directly from projections by a variety of processing procedures that have generated much theoretical discussion in the literature. In these methods the three-dimensional reconstructions are calculated by directly recombining the set of two-dimensional projections (views) of the specimen. The algebraic algorithms required to perform such reconstructions offer some computational advantages when compared to the Fourier methods, but they are generally less familiar to electron microscopists than the Fourier transform algorithms known by most crystallographers and digital image processors. The application of these methods to biological problems has mostly been limited to the study of phage tail structures and noncrystalline enzyme molecules (Table 1.IV.D.2.b,c).

3 TABLE 2: TOPICS RELATED TO IMAGE PROCESSING

Although less comprehensive than Table 1, Table 2 outlines several selected topics related to the theory and practice of structural studies by electron microscopy and image processing.

3.1 Diffraction and Symmetry

The principles of symmetry and X-ray diffraction form the fundamental framework of the Fourier image-processing methods. Structure analysis by this type of image processing is completely analogous to the protein crystallographic methods. For example, a basic understanding of diffraction and symmetry greatly facilitates the ability to correctly index and interpret optical diffraction patterns. The functional significance of symmetry in biological macromolecules and aggregates has been discussed previously (679).

3.2 Specimen Preparation

Image-processing results are largely influenced by the choice of specimen preparation method. Most specimens are negatively stained because this leads to better preservation, image contrast, and resolution than the other commonly used techniques. Sometimes it is more informative to study the specimen when it is shadowed, freeze-dried (etched, or fractured), thin sectioned, or unstained. The appearance of structural features usually varies for a given specimen, depending on which method of preparation is used.

The destructive effects of dehydration on biological specimens are a major source of artifact in studying structure by electron microscopy (Table 2.II.B). The most promising methods that aim to reduce such damage are those that retain the aqueous (liquid or frozen) environment of the specimen (Table 2.II.B.2.c). Critical point drying methods have not been widely practiced in image-processing studies, nor have suitable (nondenaturing, nondestructive), chemical cross-linking agents been discovered.

Support films (Table 2.II.C), although necessary for studying specimens in the electron microscope, contribute most of the noise (Table 2.III.E) that obscures specimen details in micrographs. The use of ultrathin or perforated films may be advantageous in certain instances—for example, when the specimen is small or very thin.

Since by the usual techniques it is generally more difficult to perform the image processing of aperiodic specimens than of periodic ones, it is beneficial if the noncrystalline specimen can be induced *in vitro* to form suitable thin crystals or paracrystals. One method, successful with several viruses (helical and spherical), is the negative-stain, carbon film, mica flotation technique of Horne and Pasquali-Ronchetti (see Table 2.II.D).

3.3 Imaging Conditions

Image-processing results also depend on the imaging conditions of the microscope at the time the images are recorded (Table 2.III). Selected texts and reviews on the principles and practice of electron microscopy are cited in Table 2.III.A. The mechanisms of contrast formation in the electron microscope are described in papers cited in Table 2.III.B.

The destructive effect of the electron beam on the specimen poses a severe restriction on the study of biological structure at high resolution. Major advances have been made in recent years in the development of methods to reduce the total dose delivered to specimens during microscopy. The low-dose technique of Unwin and Henderson (1002) is successful for the study of large, periodic arrays of unstained specimens, while the minimal-dose method of Williams and Fisher (1077) is widely practiced with negatively stained specimens. Normal, high-dose microscopy, in which no precaution is taken to minimize the total dose, leads to stain migration and crystallization and results in lower resolution of specimen details (Table 2.III.C.5).

Instrumental aberrations (Table 2.III.D) are a major source of image artifacts and, therefore, they can lead to incorrect interpretation of structural features. Digital image processing (Table 1.III.B.6.b and 1.III.C.3.g) can successfully compensate for correctly diagnosed artifacts. Most instrument-induced aberrations are easily detected by optical diffraction and can be quantified by digital techniques.

The main benefit of image-processing procedures, the formation of clear specimen images, is a consequence of image noise suppression. Three main sources of noise (Table 2.III.E) in micrographs are the granularity of the support film, irregularities in the specimen, and aberrations in the electron optics of the microscope. Since most processing methods involve a step in which signal and noise are distinguished, a familiarity with the possible sources of noise is useful.

3.4 Image Interpretation

Once there is assurance that all processing manipulations have been correctly applied, the most important stage in any structural study is the final interpretation of structural features. The effects of specimen preparation, imaging, and processing conditions (including the display of results), all influence the appearance and interpretation of specimen features. Model-building studies (Table 2.IV.C) and the correlation of image-processing results with other data, such as X-ray diffraction (Table 2.V), often complement and verify the image-processing results. In many instances an expected result is more easily interpreted and believed than an unanticipated finding. For example, the results that demonstrate hexagonal packing of the morphological units in the polyheads of T-even bacteriophages (Table 3) were completely expected and consistent with the principles of viral assembly as proposed by Cooper and Klug (144). On the other hand, totally unexpected results may lead to unique and interesting revelations. A beautiful illustration of this is the demonstration by Kiselev and Klug (570, 586) that tubes of human wart virus can be wholly assembled from pentamer units.

4 TABLE 3: BIOLOGICAL APPLICATIONS OF IMAGE PROCESSING

Table 3 cites hundreds of biological structure studies that are aided by image processing. The applications for each specimen are categorized according to the types of processing method employed (see key to Table 3, footnote *a*).

This serves as a guide for those interested in studying similar specimens by established techniques. Articles that review many of the biological applications of image processing include those by Lake (622), Amos (21), Crowther and Klug (189), and Finch (277).

Some examples in Table 3 are optical diffraction studies of models rather than electron micrographs (e.g., 322, 381, 687, 759) and are included because they illustrate how model building can contribute structural information. Also, examples of structural information obtained by combined model building and image processing of micrographs may be found in articles cited in Table 2.IV.C.

5 IMAGE PROCESSING OF SPECIMENS OF DIFFERENT TYPES

The choice of which image-processing method to apply is often dictated by the type of specimen studied. Specimens may be grouped into five main structure classes according to the symmetry they possess: helical, sheet, crystal, rotationally symmetric particles (aggregates), and noncrystalline, asymmetric particles.

5.1 Helical

For helical particles, the three-dimensional structure is usually reconstructed because structural details revealed by two-dimensional processing are often confusing and uninterpretable. Therefore, optical or digital filtering have limited value in the study of helical structures. The high symmetry of most helical particles and aggregates makes them ideal subjects for image processing, since one view often suffices to determine the three-dimensional structure to 2–3 nm resolution (Table 1.III.C.1.a).

Particles with the best preserved symmetry and structure are chosen on the basis of their optical diffraction pattern quality and the helical parameters determined from indexing the diffraction pattern. Helical indexing is often nontrivial because the diffraction spots tend to be very diffuse or the patterns do not display perfect mirror symmetry owing to distortion or uneven staining of particles. The computed transform is an aid to indexing because it reveals the symmetry or antisymmetry in the phases of diffraction spots (arising from the two sides of the particle) mirrored across the meridional line (parallel to the helix axis) in the diffraction pattern (Table 1.III.B.4). This, in turn, indicates whether the pair of spots arises from an even or odd numbered set of helices (helical family). DeRosier and Moore (208) give an excellent and comprehensive description of most of the important aspects of the three-dimensional reconstruction method applied to helical particles. This paper also discusses the treatment of the data for particles that lie on the microscope grid that is tilted relative to the electron beam.

It is common and recommended practice to compare and average the reconstructions of several individual particles so that the interpretation of structural detail is not biased by one specific example. Averaging helps to reduce the influence of staining artifacts present in individual reconstructions. Computer averaging also provides a quantitative measure of the similarities and differences between the separate reconstructions. Several examples of image reconstruction averaging are cited in Table 1.III.C.3.i.

The structural handedness of helical particles may be determined in a number of ways. The appearance of a particle changes at the edges when it is tilted about a line normal to the helix axis, and this often provides unequivocal evidence for establishing the correct hand (275, 641, 761). If it is too difficult to see the changes directly, the same experiment can be performed and optical diffraction patterns obtained from the right and left halves of the tilted particle (150, 275, 761). Spot intensities increase or decrease depending on the hand of the respective helical family. The particle may also be axially tilted and the phase changes in the computer transform used to determine handedness (96). Computer-simulated models are also informative in deciding the correct hand (275, 641). It also may be possible to determine handedness by metal shadowing one side of a particle if it contains deep grooves on its outer surface (1055).

The sheath of the T4 bacteriophage tail was the first helical particle (and the first particle of any kind) to be studied by the Fourier three-dimensional reconstruction method of DeRosier and Klug (204). This method is now routinely used in the study of a wide range of helical structures including actin (732, 916, 1039), tobacco mosaic virus rods (911) and stacked disks (285, 1003), ribosome helices (621, 627, 628, 629), bacterial flagella (883), flagellar microtubules (29), sickle-cell hemoglobin fibers (237, 238), alfalfa mosaic virus (696), catalase tubes (578), glutamine synthetase cables (327, 329, 1055), hemocyanin (621), and the T4 tail sheath (30, 903). In addition, Table 3 includes numerous examples of optical diffraction analyses of helical particles.

5.2 Sheets

Specimens characterized as sheets, such as membranes, cell walls, and thin crystals, are distinguished by their prevalent two-dimensionality. Optical and digital filtering, or photographic superposition methods often provide interpretable structural information if the specimen is thin (< 50 nm) and contains only one or a few repeating units (unit cells) in the short dimension. Nevertheless, structural details are generally more reliably interpreted from a three-dimensional reconstruction. This requires the collection and combination of several independent views, usually obtained from images of the sheet systematically oriented at known angles of tilt in the microscope. Unfortunately, the restricted range of tilts ($\pm 60°$) in most microscopes results in a missing cone of diffraction data, which in turn produces a loss of structural resolution in the direction perpendicular to the plane of the sheet.

In fortuitous cases, such as the purple membrane (450), where the diffraction pattern is very weak in the unobserved region, the absence of such data does not severely alter the reconstruction results. In general, however, the unobserved data in the missing cone may contain useful structural information, and strategies should be devised to measure or generate some of the data. Embedding and sectioning the specimen at right angles to the plane was one successful solution in at least two studies (27, 999). A constrained iterative Fourier refinement method (solvent flattening technique) is another possibility under investigation (R. M. Stroud, personal communication).

Several papers describe the methods and applications of two-dimensional image processing of sheets (Tables 3 and 2.III.A.2.a and 2.III.B.10.a,b), however, only five sheets have been studied by three-dimensional reconstruction techniques: purple membrane (450), cytochrome oxidase vesicles (337, 449), membrane-bound ribosomes (999), actin filament bundles (207), and tubulin sheets (26, 27, 941).

5.3 Three-Dimensional Crystals

Crystalline specimens that extend in three dimensions pose two major problems for the image processor. First, the specimen must be made thin enough (< 50 nm) to examine in the microscope, and second, if three-dimensional structural information is required, independent views of the crystal must be photographed and combined. The specimen may be embedded in plastic and thin sectioned in several known orientations, or it may be possible to crush the crystals to produce thin fragments that reveal an adequate number of independent views. Thin sectioning is usually unsatisfactory because the preparation procedures are rather drastic and lead to lower resolution of structural details (5–10 nm). Sectioning may be necessary, though, if suitable crystals can not be obtained *in vitro* (e.g., intracellular crystalline inclusions) or by fragmentation.

Several three-dimensional crystalline specimens have been studied by optical and computer diffraction and filtering, and by photographic superposition (Tables 3 and 1.III.B.10.a,b), but few have been reconstructed in three dimensions: lipoyl transsuccinylase (203, 209), lipovitellin-phosvitin (769), nucleosomes (291), and catalase (1022). It is interesting to note that in each of these examples, structural data were obtained from X-ray diffraction experiments as well as from microscopy. The combination of diffraction amplitudes measured by X-ray or electron diffraction with phases measured from images has proved to be valuable in several structural studies (Table 1.III.B.8,9).

5.4 Rotationally Symmetric Particles

Specimens that display only rotational symmetry, such as noncrystalline, oligomeric proteins, spherical viruses, and bacteriophage base plates, have been studied by one or more of the rotational photographic superposition, rotational filtering, and three-dimensional reconstruction techniques. Digital rotational filtering and photographic superposition techniques produce qualitatively similar results, but it is usually recommended that the latter method be used with caution and, ideally, with specimens displaying obvious or well-established symmetry. The two methods are compared by Crowther and Amos (179) and Misell (720).

Three-dimensional reconstructions of spherical viruses are calculated by combining a few unique views of the particle. The detailed procedures are outlined by Crowther (175, 176), Crowther and Amos (180), and Crowther et al. (183). These methods provide a quantitative measure of the agreement between individual views and also indicate the resolution to which the 532 icosahedral symmetry is obeyed.

Examples of image-processing studies of rotationally symmetric particles are given in Tables 3, and 1.III.B.10, and 1.IV.A.2. Except for the spherical viruses, no three-dimensional reconstructions have been performed by Fourier methods. This probably reflects the difficulty in reconstructing particles much smaller and of lower symmetry than the spherical viruses. Some three-dimensional reconstructions of isolated enzyme molecules have been calculated using real-space, back-projection methods (567, 861, 985, 1008).

5.5 Asymmetric Particles

Aperiodic specimens have rarely been studied by image processing, mainly because existing techniques are not ideally suited for the study of asymmetric particles. Nevertheless, recent advances, including the use of correlation methods in the laboratories of Frank and Hoppe (Table 1.III.C.3.e) and the optical filtering of artificially produced arrays of Ottensmeyer's group (Table 1.II.G.4), represent a needed trend in the development of new methodology for studying this type of specimen.

6 PROGRESS AND PROSPECTS

6.1 Radiation Effects

The single, most significant advance in specimen preparation and imaging techniques was the recent development of very-low-dose electron microscopy of unstained, unfixed crystalline specimens by Unwin and Henderson (1002). Their remarkable study of the three-dimensional structure of the purple membrane (450) at 0.7 nm resolution revealed the α-helical, secondary structure in the protein subunits.

The main advantages of this technique are a minimization of structural changes due to beam damage, the manifestation of genuine contrast in unstained structural features, and the reduction of dehydration damage in specimens where the aqueous medium is replaced with glucose or other low molecular weight sugars. This technique offers the best opportunity to study the native structure of specimens in the electron microscope at high resolution. Unfortunately, only a few studies of this type have been successful because the most restrictive prerequisite is that the specimen be well ordered over large areas (1–10 μm^2). This severely limits the number of specimens suitable for study by these methods. Unless ways are developed for inducing *in vitro* formation of large, periodic arrays, most specimens, including helical and single, isolated asymmetric particles, cannot be studied by these powerful methods. The negative-stain, carbon film, mica flotation technique for forming crystalline arrays of viruses (501) might prove easily adaptable for this purpose.

In the immediate future, most image processing will continue to center on the study of negatively stained, periodic specimens because negative staining remains the simplest method for preparing and visualizing biological structures, and image-processing methods are best suited for the analysis of periodic specimens. Radiation damage is also a problem with stained particles, since the electron beam causes the stain to migrate and crystallize around the specimen ashes (Table 2.III.C.5). The minimal beam technique of Williams and Fisher (1077; see also 449, 773, 1002) is quite successful in reducing resolution loss caused by irradiation of stained particles. The search for stains that are more radiation resistant and offer more protection against dehydration effects has not yet proved successful.

6.2 Dehydration Damage

Specimen flattening and other structural distortions caused by dehydration are finally being recognized as important problems in specimen preservation for electron microscopy. Maintenance of the native (hydrated and unstained) specimen structure (Table 2.II.B.2.c) has been extensively studied by Parsons's group (wet specimens: eg. 513, 802) and by Taylor and Glaeser (368, 955, 961: frozen-hydrated specimens). Their techniques are important, practical advances in specimen preparation methods, but it will probably be some time before hydrated specimens can be routinely studied with commercially available microscopes.

In the meantime, we must contend with the problems of dehydration-induced distortions. Negative stains offer better structural support than can be obtained if the material is unstained (i.e., no embedding matrix such as glucose); however some studies (Table 2.II.B.1) have reported up to 30% flattening of negatively stained specimens. Chemical cross-linking agents such as glutaraldehyde and dimethylsuberimidate have not been shown to offer much protection except when protein assemblies dissociate on the grid without prior fixation. The use of fixatives is generally discouraged because they often produce irreversible structural changes. If specimens must be examined by thin sectioning, the use of dehydrating agents such as ethylene glycol has been recommended because these substances are less denaturing than most other commonly used agents (Table 2.II.A.2.a). Present thin-sectioning preparation methods usually lead to rather low resolution (5–10 nm) of structural details. Advances in thin-sectioning technology that improve resolution will be welcomed.*

The application of digital image-processing procedures to correct for dehydration-induced or inherent specimen distortions has been successful in at least two studies (191, 325). Until new preparation methods are available to successfully and routinely prevent dehydration-induced specimen distortions, greater emphasis should be placed on development of computer methods to correct for distortions. Unfortunately, this may prove difficult, if not impossible, especially in the case of specimen flattening, unless there are ways to accurately measure the exact nature of the distortions. That is, simple models such as isotropic or one-dimensional distortions are probably inadequate to describe the actual deformations occuring in specimens. Three-dimensional image reconstructions will be somewhat artifactual if the specimen is distorted or if the distortion is incorrectly compensated. In either instance, structural features may be misrepresented, and therefore misinterpreted.

6.3 Image Distortions

Detection and correction for the contrast transfer characteristics of the electron microscope (defocus, spherical aberration, beam coherence, astigmatism, etc.) are now routine steps in digital image processing, especially at higher resolutions (< 2 nm). An unresolved and perplexing problem concerns the observation that although some specimens reveal 0.3–0.4 nm resolution in low-dose electron diffraction patterns, the diffraction patterns of their low-dose-recorded images fade out at 0.7–1.0 nm. The electron diffraction results clearly indicate specimen preservation in the microscope at near-atomic resolution, but this information is somehow lost when the image is recorded. An answer to this dilemma will represent a major breakthrough because the study of some biological structures at near-atomic resolution may then be possible in the electron microscope.

6.4 Aperiodic Specimens

Isolated asymmetric particles are the most neglected type of specimen studied by image processing. Most processing methods are, in principle, applicable to nonperiodic structures but in practice they are inadequate and inefficient. Consequently, most image-processing studies have involved the structure analysis of periodic specimens. The main disadvantages of aperiodic objects derive from the inefficiency of collecting and combining

*See Akey et al., *J. Mol. Biol.* **136:**19–43 (1980).

their images. A three-dimensional reconstruction of a nonperiodic specimen requires many more views than for a similar-sized, symmetric particle, and the radiation sensitivity of biological specimens makes it difficult to record several exposures of a single particle that is tilted to reveal different views. The combination of different views from the same or separate particles is nontrivial owing to the absence of symmetry axes, which are a useful reference frame. Preliminary results on the structural studies of single ribosomes and fatty acid synthetase molecules have been reported by Hoppe's group (478, 480, 485). Frank et al. (316–318) have used correlation methods to study single molecules of glutamine synthetase (with 622 symmetry), and these techniques may eventually be applicable with asymmetric particles.

6.5 Missing Three-Dimensional Data

A serious problem in three-dimensional data collection for sheet-type specimens arises because the limited range of tilts leads to a missing cone of diffraction data (see Section 5). If the missing data are not measured in some way (e.g., by thin sectioning) or generated by iterative refinement, the reconstruction will be poorly resolved in the direction perpendicular to the plane of the sheet. The three-dimensional structural study of the purple membrane (450) was successful because it was known from X-ray diffraction measurements that the missing data are very weak and can be ignored because they would have little effect on the overall quality of the reconstruction.

6.6 Treatment of the Constant Term of Diffraction Data

In the study of helical particles by the three-dimensional reconstruction technique, the treatment of equatorial diffraction data is not always straightforward, and such data are often completely left out of the reconstruction. It is common practice to assume that the particle is circularly symmetric (i.e., undistorted) and therefore that the equatorial data are all real (i.e., the phases are either 0 or 180 degrees). Also, because the digitized image is floated, the amplitudes of the equatorial data are likely to differ from the true values. Therefore, the constant term (average density) at each radius in the reconstruction is probably in error. At present there is no simple way to restore the correct amplitudes and also measure the nature and extent of particle distortion.

Measurement of the zero-order term (magnitude of the central diffraction spot) is also a problem in two-dimensional, digital reconstruction studies. The zero-order term specifies the average density in the image and is important for defining the boundaries between stain-accepting and the stain-excluding regions. One usually adds enough constant term (i.e., increases the amplitude of the central reflection) to produce a reconstruction with reasonable connectivity between the stain-excluding regions. Thus the boundary between stain and specimen is never exactly determined, so absolute measures of intrasubunit extents are not always reliable.

6.7 Correlation with Other Data

Combination and correlation of electron microscope data with measurements from other physical techniques such as X-ray and electron diffraction are likely to become more common in the future. For example, the combination of phases measured from micrographs and amplitudes measured from electron diffraction patterns was instrumental in the determination of the purple membrane structure (450, 1002). Amplitudes measured from electron diffraction patterns are more accurate than those measured from digitally processed images because the diffraction amplitudes from images are modulated by the contrast transfer function of the electron microscope. The effect of the transfer function on the pattern can only be partially corrected by digital methods.

Several studies (Table 1.III.B.8,9) include results based on the combination of phases measured from images and amplitudes determined by X-ray diffraction. Because the X-ray experiments are performed with specimens in their native, hydrated state, it is believed that these measurements better reflect the true amplitudes than those measured from the diffraction patterns of negatively stained specimens. At low resolution, the phases from stained particles are believed to agree reasonably well with the phases of particles in their native state.

6.8 Technology

Except for the unparalleled growth of computer technology, there have been no recent, major advances in the development of image processing procedures. The availability of optical diffractometry on-line with the electron microscope (Table 1.I.C.1) may be useful for rapid specimen screening and adjustment of microscope imaging conditions (defocus and astigmatism correction), as well as for proividing a quick method to simultaneously record images and their diffraction patterns. Frank (313) has noted that the advent of array processors and fast mass storage will likely lead to a breakthrough in the speed and versatility of image processing. Thus, real-time modes of operation and interactive problem solving will become not only possible, but also fast and inexpensive.

6.9 Interpretation of Image Reconstructions

Perhaps the most important but understated topic is the interpretation of structural features revealed in image reconstructions. What are the structural and functional significance of details revealed by reconstructions? Crowther (178) states that the problem of interpretation is really one of pattern recognition. He advocates the combination and correlation of structural information from micrographs with other types of experiment (i.e., biochemical, genetic, immunological, or other complementary structural information) to improve our understanding of biological structure-function relationships. The work of Crowther et al. (190) on the molecular reorganization in the base plate of bacteriophage T4 is an excellent example of how data of different types are combined and interpreted.

Perhaps the most appropriate statement is that given by DeRosier:

> *Even the clearest and most detailed images, however, have little meaning without the proper conceptual framework. The interpretation of such images in terms of molecular organization requires the coordinated application of mind and eye, that is the ability to recognize in images the consequences of molecular interactions.**

Table 1 Theory and Methods of Image Processing

I Optical diffraction

- **A** Basic theory and methods: **500,533,**588,**589,**622,720,740
- **B** Applications of optical diffraction
  - **1** Objective means to reveal and assess periodic structural detail: 11,21, 178,**263,**289,390,**588,589,767**
    - **a** Accurate measurement of lattice dimensions (unit cell size and shape): 256,**289,**533,569
      - Spacings and orientation are used to correctly apply translational photographic superposition: 670,720
      - Dimensions may aid in molecular weight estimates and the identification of chemical species: 81
    - **b** Detection of rotational and translational symmetry elements: 533,624,636,637,947
    - **c** Determination of the relationship of layers in multilayered specimens: 178
    - **d** Detection of specimen preservation (distortion, resolution, staining artifacts) This is a good criterion for selecting the best images for reconstructions: 178,**263,**622,1097
    - **e** Assess long- and short-range order
      - Spots are sharper, the more regular or more extensive the specimen: **263,**415,**500,533,**739,1060
      - The narrowness and curvature of helical specimens produces elongated spots: 256,263
    - **f** Identify signal and noise components in the image: 263,533,588
  - **2** Examine very small specimen areas

*D. J. DeRosier, in *Topics in the Study of Life: The BIO Source Book,* A. Kramer, Ed., Harper & Row, New York, 1971, p. 24.

3 Display imaging characteristics of the electron microscope
- **a** Contrast transfer function: 198,305,**500,533,**588,602,720,740
- **b** Astigmatism: 198,**500,533,**670,720
- **c** Resolution in micrograph (and specimen): 263,490,**500,533,**720
- **d** Beam coherence: 310,500,**530,**533,**670,**720
- **e** Specimen (stage) drift: 500,533,670,720
- **f** Focal drift: 500,**530,**533,**670**
- **g** Miscentering or electrical charging of the objective aperture: 720

4 Determine handedness of structures: **150,275,466,637,**761,1097

5 Measure specimen thickness: 81

6 Determine or verify chemical stoichiometries: 637

7 Aids understanding of diffraction theory and methods: **114,419, 645,**946,947,951

8 Represent electron diffraction patterns: 559

9 Examples of optical diffraction applications: (see entries (1a) in Table 3 for numerous examples)

C Apparatus: The optical diffractometer

1 Design: Descriptions and schematic diagrams

	Linear	Folded
Lensless	85,527,**740,**951	**249,**951
One lens	37,84,**263,**490,500,527, 720,740	6,206
Two lens	83,108,109,121,418,481, 490,**500,**530,533,645,670, 695,**740,**822,931,951,971, 1093	**11,45,**206,**263,**326,407, 408,417,458,**500,510,**533, 601,643,644,646,740,946, 947,948,949,951,**953,**967, 1100

Advantages/disadvantages of different designs: 84,**206,263,**636,637,1093
Advantages/disadvantages of different object positions (i.e., in front or behind the diffractometer lens): 263,533,966
Commercial designs: **500,533,**951
Other designs: Conversion of light microscope: **344,**500,822
Use of ordinary spectrophotometer: 946
Diffractometer on-line with electron microscope: 99,457, 853

2 Components
- **a** Optical benches, support and mounts for components: 263,500,533, 646,670,953
- **b** Light source
  - Mercury arc: 107,500,**510,646,1093**
  - Introduction of laser: **84,421,**500,740,1015
  - Tungsten bulb: 344
  - Coherence of source: 500,646,951,**953,967**
    - Coherent source reveals system flaws (e.g., dust): 592

- **c** Expansion and filtration system
  - Expansion lens (converging or diverging): 500,533
  - Pinhole filtering and hole size determination: 83,417,533,951
  - Alteration of illumination profile: **417**,533,947
  - Modified condenser to produce circularly polarized light: 418
- **d** Collimating and diffraction lenses
  - One-lens or two-lens design: 84,263
  - Use of corrected doublets: 405,500
  - Arrangement to reduce spherical aberration: 530,1093
  - Use of central part of lens: 83,263
  - Antireflection coating: 206
- **e** Liquid gate (reduces inhomogeneities of the micrograph emulsion)
  - Sandwich micrograph in oil between optical flats: **82**,326,382,**416**,533,589
  - Optical gate (combination of two main lenses and micrograph into a single optical component): **82**,83,533
  - Reduction of phase effects without oil immersion by using small areas of the micrographs: 206,263,670
  - Glass plates vs. film: 11,263
- **f** Other components (mirrors, camera, film polarizing filter, shielding, etc.): refer to references in I.C.1

3 Alignment

- **a** Object is to position all components so that they share a common optical axis: 533,953
- **b** Detailed alignment procedure: **45**,82,417,500,**530**,533,**670**,951,**953**
  - Use of Boy's point: 417,530,533,646,**670**,951,**953**

4 Calibration

- **a** Diffractometer constant: **11**,45,206,500,**530**,**533**,740
  - Calibration specimens: 45,500,533,740
- **b** Transform focal plane position: 45,500,533,953
- **c** Instrument resolution: 500,510
- **d** Pattern distortion (pincushion or barrel): **530**,533

D Pattern indexing (decide which spots are noise and which are from the specimen): 720

1 Indexing aids

- **a** Spots lie on a lattice if specimen is periodic: 592,720
- **b** Uneven stain penetration of two-sided specimen (leads to one-sided image, which may be easier to index): 263,592,622 (or more difficult to index: 500,533)
- **c** One-sided image of shadowed material: 1097
- **d** Friedel symmetry (pattern should have minimum inversion symmetry): 533
- **e** Uneven particle distortion: 622
- **f** Mirror line in pattern for undistorted helical particles: 206,592,637 or collapsed tubes: 206,263
- **g** Systematic absences (information about symmetry): 48,720

2 Indexing of helical particles (also see 8)
  a Spots are broad and more difficult to index: 570,622
  b Computed transform is more informative, since the phases of mirror-related spots may be calculated: 201
  c Mirror or glide symmetry in the image indicates even or odd symmetry: 726,729
3 Indexing is necessary step for reconstruction work: 263,500,592,952
4 Inspect several patterns from different images of a specimen to determine constant features: 206,263,592
5 Indexing should be consistent with features visible in the image: 206,592
6 At low resolution, pseudo-symmetry may be misleading: 50,720
7 Non-indexible spots resulting from multiple scattering in multilayered specimens: 288; most specimens obey first-order theory, so these effects can be ignored: 588 (but see 645). Optical diffraction from multilayered specimens is, to a first approximation, equal to the sum of the optical diffraction from the individual layers: 206
8 Examples of diffraction pattern indexing
  a Plane layers (two-dimensional sheets, collapsed tubes, etc.): 9, 13, 28,**48,**170,171,**206,**218,**256,288,**462,490,500,**555,**569,**570,571,**572, 573,577,**636,**664,**720,**844,846,847,**926,**941,**1004,**1095,1096,**1097,** 1098
  b Helical particles: 3,23,24,**29,**168,219,**286,**300,329,549,568,569,**571,** 572,**573,**589,592,600,**622,**629,**637,**698,**701,705,707,**708,711,**720,** **725,**728,729,**763,**819,903,910,911,980,**986,**1014,1018,1019,1020, 1039

E Artifacts of optical diffraction
1 Absence of Friedel symmetry (non-ideal system)
  a Lens quality: 83,263,533,947,953
  b Property of illumination beam: 417,533,947,953
  c Diffractometer misalignment: 83,533,953
  d Transform plane focus: 533,953
  e Mirror quality (diffraction rings from dust): 592
  f Uneven micrograph emulsion thickness (phase effects): 83,**206,**500
2 High-order, non-indexible spots from non-linear transfer function of photographic process: 500
3 Multiple scattering in multilayered specimens: 206,440

F Related topics
1 Comparison of optical diffraction with computed transforms: see III.D
2 Comparison to X-ray diffraction: 408,500,588,589,**644**
3 Comparison to electron diffraction: 500,589
4 Record of pattern contains amplitude but not phase information: 263,500,588
5 Pattern is characteristic of the micrograph, not the specimen, because it

contains information not necessarily related to the specimen structure (e.g., astigmatism, specimen drift, focus, nonlinearity of the photographic process, beam coherence, focal drift, stain distribution, and the shape of the window limiting the area for diffraction): **500**,530,**533**,670

6 Use in X-ray crystal structure determinations

- **a** Reviews: 151,410,**644**,946,947,**951**
- **b** Optical Fourier summation: **106,107**,118,119,120,121,**151**,407,409, **414**,644,**951**
- **c** Optical transform: 85,**108**,211,**410**,644,739,932,946,947
- **d** Patterson synthesis: 107,118,119,951
- **e** Study of specimen imperfections: 109,948,**949,1081**
- **f** Alteration of phases of diffracted rays: 107,118,119,**120**,121,407, 409,414,418
- **g** Methods to determine phases of diffracted rays: **151**,407,**408,414**, 642,**823**,945,1093
- **h** Use of the fly's eye method: **109**,110,211,406,**931**,951
- **i** Three-dimensional optical transform: 151,**418**,951

II Optical filtration

A Theory and methods: 6,**11**,45,55,83,**206,263**,277,**326**,350,353,**500**,533,**592**, **622**,720,894,924,1009

B Applications of optical filtration

1 Reduce noise and present in a useful way the information contained in the image: 263,533,**592,622**

2 Determine the average size and shape of subunits: 12,178,**263**,326,588,**592**

3 Separate moiré images of multilayered specimens: 263,622

Examples: 8,**11,206**,218,259,**263**,277,**288**,490,**500,570**,571,573,581,622, **636**,924,**952**,1096,1098

4 Determine or verify chemical stoichiometries: 10

5 Detect conformational changes in macromolecules: 12,581

6 Study macromolecular assembly: 178,206,586,622, and molecular events leading to movement: 622

7 Adjust image contrast by attenuating the undiffracted beam: 45,263, 720,924

8 Correct for instrumental aberrations: 968,971

9 Examples of optical filtration applications: (see entries (1b) in Table 3 for numerous examples)

C Reconstruction apparatus

1 Linear design: 490,**500**,533,592,740,770,952,968,972

2 Folded design: 6,45,206,**263**,326,500

(See I.C for appropriate references concerning components, alignment and calibration)

D Image selection and indexing

- **1** Images selected on the basis of optical diffraction patterns (number and sharpness of spots, resolution, high signal-to-noise ratio) that indicate specimen preservation: 263,720
- **2** Indexing (also see I.D)
  - **a** Particle symmetry must be solved: 263,592,720
  - **b** Distinguish sources of signal and noise for correct indexing: 263,952
  - **c** Proper mask design depends on correct indexing: 500
  - **d** Multilayered specimens: need to identify which spots belong to which lattice.
    - Success of filtering depends on the spatial separation of diffracted rays associated with each side: 490,592,952

E Filter mask design and preparation

- **1** Design
  - **a** Position of holes determined by indexing: see II.D.2
  - **b** Size of holes determines the amount of averaging and noise reduction: **11,45,**206,255,**263,326,**782,924
- **2** Methods to prepare masks for diffraction or filtering experiments
  - **a** Direct methods (quick and simple)
    - Pantographic punch: 206,**500,510,**570,642,644,739,947,**951**
    - Drilled holes: 55,106,107,500,949,**951**
    - Punched or cut holes: **263,**419,500,533,571,642,782,950,1093
  - **b** Photographic reduction of drawing (slow, but more exact and allows for complex patterns)
    - Direct use of photoreduced copy: 85,117,235,**420,739,946,**947, **1093**
    - Etching methods
      - Film: 37
      - Metal sheet: 11,**45,**236,**263,**414,419,420,459,**471,**636,720,952
    - Geometrical distortion during photoreduction: 947
    - Grating machine (high positional accuracy): 1081
  - **c** Computer-produced masks: 380,**415,**419,558,739,946,948

F Filtering artifacts

- **1** Incorrect design or positioning of the mask
  - **a** Misindexed pattern leads to incorrect hole positions and an erroneous reconstruction: 592,952
  - **b** Mask imposes an assumed model (reconstruction cannot be proof of the existence of a particular symmetry): 83,441,952
  - **c** Partial or total blocking of genuine spots by the mask leads to an erroneous reconstruction: 83,**440,500,**533,670,894
  - **d** Loss of information if holes are made too small: 326,740
  - **e** Spurious fringes appear in the reconstruction if diffraction from the box which limits the area investigated is allowed to pass through the holes: **45,**263

2 Multiple scattering from multilayered specimens (some spots from different layers overlap and interfere constructively or destructively so artifacts cannot be avoided): 500,592,625,952
(e.g., in helical particles there may be overlap of subsidiary maxima between different spots: 500,592)

3 Lattice defects may cause artifacts: 326,952

4 Periodic noise (cannot be filtered): 6,55,263

5 Frequency doubling (over-attenuation of the unscattered beam): **45**,263, 720,924,966
Examples of central beam attenuation: **45**,116,192,222,263, 267,445,**924**, 966

6 Artifacts caused by flaws in the optics of the apparatus: 45,740

G Related topics

1 Optical filtering is suited only for objects with translational symmetry: 263

2 Comparison of optical filtering with digital methods: see III.D

3 Comparison of optical filtering with translational photographic superposition method: 45,**206**,263,326,588

4 FAIRS method (filtering of arrays in reciprocal space): 720,**782**
Examples of biological applications: 574,782,783,912

III Digital Fourier image processing

A Theory and methods

1 General principles of digital processing: 6,91,**306**,**720**,752,**866**

2 Two-dimensional processing

a Translational symmetry: **11**,**48**,**171**,**191**,208,**255**,**256**,306,**449**,605,618, 624,**625**,**720**,752,805,**850**,872,**904**,**927**,**1002**

b Rotational symmetry: 21,**179**,189,263,**720**,1009
Comparison of rotational filtering with the method of rotational photographic superposition: 21,179,189,263,277,720,740

c Limitations of two-dimensional processing: 720

3 Three-dimensional reconstruction: 21,177,181,**183**,189,**201**,**204**,**208**,306, 588,620,**622**,720,1009,1053 (stereomicroscopy is less informative: 175, 189,533,622)

a Helical: 3,22,24,**29**,**30**,189,200,201,202,204,205,208,237,238,**285**,**621**, 622,**628**,696,**698**,720,732,883,**903**,**1003**,1008,**1039**

b Spherical: **175**,176,180,**181**,**183**,186,189,201,622,1008,1009

c Sheet: 27,207,337,449,450,941,999

d Three-dimensional crystal: 769

4 Other processing of micrographs: 52,308,475,487,607,751,970,1062, 1064,1067,1068

B Applications of digital Fourier image processing

1 All applications of optical processing also apply for digital methods (see I.B and II.B)

2 Recover three-dimensional information from two-dimensional projections: 204

3 Quantitatively analyze structure

- **a** Assess specimen symmetry: **48,179,**189,277,624,**626**
- **b** Assess and combine data from different images of the same projection: 3,22,**29,30,48,171,**696,883,**903,**911,925,**1003,1039**
- **c** Assess and combine data from different projections: 27,175,178, 182,337,449,941

4 Aid helical indexing: 3,201,204,237,277,285,622

5 Determine specimen handedness: 96,275,**641**

6 Correct for distortions

- **a** Deformed specimen: **191,325,**580,581,720
- **b** Electron microscope imaging distortions: 127,**260,**261,262,306,315, 476,927,969,**1002,1003,**1022,1065,1069,1070

7 Detect multiple scattering: 1038

8 Obtain reconstructions by combining amplitudes from electron diffraction and phase from images: 444,450,704,720,1001,**1002,**1009

9 Obtain reconstructions by combining amplitudes from X-ray diffraction and phases from images: 26,203,**209,291,**400,**525,**517,525,660, 1011,1012

10 Examples of digital processing of biological images

- **a** Two-dimensional transforms and filtering
  - Translational symmetry (see (2a) in Table 3 for additional examples)
    - Helical: 167,228,280,493,**641,696,**699,**764,805**
    - Sheet: **11,**13,**28,**48,67,171,**191,**255,**256,**327,449,605,618,**625,**663, 699,741,850,**904,915,**925,927,928,955,**998,**1000,**1002,**1004,1102
  - Rotational symmetry: (see (2b) in Table 3 for additional examples) 521,**179,185,**189,**190,**191,339,**698,**720,772
- **b** Three-dimensional reconstruction
  - Helical: **3,**22,23,24,**29,30,31,**165,167,189,199,201,202,**204,**205,237, **238,**245,**285,**327,329,622,627,**628,629,**696,**698,**699,720,**732,**735, 764,772,883,902,**903,**911,916,**1003,**1008,**1039**
  - Spherical: **175,176,**178,**180,181,**182,**186,**189,201,276,278,279,622, 697,1008,1009
  - Sheet: 26,**207,337,449,450,**720,941,999,1009
  - Three-dimensional crystal: 203,209,291,**769,**1022

C Data collection and processing

1 Collection of three-dimensional data

- **a** Number of views required: 175,176,177,178,**183,204,**277,533,588, 622,720,**941**
- **b** Methods to collect views
  - Use of symmetry: 176,183,201,**204,**306 (one view may suffice if specimen has high symmetry: 175,181,**204,**277,588,622)

Record systematically tilted views of the same or different particles: 24,26,**27**,175,181,201,204,245,**337**,**449**,**450**,588,622,**941**,999

Record images of variously oriented particles: 175,181,182,201,**769**

c Methods to obtain missing data from limited tilts

If most diffraction from the specimen lies in a plane normal to the electron beam, the extra data are not required: 450

Embed and section the specimen at right angles: 26,**27**,**999**

d Use shadowed material (three-dimensional data about surface structure): 579,**904**

2 Densitometry

a Selection of micrographs: **3**,30,**48**,190,204,**208**,**285**,444,883,903,1003,1039

b Densitometer: 208,256,306,702,**720**,**866**

c Scanning conditions

Orientation of image with respect to the scan direction: 6,**9**,**11**,204,**208**,720,903,927,928

Sampling frequency: 6,9,204,**208**,306,349,720,866,**903**

Aliasing errors: 204,866

Scanning aperture size: 204,306

3 Digital processing

a Software systems: 7,238,**320**,864,865,866,**868**,**900**

b Image boundary—floating: 11,204,**208**,306,720,866,903
—need for integral number of unit cells: 9,204,720

c Fast Fourier transform: 204,**208**,349,**720**,850,866

d Data interpolation: 9,**11**,175,181,**183**,204,**208**,306,720,769,866,928,1009

e Determination of particle orientation and phase origin position: 3,27,**29**,**30**,**48**,171,176,179,182,201,204,**208**,285,306,**449**,**624**,**625**,696,**769**,**805**,866,883,**903**,**941**,1004,**1039**

Use of auto- and cross-correlation procedures

Principles: 306,307,**309**,**312**,313,**314**,398,476,479,483,486,**720**,865,866,**867**

Applications: 253,312,313,**316**,317,**318**,524,846

f Mask design for two-dimensional filtering: 28,256,263

Hole size and shape: 618

Measurement of integrated spot intensity: 48,850

Subtraction of background surrounding spots: 337

g Correction or detection of specimen and image defects

Symmetrization: 48,228,625,927,941

Contrast transfer function (focusing effects): **260**,261,**262**,449,**450**,**927**,1001,**1002**,1022,1069

Lattice distortions: 191,325,850,903

Image flaws (e.g., scratches or dust on the micrograph): 605

Temperature factor correction for short-range disorder: 1004

Correction for second-order effects (multiple scattering): 1038
Contrast stretching (histogram equalization): 376,720

h Image subtraction: 476,720,903,1063

i Averaging of separate reconstructions: **3**,**22**,29,48,171,182,883,903, 911,916,**1003**,1039

j Display of reconstruction results: see Table 2.IV.B

D Comparison of optical and digital processing by Fourier methods: **11**,127,255,263,**720**,866,970

1 Data and transforms differ: 11 (optical is continuous; digital is sampled and discrete)

2 Advantages of optical filtering

a Transforms can be obtained quickly (speed of light!), easily and inexpensively from any size area. This is useful for screening and selecting images for reconstruction: 263

(To process images by computer, optical densities on the micrograph must first be digitized by a scanning microdensitometer. This requires expensive equipment and data storage handling capabilities: 255,256)

b Results are qualitatively similar to those obtained digitally: 11,75,263

c Reconstructions are done photographically (digital methods require good display facilities): 720

3 Advantages of digital processing

a Only way to do three-dimensional reconstruction or rotational filtering: 769

b More precise, flexible and quantitative: see B.3

Phases as well as amplitudes are obtained: 203,277,308,567,625

Useful or essential for indexing helical particles: 201,204,277, 622,**629**,637

Locate symmetry elements for orientation: 3,11,263,500,625

Assess specimen symmetry and preservation: 48,179,277,567, 625,626,850

Assess agreement between different images or reconstructions: 22,903

Manipulate phase and amplitude information: 28,500,625,740, 1003 (e.g., enhance certain features: 28,641; or control average density by setting the amplitude of the zero-order: 625. This may be done by optical filtering, but caution is required because of the possibility of frequency doubling effects: see II.F.5)

c Easier to correct for distortions in the specimen: 11,28,325,625, 720,740; or caused by the microscope: 127,720

d Easier to separate moiré images: 625

e Allows infinite control over mask design for filtering

Variable hole size, shape and weight: 28,48,618

Complete averaging can be obtained (equivalent to infinitely small holes in optical filtering): 255,256,263

(Mask design and production is more difficult and tedious for optical filtering experiments: 720)

f Easier to symmetrize an image (idealization of rotational or translational symmetry): 11,**625**,720,**850**

g Allows addition or averaging of separate reconstructions: see III.C.3.i

h Easier for processing images with low signal-to-noise ratio or with closely spaced spots: **28**,189; (digital methods are essential for processing low-dose images: 263,**609**,1001,**1002**; or other weak images: 625)

IV Real space and other reconstruction methods

A Rotational photographic superposition (enhancement of rotational symmetry)

1 Method: 18,263,**342**,490,**500**,**671**,720,762

a Test for artifacts: 18,**332**,**500**,**671**,720,**737**,740,**762**

b Calculation of reduction in noise: 671,762

c Comparison with digital, rotational filtering: see Table 1.III.A.2.b

2 Applications: see entries (3a) in Table 3 for numerous examples

Selected examples: 101,195,234,**334**,**335**,**342**,452,**458**,468,**469**,500,651, **671**,754,766,**790**,819,843,1032,1075

3 Apparatus

a Rotational and stroboscopic integrators: 490,500,671

Use of dove prism: 500,762

B Translational photographic superposition (enhancement of translational symmetry)

1 Method: 263,490,**500**,**672**,691,720,740,**1050**

a Artifacts

Correct translation is determined by trial and error: 589,1050 (optical diffraction may be used to determine orientation and translation: 21,466,670,720)

The reconstruction may be erroneous if the strongest visible periodicity in the image is not the principal repeat.

b Analogy with optical filtering: 45,263,588

c Calculation of reduction in noise: 1050

2 Applications: see entries (3b) in Table 3 for numerous examples

Selected examples: 340,465,**466**,500,506,**614**,619,**672**,720,920,**1046**, **1048**,1049,**1050**,**1051**

3 Apparatus

a Linear integrator: 490,500,**672**,720

b Shadow linear integrator: 490,500,740,**1050**

c Grating periodograph: 740

C Other optical methods and their applications

1 Photographic differentiation: 38,**671**

2 Multiple image superposition: 290,567,**776,841,851,**918,930,1029

3 Optical self-convolution (autocorrelation): 246,**247,**296,**297,298,**299,794

D Reconstruction from projections: 41,351,352,**387,453,**591,905,1103,**1107**

1 Iterative algebraic reconstruction

a ART (algebraic reconstruction technique): 77,78,187,306,331,351, **385,386,**453,454,455,**456,**1008,1009,1107

b SIRT (simultaneous iterative reconstruction technique): **351,352,** 456,630,1008

c EFIRT (extended field iterative reconstruction technique): 188

2 Back-projection

a Exact solution: **183,**306,352,453,540,588,779,780,**1006,**1008,**1009,** 1021

b SPF (synthesis by projecting functions): 306,1006,**1007,**1008,**1016,** 1021

Biological applications: 711,712,861,985,1008,1017

c SMPF (synthesis by modified projecting functions): **1009,**1021, Biological applications: 241,242,567,706,708,709,710,1008,1018, 1019,1020

3 Three-dimensional reconstructions of non-periodic structures

a Principles: **476,**477,**479,**480,482,**483,**484,989

b Biological applications: 478,480,**485**

4 Convolution method: 456,**828,**1107

5 Polytrophic montage: **426,**427

E Other methods

1 Holographic: 223,224,348,854,**936,**937

2 One-dimensional averaging with a cylindrical lens: 222

Table 2 Topics Related to Image Processing

I Principles of diffraction and symmetry

A Diffraction theory

1 Diffraction from crystals: 98,122,214,**243,375,**470,881,935,1083

2 Diffraction from helical and cylindrical structures: 156,214,**470,590,** 720,881,1071,1072,1073,1083

3 Diffraction symmetry: 122,214,243,375,470,**881,**1083

B Symmetry in biological molecules: 140,**144,**214,**243,**412,432,470,503,**585,** 587,**679,**1026,1083

C Fourier optics: 57,**382,**445,528,**646,**792,822

D Fourier mathematics: **102,**103,122,**643,**791,881,935

E Protein crystallography: **98,**214,243,470,**881,**1083

II Specimen preparation for electron microscopy

A Preparation methods

1 Negative staining: 277,343,370,403,429,**433**,658,675,700,**774**,1027,1028, 1086

2 Fixation, embedding, sectioning, and staining of sections:

a Techniques: **372**,**437**,639,652,695,716,**835**,842,890,1086

b Image processing of sectioned material (selected examples):

Optical diffraction: **5**,27,32,51,378,379,**547**,632,**647**,654,689,735, 736,749,**809**,833,**838**,**891**,**919**,922,923,1106

Optical filtration: 326,353,379,654,717,**749**

Three-dimensional reconstruction: 288,**621**,735

Photographic superposition: 19,196,**334**,335,341,**342**,561,613,614, 615,**616**,617,665,680,**820**,**843**,846,849,857,**862**,887,944,1043, 1044,1045,1047,1048

3 Shadowing

a Techniques: 2,403,451,695,893

b Selected image-processing examples (see also 4.b): 71,462,495,756, 757,845,847,1096,1097

4 Freeze-dry, fracture, etch

a Techniques: 124,**125**,393,**580**,599,668,755,832,893,896,933

b Selected image-processing examples: 9,13,32,95,218,**294**,392,393, 394,395,560,**579**,**580**,581,582,597,605,**608**,**668**,669,770,777,874, 875,879,**880**,895,**904**,917,928,1056,1082,**1092**

5 Image processing of unstained specimens: 152,170,297,327,426,**449**, **450**,547,704,771,782,783,929,954,955,961,**998**,1001,**1002**

B Dehydration damage

1 Specimen flattening: 21,75,178,189,256,277,**282**,670,**878**
Examples: 11,55,175,178,201,467,589,592,637,672,1076,1097

2 Methods to reduce damage

a Critical point drying: 75,403,438,893

b Freeze-drying or -etching: 75,**125**,392,393,**579**,**580**,581,**583**,668,669, 893

c Hydration chamber: 75,148,216,511,512,513,550,676,677,734,742, 797,799,800,801,802,956,1042

Frozen-hydrated specimens: 363,364,367,**368**,444,954,**955**,958, 959,960,**961**

d Substitution for aqueous medium

Negative stains: 75,277,941

Sugars: 152,153,154,170,449,450,704,998,1001,**1002**

e Chemical cross-linking: 11,51,157,193,581,636,647,774,842,962,1097

C Support films

1 Normal films (non-perforated): 64,403,404,430,433,437,695,774,863,893

2 Perforated films: 68,226,250,277,336,346,423,518,546,774,807,834,889, 890,942

3 Ultrathin films: 221,532,535,537,738,957,1079,1080

4 Distortion of film may distort specimen: 549

D Preparation of crystalline arrays on mica: 490,**491**,492,**493**,498,**501**

III Imaging conditions

A General texts and reviews on the principles and practice of electron microscopy: **15**,403,695,890,894,**1086**,1087

B Contrast enhancement

1 Elastic and inelastic scattering: 15,126,161,403,695,719,721,894,1086,1104

2 Phase and amplitude contrast (contrast transfer function): 33,126,161, 162,**260**,261,391,**439**,**442**,**531**,542,**543**,544,545,**968**,995

3 Dynamical (multiple) scattering: 254,262,720,1038

4 Dark-field imaging: 75,197,**230**,**231**,233,273,403,720,771,781,782,786
Image-processing applications: 73,224,230,252,299,753,**782**,783,787,788

5 Phase plates/zonal correction plates: 472,473,**474**,**481**,603,724,968,971, **972**,973,974,977,990,991,**992**,993,**994**,995,997

6 Coherence of illumination: 435,720,721

C Radiation effects

1 Review articles: 159,357,359,368,921

2 Types and mechanisms of specimen damage: 44,62,63,69,75,112,152, 158,**159**,178,234,277,368,397,402,425,446,460,519,522,598,745,783,796, **836**,837,886,**921**

3 Measurement of damage: 44,47,61,63,65,66,75,152,158,**159**,173,**232**,319, 356,**357**,359,360,362,363,364,365,**368**,369,396,402,**443**,507,508,509,520, **521**,536,563,598,607,**610**,720,744,785,789,796,830,**836**,858,859,860,886, 921,964,**965**,988,**996**,1001,**1002**,**1082**

4 Methods to reduce damage: 15,75,152,**159**,178,215,232,263,277,**359**,363, 364,**368**,428,443,447,449,536,538,562,596,**631**,720,**773**,**789**,795,798,829, 830,837,859,860,869,941,**1002**,**1077**,1082

5 Effect on stain distribution: **47**,75,178,189,277,500,536,720,**996**

6 Image processing of low-dose images: 152,311,327,359,366,449,486, 609,**610**,1001,**1022**

D Instrumental aberrations

1 Astigmatism: 15,403,530,533,670,695,720,740,890,1086

2 Contrast transfer function (spherical aberration and focus): 15,**254**,260, 261,**262**,**413**,473,474,533,542,720,727,**876**,1066

3 Focal drift: 533,695,890

4 Image drift: 15,530,695,720,740,890,1086

5 Beam coherence: 15,403,533,695,720

E Sources of noise in electron micrograph images

1 Variability in the specimen support film: 55,178,222,255,263,308,326, 424,500,588,635,782,783

2 Specimen irregularities

a Inherent in the specimen: 55,178,206,256,588

b Caused by preparation and imaging procedures

Dehydration and beam-induced damage: 206,222,326,500,740

Unfavorable interaction with stains: 740

Variability and granularity of stains: 55,83,178,206,222,255,256, 263,326,588

Contamination buildup from beam-induced fixation of volatile molecules in the microscope column to the specimen: 222,500

Presence of impurities such as extra protein in the specimen sample: 83,256

3 Microscope imaging conditions

a Astigmatism and spherical aberration in the objective lens: 500

b Chromatic aberration and electron charge effects in thick specimens (usually > 50 nm): 500

c Statistics in the random arrival of electrons at the specimen: 222,308,500,635,782,783

d Random scattering of electrons from the internal parts of the microscope: 500

e Defocusing of the objective lens leading to phase contrast effects: 439,500

4 Photographic effects

a Emulsion granularity: 222,308,500,635,966

b Non-linear exposure and development conditions: 136,**268**,269,589, 1024,**1025**

F Other topics

1 Double-exposure method: 534,720,1105

2 Electron diffraction

a Reviews: 271,272,369,747

b Correlated with image processing: 367,444,450,704,746,961,998, 1001,1002,1009,1022

IV Interpretation of images and reconstructions

A Interpretation of structural features revealed by negative staining: 21,**141**, 175,**178**,189,190,201,**277**,282,429,433,504,505,**720**,774,1027,1028

1 Staining artifacts: 50,**206**,533,589,592,622,670,1027

2 Interpretation of images of multilayered specimens: 206,500,592,670

B Examples of different image reconstruction displays

1 Two-dimensional: 11,48,170,190,255,256,262,316,625,696,720,752,813, 850,998

2 Three-dimensional: 30,165,175,180,181,720,903,904,996

C Model building as an aid to structure analyses

1 Modeling of structures: 51,70,141,184,275,277,282,283,284,347,491,516,

570,592,594,604,616,641,687,765,815,816,817,818,980,1034,1057,1058, 1059

2 Modeling of transforms: 901

V Comparison and correlation of X-ray diffraction and electron microscopy in structure determination

A Complementary nature: 189,588,647

B Resolution: 175,201,277

C Specimen requirements: 75,175,201,533,622

D Image reconstructions from combined X-ray and electron microscope data: see Table 1.III.B.9

E Comparison of X-ray diffraction and diffraction from electron micrographs: 5,26,**50**,**51**,**96**,142,147,203,**209**,210,256,274,291,300,400,622,**647**,660,730, 768,883,998,1094

Table 3 Biological Applications of Image Processing[a,b]

SPECIMEN	
BACTERIAL FLAGELLA	(1a) 147,286,381,553,600,763,883; (1b) 264,600,763; (2a,c) 883; (3a,b) 264
CELL MEMBRANE–WALL	
ATPase in artificial membrane	(3a) 1031
Bacillus brevis T layer	(1a) 6,**11**,230,579,580,720,**904;** (1b) 6,**11**,579,580,720; (2a) 6,**11**,720,**904**,1013; (2b) 904; (3a) 6,11
Bacillus polymyxa wall	(1a,b) 288
Bacterial wall peptidoglycan	(1a) 128
Bladder luminal membrane	(1a) 597,740,880,917,1034,1049,1050,1051; (1b) 597,879,880,917,1034,1051,1052; (3b) 740,1049,1050,1051
Capillary endothelium pores	(3a) 683
Chlamydomonas reinhardi wall	(1a) 145,462,463,494,495,**844**,845,847; (1b) 462,**844**,847; (3b) 462,844
Chlorogonium elongatum wall	(1a,b,3b) 846
Cholinergic receptor	(1a) 115,139,850; (1b) 115; (2a) 850; (3a) 139
Clostridium	(1a) 191,653,720; (2a,b) 191,720
Coated vesicle	(3a) 1090
Complex tubule of gymnostome	(3a) 827
Cytochrome *c* reductase vesicle	(1a,2a) 1085
Cytochrome oxidase membrane	(1a) **327**,328,**337**,**449**,667; (1b) 327,328,667; (2a) 327,337,449; (2e) 337,449
E. coli envelope	(1a) 72,720,**928**,1056; (1b) 928,1056; (2a) 720,928; (3a) 92
Flagellar plasma membrane	(1a,b) 777
Gap junction	(1a) 32,**143**,355,448,720,**1102;** (1b) 448,720; (2a) 143,720,1102; (3a) 811,812
Halobacterium halobium gas vesicle	(1a) 367,954
Ileum endocytic membrane complex	(1a,2a,3b) 848
Lipid bilayer-micelle	(1a) 160,514; (3a) 650
Lipovitellin-phosvitin complex	(1a) 768; (2e) 769
Lysolecithin-water mixture	(1a) 551

Micrococcus radiodurans HPI layer	(1a) **606**; (2a) 67,606
Microcystis marginata wall	(1a,b,3d) 560
Milk fat globule	(2a) 575
Miscellaneous bacterial wall–membrane	(1a) 87,301,373; (2a) 373,874
Mitochondrial cristae	(3a) 334
Nuclear pore complex	(3a) 1,97,213,293,302,304,345,681,682,870
Plasma membrane	(1a) 42,303; (1b) 303; (3a) 79
Purple membrane	(1a) 100,294,295,**444**,565,608,720,756,757,**1002**,1009; (1b) 100,294,720; (2a) **444**,**450**,608,702,704,1001,**1002**,1009; (2e) **450**,720,1009
Retinal receptor	(1a) 93,94
Spirillum serpens	(1a) 152,367,955; (2a) 361
Spirillum putridiconchylium	(1a,b) 89
Sporosarcina urea wall	(1a,b) 88
Synaptic vesicle	(1a) 1101; (3a) 334
Thylakoid	(1a) 669,688,714; (2a) 714
Yeast plasmalemma	(1a) 392,393,395,**605**,895; (2a) 392,394,**605**
CHROMATIN-HISTONE–NUCLEIC ACID	
DNA	(1a) 633,759,939; (2a) 306,939,1064
Histone fiber	(1a) 909,910
Nucleosomes	(1a,b) 234; (2a) **291**,595; (3a) 234
Nucleoprotamine	(1a) 73
Nucleotide	(2a) 223; (3f) 298
ENZYMES	
Acetyl-CoA carboxylase	(1a) 655
Adenosine triphosphatase	(1a) 913,1040; (2a) 1040; (3a) 1031
Amylase	(1a) 692,693
Argininosuccinase	(1a) 728
Aspartate transcarbamylase	(3c) 277,359,431,841

Table 3 (Continued)

SPECIMEN	
Enzymes (continued)	
Canavalin	(1a,3b) 694
Catalase	(1a) 47,84,130,201,**255**,260,261,262,270,306,344,367,401,500,**569**,**578**,589,609,610, 622,**647**,720,746,752,954,**955**,960,**961**,976,**998**,**1002**,**1008**,1010,1011,1015,**1022**,1038; (1b) 277,353,371,**569**,**578**; (2a) **60**,130,255,260,261,262,270,306,720,751,752,872,**955**, **998**,**1002**,1009,1011,1012,**1022**,**1038**; (2c) 200,201,202,**578**; (3d) 60; (3e) 578, 1008,1009; (3f) 1100; (3g) 348,1100
Citrate synthase	(3c) 851
Coupling factor 1	(1a) 793
Dihydrolipoyl transsuccinylase	(1a) 203,**209**,210,774
Endonuclease EndoR.Bgl I	(3a) 529
Fatty acid synthetase	(1a) 485; (3f,j) 478,480,**485**
Glucose oxidase	(1a) 578,**1014**; (1b) 567,578,1009,**1014**; (2c) 567; (3e) 567,578,1009,**1014**
Glutamate decarboxylase	(3a) 978
Glutamate dehydrogenase	(1a) 548,549,1074; (3f) 296,297
Glutaminase	(3c) 776
Glutamine synthetase	(1a) 327,329,330; (1b) 329; (2c) 327,329,1055; (3f) 312,313,**316**,317,**318**
Lactate dehydrogenase	(3b) 617
Leucine aminopeptidase	(1a) 567,577; (1b) 567,577,1009; (3a) 567,944; (3b) 944; (3c) 567; (3e) 567,985
Malate hydrolase	(2a) 699
Muconate cycloisomerase	(1a) 40
Neuraminidase	(1a,2a) 720
Nitrogenase	(1a,b) 722
Parahydroxybenzoate hydrolase	(1a) 871
Phosphorylase *a*	(1a) 573,986; (1b) 573
Phosphorylase *b*	(1a) 239,277,**571**,572,601,986,1008,1009; (1b) 239,277,431,**571**,572,601,1008,1009

Protocollagen proline hydroxylase (3a) 775
Pyruvate carboxylase (3c) 1029
Pyruvate dehydrogenase (3b) 1075
Pyruvate kinase (1a,b) 908
Ribulose bisphosphate carboxylase (1a) 45,**50**,51,1106; (1b) 45,50; (3a) 101,567; (3b) 1082; (3e) 567,861
Tryptophan-tRNA synthetase (1a) 566

FIBROUS MOLECULES
Amyloid fibril (3a) 374,882
Chitin (1a) 138,758
Collagen (1a) 74,214,**225**,**821**,840,940; (3f) 253; (3i) 222
Elastin (1a) 388,389,803,877,1035,1036,1037
Fibrinogen (1a) 981,1054; (1b) 1054
Keratin (1a) 322,323,324,**326**,533; (1b) **326**,533,**717**
P Protein filament (3a) 174
Poly-L-alanine (1a) 248
Purkinje fiber glycogen (1a) 975

IMMUNOGLOBULIN (3b) 611,612,613,**614**,**616**,**862**

MICROTUBILE–FLAGELLA–CILIA–CENTRIOLE
Acrosome (3a) 943
Axoneme/cilia (1a) **24**,**31**,**149**,741,986; (1b) 149; (2a) 24,31,986; (3a) 19,355,887,1044 (3b) 808,857,979,1044,1045,**1046**,1047,1048,1088
Axostyle (1a) 95,689
Cardiac muscle microtubule (3a) 377
Central pair tubules (1a,b) 640,778
Centriole (3a) 34,43,196,342,723,814,849
Chloroplast microtubule (3a) 856
Duplex tubule (1a) 526
Epithelial cell microtubule (3a) 332

Table 3 (Continued)

SPECIMEN	
Microtubile–Flagella–Cilia–Centriole (continued)	
Flagellar tubules	(1a) **23,29,**31,59,133,134,**150,**344,390,690,720,1094; (1b) **23,29,**31,720,1009; (2a) **29,**31,641; (2c) 23,**29,**31,258,720,1009; (3a) 39,132,174,556,820,843,887; (3b) 58,133,342,690,887
Macrotubule	(1a,b) 259
Mitotic spindle	(1a) 687
Neurofilament association with microtubule	(3a,b) 86
Neurotubules and walls	(1a) 25,**255,256,263,**660,662,720,986,**987;** (1b) 255,263,987; (2a) 255,**256,**257,258, 259,263,720,986; (3a) 334; (3b) 561; (3e) **987,**1009
Root tip cell microtubule	(3a) 634
Tubulin aggregates	(1a) 27,47,**48,49,170,**661,663,**664,**919,941; (1b) 1009; (2a) 26,27,48,49,170,171,661, 663,941; (2e) 26,27,941; (3a) 333
MUSCLE	
A-Band of skeletal muscle	(1a) 163,411,891
Actin and actin complexes	(1a) 111,**207,354,**589,657,**732,**748,764,**765,**914,**915,**916,1005,**1039,1095;** (1b) 764,**765, 1095;** (2a) **732,**764,**915,**1009,**1039;** (2c) 189,199,200,207,277,**732,**764,**916,**1008,1009, **1039**
Adductor	(1a) 715,750,760; (2c) 245
I-Band	(3a) 266
Insect flight muscle	(1a) 123,547,673,831,**833**
M-Band	(1a,b) 654
Myosin	(1a) **80,**123,135,164,244,**246,**733,**809,897,**906; (3a) 810; (3f) 246,247
Retractor muscle	(1a) 713,907
Smooth muscle	(1a) 164,717,839,**897,**898,**899,**906
Striated muscle	(1a) 36,212,**517,**892,1099; (2a) 517

Tropomyosin	(1a) **142,**277,326,622,**648,649,**929; (1b) 326; (2a) 929; (3f) 794
Z-Band	(1a) 378,379; (1b) 353,379; (3a) 321; (3b) 852
PILI	(1a) 252,300
RESPIRATORY PROTEINS	
Chlorocruorin	(3a) 399
Erythrocruorin	(1a) 632; (3a) 195
Hemocyanin	(1a) 21,**698,**976,1030,1074; (2a) 698; (2b) 189,**698;** (2c) 21,189,277,**698,**1009; (3a) 342
Sickle-cell hemoglobin fibers	(1a) 168,238,292,772,1061; (2a) 169,237,238; (2b) 772; (2c) 237,238,772; (3a) 194,342; (3b) 342
RIBOSOME	
Bacterial helices	(1a) 564,665,666; (3a) 665,666
Chromatoid body helix	(1a) 622,627,628,**629,**735,736; (2a) 622,628,**629,**1008; (2c) 205,306,**621,**622,627,628, **629,**735,1008
In vitro helical array	(1a) 155
Membrane crystal	(1a) 1004; (2a) 623,999,1000,1004; (2e) 999
Nonaggregated particles and subunits	(1a) 574,576,963; (1c) 574,912; (2e) 1008; (3d) 78,306; (3h) 427
Ribosome precursor	(3a) 105
VIRUS: BACTERIOPHAGE	
AR9	(1a,3e) 1020
A. eutrophus polysheath	(1a,b,3a) 1041
Butyricum	(1a) 710,1017; (1b) 710; (2a) 1017; (3e) 710,1009
Caulobacter crescentus ϕ-CbK	(1a) 624,625,626,636,637; (1b) 625,636; (2a) 624,625,626,637
DD-6	(1a) **705,**824,1017,1019; (1b) 705,1019; (2a) 1017; (3e) 705,712,1008,1009,**1019**
E. coli capsule	(3a) 930
f2	(2d) 182; (3c) 433
fd	(3g) 854,**936,**937
G	(1a) 218,219,220; (1b) 219,220
H17	(1a,3e) 241,242; (3f) 524

Table 3 (Continued)

SPECIMEN	
Virus: Bacteriophage (continued)	
INCO particles	(1a,3a) 819
λ	(1a) 555,1092; (1b) 555,1091,1092; (3a) 684; (3b) 506
μ	(1a) 3; (2a,c) 3,699
N1	(1a,2c) 1017
PBS phage	(1a,2a,c) 165,166,167
ϕ	(1a,b) 707,708; (3e) 708
ϕ-5	(1a,3e) 706,1018
T2	(1a) 8,533,670,**705**,730,**825**,826; (1b) **8**,10,12,251,720; (3e) 712,1008
T4 head	(1a) 6,**8**,9,10,**13**,21,71,**90**,137,189,**206**,**289**,**523**,**618**,668,720,875,**924**,925,**926**,**1096**,**1097**,1098; (1b) 6,**8**,**9**,10,12,**13**,14,21,**90**,189,**206**,264,277,**523**,557,581,582,**618**,668,720,**924**,**926**,**1096**,1098; (2a) 6,9,13,**28**,189,**618**,805,**925**; (3a) 6; (3b) 9,90,619,924
T4 Tail	(1a) 30,35,189,204,208,277,306,**592**,**604**,622,720,**725**,**729**,731,**903**,983; (1b) 189,**204**,277,**592**,720,972; (2a) **30**,**204**,**208**,720,**903**,1008,1009; (2b) 189; (2c) 22,**30**,189,**204**,**208**,277,306,622,720,902,**903**,997
T6 Tail	(1a) 711,1008,1009; (1b) 1009; (3e) 711,1008,1009
T-Even base plate	(2b) 178,**179**,**190**,720,1009; (3a) 104
Temperate phage no. 1	(1a,3e) 709
Unknown phage (tail)	(3a) 105
VIRUS: HELICAL	
Alfalfa mosaic	(1a) 229,515,516,**696**,**701**; (2a) 227,228,**696**,**699**,**701**; (2c) 696
Barley stripe mosaic	(1a,b) 568
Broccoli necrotic yellow	(1a) 461
Carnation yellow fleck	(1a) 56
Chlorotic leafspot	(1a) 384,490,498
Clover yellow mosaic	(1a,b) 54

Hemagglutinating virus capsid	(1a) 761
Herpes simplex type 2	(1a) 584
Influenza A2 nucleocapsid thread	(3a) 467
Papaya mosaic	(1a,b) 980
Potato virus T	(1a) 855
Potato virus X	(1a) **383,384,**492,493,498,501,552; (2a) 492,493; (3a) 1032
Potato virus Y	(1a) 685,686; (1b) 686
Rattle	(1a) 670
Red clover vein mosaic	(1a) 1032
Sendai	(1a) 280,531; (2a) 280; (3b) 342
Sugar beet yellow	(1a) 498
Tobacco mosaic PM2 protein	(1a) 670,885
Tobacco mosaic rod	(1a) 47,230,**274,275,**281,344,441,458,489,490,492,493,498,**589,**592,771,911; (1b) 281,441,592; (2a) 275,911,1069; (2c) 480,911; (3b) 493,**497,**505,**672;** (3f) 486; (3h) 426
Tobacco mosaic stacked disk	(1a) 96,252,287,458,490,**536,**537,538,720,991,994,**996;** (2a) **96,285,**994,**996,1003,**1009; (2b) 21,179,593; (2c) 189,277,**285,**593,994,995,**996,1003,**1009; (3a) 290,**458,**488,490, 504,656,**671;** (3b) 490,672; (3c) 290,671
Tobacco mosaic X protein	(3a) 1084
White clover mosaic	(1a) 1084
VIRUS: ISOMETRIC	
Adenovirus	(1a) 490,493,498,501,**502,**530,531,533,720; (2b) 21,**185,**277,720; (3a) 720,754; (3b,c) 720
Arabis mosaic	(3a) 17
Brome mosaic	(1a) 490,498,501,1023
Carnation mottle	(3a) 671
Cowpea chlorotic mottle	(1a) **46,**47,53,55,490,491,492,493,498,**499,501,**670,695,**927;** (1b) **53,55,**670,695; (2a) 4, **46,**493,505,**927;** (3b) 491,499; (3f) 867
Cowpea mosaic	(2d) 178,186,189,1009
Cucumber mosaic	(3a) 743
Cytoplasmic deoxyribovirus	(3a) 934
Fiji disease	(3a) 434

Table 3 (Continued)

SPECIMEN	
Virus: Isometric (continued)	
Foot and mouth disease	(3a) 113
Gladiolus	(3a) 146
H-Viruses	(3a) 554
Herpesvirus nucleocapsid	(2a) 338,339
Infectious bursal disease virus	(3a) 464
Necrosis	(3a) 659
Nudaurelia capensis β	(2d) 189,**279,**1009
Papilloma-polyoma type	(1a) **570,**586,622,670; (1b) 277,501,**570,**622; (2d) 21,**175,180,**181,**276,**277,278,1008, 1009; (3a) 678
ϕ-X174	(3a) 240
Plant virus (unknown name)	(3a) 737
R17	(2d) 182; (3a) 718
Rabbit kidney vacuolating	(1a) 55
Reovirus	(3a) 20,651,790,1033
Reo-like virus	(3a) 703
Rotavirus	(3a) 265; (3c) 918
Sinbis	(3a) 888
Southern bean mosaic	(1a) 492,493
Tomato bushy stunt	(2a) 176,1008; (2d) 175,176,**180,181,**201,277,**525,**622,1008,1009
Turnip crinkle	(1a) 492,493; (3a,c) 671
Turnip rosette	(1a) 492,493,496
Turnip yellow mosaic	(1a) 465,**466,**500,670; (1b) 500; (2d) 277,**697,**1009; (3a) 465,**466,**671; (3b) 465,**466**
Vesicular exanthema	(3a) 131
Wound tumor	(3c) 671

OTHER SPECIMENS	
Antibiotic/hormone	(1a) 782; (1b,c) 720,**782,783**,784; (3f) 299
Bacteriochlorophyll protein	(3b) 615,680
DNA helix destabilizing protein	(1a) 152
Edestin	(1a) 873
Egg shell protein	(3b) 340,341
Elastoidin	(1a) 1089
Erythrocyte ghost protein	(3a) 422
Excelsin	(1a) 873
Ferritin	(1a) 127,534,538; (1b) 127; (2a) 52,127,129; (3a) 342,436
groE Protein	(2a,b) 468; (3a) 245,468,469
Hepatocyte microbody	(3a) 984
High-density lipoprotein	(1a,b) 770
Intracellular crystal inclusions	(1a) 217,530,674,749,838,922,923,1050; (1b) 749; (3b) 1050
Lac repressor	(1a,3b) 920
Rhapidosome	(3a) 804
Staphylococcal α-toxin	(3c) 38
Termination factor rho	(3a) 766
Trophoblast	(1a) 638
Urinary bladder tubule structure	(3a) 982

[a] Key—1. optical Fourier methods: a, optical diffraction; b, optical filtration; c, FAIRS method. 2. Fourier image processing by computer: a, diffraction and/or filtering; b, rotational filtering; c, three-dimensional reconstruction (helical); d, three-dimensional reconstruction (spherical); e, three-dimensional reconstruction (planar). 3. Real-space and other methods: a, rotational photographic superposition; b, translational photographic superposition; c, photographic differentiation and superposition; d, algebraic iteration methods; e, synthesis by projecting functions; f, auto- and cross-correlation methods; g, holographic methods; h, polytrophic montage; i, one-dimensional averaging by cylindrical lens; j, three-dimensional reconstruction of non-periodic structures.

[b] Review articles of biological applications: **21**,175,**189**,201,277,**622**,720,**1009.**

ACKNOWLEDGMENTS

D. J. DeRosier and T. Wagenknecht are thanked for their critical reading of the manuscript and for helpful discussions and suggestions. Apologies are extended to the authors whose articles were inadvertently omitted or inappropriately categorized. All reference information, including subject key terms, is stored on the Rosenstiel PDP 11/40 computer, and FORTRAN programs have been written to allow input of new articles and formatted output of lists, which may be designed by specifying any or all of the following criteria: subject, author, journal, and year. I welcome correspondence about the retrieval of bibliographies and also about references that are missing from this review. This survey was funded in part by a postdoctoral fellowship to the author from the National Institutes of Health

REFERENCES AND BIBLIOGRAPHY

1 Abelson, H. T. and Smith, G. H. (1978) "Nuclear Pores: The Pore-Annulus Relationship in Thin Section" *J. Ultrastruct. Res.* **30**:558–588.

2 Abermann, R., Salpeter, M. M., and Bachmann, L. (1972) "High Resolution Shadowing," in *Principles and Techniques in Electronic Microscopy*, Vol. 3 (M. A. Hayat, Ed.), Van Nostrand Reinhold, New York, pp. 195–217.

3 Admiraal, G. and Mellema, J. E. (1976) "The Structure of the Contractile Sheath of Bacteriophage μ" *J. Ultrastruct. Res.* **56**:48–64.

4 Adolph, K. W. and Butler, P. J. G. (1974) "Studies on the Assembly of a Spherical Plant Virus. I. States of Aggregation of the Isolated Protein" *J. Mol. Biol.* **88**:327–341.

5 Adolph, K. W., Caspar, D. L. D., Hollingshead, C. J., Lattman, E. E., Phillips, W. C., and Murakami, W. T. (1979) "Polyoma Virion and Capsid Crystal Structures" *Science* **203**:1117–1119.

6 Aebi, U. (1977) "Analysis and Biological Significance of the Structural State of Ordered Macromolecular Assemblies When Visualized in the Electron Microscope," Ph.D. Thesis, University of Basel, Switzerland.

7 Aebi, U. (1978) "Image Processing of Electron Micrographs of Ordered Biomacromolecular Assemblies" *Proc. Ninth Intl. Congr. Electron Microsc. (Toronto)* **3**:81–86.

8 Aebi, U., Bijlenga, R., Broek, J. V. D., Broek, R. V. D., Eiserling, F., Kellenberger, C., Kellenberger, E., Mesyanzhinov, V., Muller, L., Showe, M., Smith, R. and Steven, A. (1974) "The Transformation of τ Particles into T4 Heads II. Transformations of the Surface Lattice and Related Observations on Form Determination" *J. Supramol. Struct.* **2**:253–275.

9 Aebi, U., Bijlenga, R. K. L., ten Heggeler, B., Kistler, J., Steven, A. C., and Smith, P. R. (1976) "Comparison of the Structural and Chemical Composition of Giant T-Even Phage Heads (Appendix: The Computer Filtration of Hexagonal Lattices, by P. R. Smith and U. Aebi)" *J. Supramol. Struct.* **5**:475–495.

10 Aebi, U., Bijlinga, R. K. L., van Driel, R., ten Heggeler, B., van den Broek, R., Steven, A. C., and Smith, P. R. (1976) "Protein Lattice Engineering as Seen in the Electron Microscope When Combined with a Posteriori Image Processing" *Sixth Europ. Reg. Conf. Electron Microsc. (Jerusalem)* **2**:498–501.

11 Aebi, U., Smith, P. R., Dubochet, J., Henry, C., and Kellenberger, E. (1973) "A Study of the Structure of the T-Layer of *Bacillus brevis*" *J. Supramol. Struct.* **1**:498–522.

12 Aebi, U., ten Heggeler, B., Onorato, L., Kistler, J., and Showe, M. K. (1977) "New Method for Localizing Proteins in Periodic Structures: Fab Fragment Labeling Combined with Image Processing of Electron Micrographs" *Proc. Natl. Acad. Sci. U.S.A.* **74**:5514–5518.

13 Aebi, U., van den Broek, R., Smith, P. R., ten Heggler, B., Dubochet, J., Mesyanzhinov, V. V., Tsugita, A., and Kistler, J. (1979) "Crystalline Aggregation of a Proteolytic Fragment of the Major Head Protein of Bacteriophage T4" *J. Mol. Biol.* **130**:255–272.

14 Aebi, U., van Driel, R., Bijlenga, R. K. L., ten Heggeler, B., van den Broek, R., Steven, A. C., and Smith, P. R. (1977) "Capsid Fine Structure of T-Even Bacteriophages. Binding and Localization of Two Dispensable Capsid Proteins into the P23 Surface Lattice" *J. Mol. Biol.* **110**:687–698.

15 Agar, A. W., Alderson, R. H., and Chescoe, D. (1974) "Principles and Practice of Electron Microscope Operation," in *Practical Methods of Electron Microscopy*, Vol. 2 (A. M. Glauert, Ed.), North Holland, Amsterdam, pp. 1–345.

16 Agar, A. W., Frank, F. C., and Keller, A. (1959) "Crystallinity Effects in the Electron Microscopy of Polyethylene" *Phil. Mag.* **4**:32–55.

17 Agrawal, H. O. (1967) "The Morphology of Arabis Mosaic Virus" *J. Ultrastruct. Res.* **17**:84–90.

18 Agrawal, H. O., Kent, J. W., and MacKay, D. M. (1965) "Rotation Technique in Electron Microscopy of Viruses" *Science* **148**:638–640.

19 Allen, R. D. (1968) "A Reinvestigation of Cross Sections of Cilia" *J. Cell Biol.* **37**:825–831.

20 Amano, Y., Katagiri, S., Ishida, N., and Watanabe, Y. (1971) "Spontaneous Degradation of Reovirus Capsid into Subunits" *J. Virol.* **8**:805–808.

21 Amos, L. A. (1974) "Image Analysis of Macromolecular Structures" *J. Microsc.* **100**:143–152.

22 Amos, L. A. (1975) "Improved Resolution in Three-Dimensional Image Reconstruction of Helical Structures" *Electron Microsc. Soc. Am. Proc.* **33**:290–291.

23 Amos, L. A. (1975) "Substructure and Symmetry of Microtubules," in *Microtubules and Microtubule Inhibitors* (M. Borgers and M. de Brabander, Eds.), North Holland, Amsterdam, pp. 21–34.

24 Amos, L. A. (1976) "Three-Dimensional Image Reconstruction of Intact Flagellar Axonemes from a Tilt Series of Electron Micrographs" *Sixth Eur. Reg. Conf. Electron Microsc. (Jerusalem)* **1**:14–19.

25 Amos, L. A. (1977) "Arrangement of High Molecular Weight Associated Proteins on Purified Mammalian Brain Microtubules" *J. Cell Biol.* **72**:642–654.

26 Amos, L. A. and Baker, T. S. (1979) "The Three-Dimensional Structure of Tubulin Protofilaments" *Nature (London)* **279**:607–612.

27 Amos, L. A. and Baker, T. S. (1979) "Three-Dimensional Image of Tubulin in Zinc-Induced Sheets, Reconstructed from Electron Micrographs" *Intl. J. Biol. Macromol.* **1**:147–156.

28 Amos, L. A. and Klug, A. (1972) "Image Filtering by Computer" *Fifth Eur. Reg. Conf. Electron Microsc. (Manchester)* **1**:580–581.

29 Amos, L. A. and Klug, A. (1974) "Arrangement of Subunits in Flagellar Microtubules" *J. Cell Sci.* **14**:523–549.

30 Amos, L. A. and Klug, A. (1975) "Three-Dimensional Image Reconstructions of the Contractile Tail of T4 Bacteriophage (Appendix: Combination of Data from Helical Particles: Correlation and Selection, L. A. Amos)" *J. Mol. Biol.* **99**:51–73.

31 Amos, L. A., Linck, R. W., and Klug, A. (1976) "Molecular Structure of Flagellar

Microtubules," in *Cell Motility* (R. Goldman, T. Pollard, and J. Rosenbaum, Eds.), Cold Spring Harbor Laboratory, Cold Spring Harbor, NY, pp. 847–867.

32 Amsterdam, A., Josephs, R., Lieberman, M. E., and Lindner, H. R. (1976) "Organization of Intramembrane Particles in Freeze-Cleaved Gap Junctions of Rat Graafian Follicles: Optical-Diffraction Analysis" *J. Cell Sci.* **21**:93–105.

33 Andersen, W. H. J. (1972) "Phase Contrast Enhancement by Single Sideband Modulation Transfer" *Electron Microsc. Soc. Am. Proc.* **30**:616–617.

34 Anderson, R. G. W. and Brenner, R. M. (1971) "The Formation of Basal Bodies (Centrioles) in the Rhesus Monkey Oviduct" *J. Cell Biol.* **50**:10–34.

35 Anderson, T. F. and Krimm, S. (1966) "Diffraction of Light from Electron Micrographs of Helical Structures in Bacteriophage Tail Sheaths" *Proc. Sixth Intl. Congr. Electron Microsc. (Kyoto)* **1**:145–146.

36 April, E. W., Brandt, P. W., and Elliott, G. F. (1971) "The Myofilament Lattice: Studies on Isolated Fibers. I. The Constancy of the Unit-Cell Volume with Variation in Sarcomere Length in a Lattice in Which the Thin-to-Thick Myofilament Ratio is 6:1" *J. Cell Biol.* **51**:72–82.

37 Aravindakshan, C. (1957) "A Simple Arrangement for Obtaining Optical Transforms of Crystal Structures" *J. Sci. Instrum.* **34**:250.

38 Arbuthnott, J. P., Freer, J. H., and Bernheimer, A. W. (1967) "Physical States of Staphylococcal α-Toxin" *J. Bacteriol.* **94**:1170–1177.

39 Arnott, H. J. and Smith, H. E. (1969) "Analysis of Microtubular Structure in *Euglena granulata*" *J. Phycol.* **5**:68–75.

40 Avigad, G., Englard, S., Olsen, B. R., Wolfenstein-Todel, C., and Wiggins, R. (1974) "Molecular Properties of *cis,cis*-Muconate Cycloisomerase from *Pseudomonas putida*" *J. Mol. Biol.* **89**:651–662.

41 Baba, N. and Murata, K. (1977) "Filtering for Image Reconstruction from Projections" *J. Opt. Soc. Am.* **67**:662–668.

42 Baccetti, B., Bigliardi, E., and Rosati, F. (1971) "The Spermatozoon of Arthropoda. XIII. The Cell Surface" *J. Ultrastruct. Res.* **35**:582–605.

43 Baccetti, B., Dallai, R., and Fratello, B. (1973) "The Spermatozoon of Arthropoda. XXII. The '12+0', '14+0' or Aflagellate Sperm of *Protura*" *J. Cell Sci.* **13**:321–335.

44 Bahr, G. F., Johnson, F. B., and Zeitler, E. (1965) "The Elementary Composition of Organic Objects After Electron Irradiation" *Lab. Invest.* **14**:377–395.

45 Baker, T. S. (1976) "The Structure of Tobacco Leaf Ribulose Bisphosphate Carboxylase" Ph.D. Thesis, University of California, Los Angeles.

46 Baker, T. S. (1978) "The Packing of Cowpea Chlorotic Mottle Virus in Crystalline Monolayers" *Proc. Ninth Intl. Congr. Electron Microsc. (Toronto)* **2**:24–25.

47 Baker, T. S. (1978) "Preirradiation and Minimum Beam Microscopy of Periodic Biological Specimens" *Proc. Ninth Intl. Congr. Electron Microsc. (Toronto)* **2**:2–3.

48 Baker, T. S. and Amos, L. A. (1978) "Structure of the Tubulin Dimer in Zinc-Induced Sheets" *J. Mol. Biol.* **123**:89–106.

49 Baker, T. S. and Amos, L. A. (1978) "The Structure of the Tubulin Heterodimer in Zinc-Induced Sheets" *Proc. Ninth Intl. Congr. Electron Microsc. (Toronto)* **2**:272–273.

50 Baker, T. S., Eisenberg, D., and Eiserling, F. (1977) "Ribulose Bisphosphate Carboxylase: A Two-Layered, Square-Shaped Molecule of Symmetry 422" *Science* **196**:293–295.

51 Baker, T. S., Eisenberg, D., Eiserling, F. A., and Weissman, L. (1975) "The Structure of Form I Crystals of D-Ribulose-1, 5-Diphosphate Carboxylase" *J. Mol. Biol.* **91**:391–399.

52 Ball, F. L., Harris, W. W., and Welton, T. A. (1971) "Computer Processing of Electron Micrographs" *Electron Microsc. Soc. Am. Proc.* **29**:88–89.

53 Bancroft, J. B. (1973) "Cowpea Chlorotic Mottle Virus," in *The Generation of Subcellular Structures* (R. Markham, J. B. Bancroft, D. R. Davies, D. A. Hopwood, and R. W. Horne, Eds.), North Holland, Amsterdam, pp. 115–122.

54 Bancroft, J. B., Abouhaidar, M., and Erickson, J. W. (1979) "Assembly of Clover Yellow Mosaic Virus and Its Protein" *Virology* **98:**121–130.

55 Bancroft, J. B., Hills, G. J., and Markham, R. (1967) "A Study of the Self-Assembly Process in a Small Spherical Virus: Formation of Organized Structures from Protein Subunits *in Vitro*" *Virology* **31:**354–379.

56 Bar-Joseph, M., Josephs, R., and Cohen, J. (1977) "Carnation Yellow Fleck Virus Particles '*in vivo*,' " *Virology* **31:**144–151.

57 Barber, N. F. (1961) in *Experimental Correlograms and Fourier Transforms*, Pergamon Press, Oxford.

58 Barton, R. (1967) "Substructure in Negatively Stained Flagella of *Pteridium* Spermatozoids" *J. Ultrastruct. Res.* **20:**6–19.

59 Barton, R. (1969) "Investigation of Negatively Stained Plant Flagellar Microtubules by Optical Diffraction" *J. Cell Biol.* **41:**637–641.

60 Barynin, V. V. and Vainshtein, B. K. (1972) "X-Ray Diffraction and Electron Microscopic Study of Hexagonal Catalase Crystals. I. Electron-Microscopic Study" *Sov. Phys. Cryst.* **16:**653–661.

61 Baumeister, W., Fringeli, U. P., Hahn, M., and Seredynski, J. (1976) "Radiation Damage of Proteins in the Solid State: Changes of β-Lactoglobulin Secondary Structure" *Biochim. Biophys. Acta* **453:**289–292.

62 Baumeister, W., Fringeli, U. P., and Hahn, M. (1976) "Latent Dose Effects in Radiation Damage" *Sixth Eur. Reg. Conf. Electron Microsc.* (*Jerusalem*) **2:**507–509.

63 Baumeister, W., Fringeli, U. P., Hahn, M., Herbertz, L. M., Kopp, F., and Seredynski, J. (1976) "Radiation Damage in Proteins" *Sixth Eur. Reg. Conf. Electron Microsc.* (*Jerusalem*) **2:**510–512.

64 Baumeister, W. and Hahn, M. (1976) "An Improved Method for Preparing Single Crystal Specimen Supports: H(2)O(2) Exfoliation of Vermiculite" *Micron* **7:**247–251.

65 Baumeister, W., Hahn, M., and Fringeli, U. P. (1976) "Electron-Beam-Induced Conformational Changes in Polypeptide Layers: An Infrared Study" *Z. Naturforsch., C.* **31:**746–747.

66 Baumeister, W., Hahn, M., Seredynski, J., and Herbertz, L. M. (1976) "Radiation Damage of Proteins in the Solid State: Changes of Amino Acid Composition in Catalase" *Ultramicroscopy* **1:**377–382.

67 Baumeister, W. and Kubler, O. (1978) "Topographic Study of the Cell Surface of *Micrococcus radiodurans*" *Proc. Natl. Acad. Sci. U.S.A.* **75:**5525–5528.

68 Baumeister, W. and Seredynski, J. (1976) "Preparation of Perforated Films with Predeterminable Hole Size Distributions" *Micron* **7:**49–54.

69 Baumeister, W. and Seredynski, J. (1978) "Radiation Damage to Proteins: Changes on the Primary and Secondary Structure Level" *Proc. Ninth Int. Congr. Electron Microsc.* (*Toronto*) **3:**40–48.

70 Bayer, M. E. and Bocharov, A. F. (1973) "The Capsid Structure of Bacteriophage λ" *Virology* **54:**465–475.

71 Bayer, M. E. and Cummings, D. J. (1977) "Structural Aberrations in T-Even Bacteriophage. VIII. Surface Morphology of T4 Lollipops" *Virology* **76:**767–780.

72 Bayer, M. E. and Leive, L. (1977) "Effect of Ethylenediaminetetraacetate Upon the Surface of *Escherichia coli*" *J. Bacteriol.* **130:**1364–1381.

73 Bazett-Jones, D. P. and Ottensmeyer, F. P. (1979) "A Model Structure of Nucleoprotamine" *J. Ultrastruct. Res.* **67:**255–266.

74 Bear, R. S. (1955) "Configuration of Collagen and Gelatin Molecules in Condensed and Dispersed States" *Symp. Soc. Exp. Biol.* **9**:97–114.

75 Beer, M., Frank, J., Hanszen, K. J., Kellenberger, E., and Williams, R. C. (1975) "The Possibilities and Prospects of Obtaining High-Resolution Information (below 30 angstroms) on Biological Material Using the Electron Microscope" *Q. Rev. Biophys.* **7**:211–238.

76 Beeston, B. E. P. (1973) "An Introduction to Electron Diffraction," in *Practical Methods of Electron Microscopy*, Vol. 1 (A. M. Glauert, Ed.), North Holland, Amsterdam, pp. 193–323.

77 Bellman, S. H., Bender, R., Gordon, R., and Rowe, J. E. Jr. (1971) "ART is Science. Being a Defense of Algebraic Reconstruction Techniques for Three-Dimensional Electron Microscopy" *J. Theor. Biol.* **32**:205–216.

78 Bender, R., Bellman, S. H., and Gordon, R. (1970) "ART and the Ribosome: A Preliminary Report on the Three-Dimensional Structure of Individual Ribosomes Determined by an Algebraic Reconstruction Technique" *J. Theor. Biol.* **29**:483–487.

79 Benedetti, E. L., and Emmelot, P. (1965) "Electron Microscope Observations on Negatively Stained Plasma Membranes Isolated from Rat Liver" *J. Cell Biol.* **26**:299–305.

80 Bennett, P. M. (1976) "The Molecular Packing in Light Meromyosin Paracrystals" *Sixth Eur. Reg. Conf. Electron Microsc. (Jerusalem)* **2**:517–519.

81 Berger, J. E. (1969) "Optical Diffraction Studies of Crystalline Structures in Electron Micrographs" *J. Cell Biol.* **43**:442–447.

82 Berger, J. E. and Harker, D. (1967) "Optical Diffractometer for Production of Fourier Transforms of Electron Micrographs" *Rev. Sci. Instrum.* **38**:292–293.

83 Berger, J. E., Taylor, C. A., Shechtman, D., and Lipson, H. (1972) "Miscellaneous Applications" in *Optical Transforms* (H. Lipson, Ed.), Academic Press, New York, pp. 401–422.

84 Berger, J. E., Zobel, C. R., and Engler, P. E. (1966) "Laser as Light Source for Optical Diffractometers; Fourier Analysis of Electron Micrographs" *Science* **153**:168–170.

85 Berry, C. R. (1950) "Optical Evaluation of Molecular Structure Factors" *Am. J. Phys.* **18**:269–273.

86 Bertolini, B., Monaco, G., and Rossi, A. (1970) "Ultrastructure of a Regular Arrangement of Microtubules and Neurofilaments" *J. Ultrastruct. Res.* **33**:173–186.

87 Beveridge, T. J. (1978) "Structure of the Macromolecular Surface Arrays on the Bacterium *Sporosarcina ureae*: Isolation, Purification and Reassembly" *Proc. Ninth Int. Congr. Electron Microsc. (Toronto)* **2**:348–349.

88 Beveridge, T. J. (1979) "Surface Arrays on the Wall of *Sporosarcina ureae*" *J. Bacteriol.* **139**:1039–1048.

89 Beveridge, T. J. and Murray, R. G. E. (1976) "Reassembly *in Vitro* of the Superficial Cell Wall Components of *Spirillum putridiconchulium*" *J. Ultrastruct. Res.* **55**:105–118.

90 Bijlenga, R. K. L., Aebi, U., and Kellenberger, E. (1976) "Properties and Structure of a Gene 24-Controlled T4 Giant Phage" *J. Mol. Biol.* **103**:469–498.

91 Billingsley, F. C. (1971) "Image Processing for Electron Microscopy: A Digital System" *Adv. Opt. Electron Microsc.* **4**:127–159.

92 Bladen, H. A., Evans, R. T., and Mergenhagen, S. E. (1966) "Lesions in *Escherichia Coli* Membranes After Action of Antibody and Complement" *J. Bacteriol.* **91**:2377–2381.

93 Blasie, J. K., Dewey, M. M., Blaurock, A. E., and Worthington, C. R. (1965) "Electron Microscope and Low-Angle X-ray Diffraction Studies on Outer Segment Membranes from the Retina of the Frog" *J. Mol. Biol.* **14**:143–152.

94 Blasie, J. K., Worthington, C. R., and Dewey, M. M. (1969) "Molecular Localization of

Frog Retinal Receptor Photopigment by Electron Microscopy and Low-Angle X-Ray Diffraction'' *J. Mol. Biol.* **39:**407–416.

95 Bloodgood, R. A. and Miller, K. R. (1974) ''Freeze-Fracture of Microtubules and Bridges in Motile Axostyles'' *J. Cell Biol.* **62:**660–671.

96 Bloomer, A. C., Champness, J. N., and Unwin, P. N. T. (1976) ''The Hand of the Stacked-Disk Aggregate of Tobacco Mosaic Virus Protein'' *J. Mol. Biol.* **105:**453–457.

97 Bloom, S. (1970) ''Structural Changes in Nuclear Envelopes During Elongation of Heart Muscle Cells'' *J. Cell Biol.* **44:**218–223.

98 Blundell, T. L. and Johnson, L. N. (1976) In *Protein Crystallography*, Academic Press, New York.

99 Bonhomme, P. and Beorchia, A. (1978) ''A Light Optical Diffractometer for Electron Microscopical Images Operating in Line'' *Proc. Ninth Int. Congr. Electron Microsc.* (*Toronto*) **1:**86–87.

100 Borovyagin, V. L., Plakunova, V. G., and Sherman, M. B. (1976) ''Microstructure of *Halobacterium halobium* Membranes'' *Dok. Biophys.* **228:**91–93.

101 Bowien, B., Mayer, F., Codd, G. A., and Schlegel, H. G. (1976) ''Purification, Some Properties and Quaternary Structure of the D-Ribulose 1,5-Diphosphate Carboxylase of *Alcaligenes eutrophus*'' *Arch. Microbiol.* **110:**157–166.

102 Bracewell, R. M. (1965) In *The Fourier Transform and Its Applications*, McGraw-Hill, New York.

103 Braddick, H. J. J. (1965) In *Vibrations, Waves and Diffraction*, McGraw-Hill, New York.

104 Bradley, D. E. (1965) ''The Structure of the Head, Collar and Base-Plate of 'T-Even' Type Bacteriophages'' *J. Gen. Microbiol.* **38:**395–408.

105 Bradley, D. E. (1966) ''The Structure of Protein Particles Released from *Pseudomonas aeruginosa* by Mitomycin C'' *Proc. Sixth Int. Congr. Electron Microsc.* (*Kyoto*), 115–116.

106 Bragg, W. L. (1939) ''A New Type of 'X-ray Microscope' '' *Nature* (*London*) **143:**678.

107 Bragg, W. L. (1942) ''The X-ray Microscope'' *Nature* (*London*) **149:**470–471.

108 Bragg, W. L. (1944) ''Lightning Calculations with Light'' *Nature* (*London*) **154:**69–72.

109 Bragg, W. L. and Lipson, H. (1943) ''A Simple Method of Demonstrating Diffraction Grating Effects'' *J. Sci. Instrum.* **20:**110–113.

110 Bragg, W. L. and Stokes, A. R. (1945) ''X-ray Analysis with the Aid of the 'Fly's Eye,' '' *Nature* (*London*) **156:**332–333.

111 Bray, D. and Thomas, C. (1976) ''Unpolymerized Actin in Fibroblasts and Brain'' *J. Mol. Biol.* **105:**527–544.

112 Breedlove, J. R., Jr. and Trammell, G. T. (1970) ''Molecular Microscopy: Fundamental Limitations'' *Science* **170:**1310–1313.

113 Breese, S. S. Jr., Trautman, R., and Bachrach, H. L. (1965) ''Rotational Symmetry in Foot-and-Mouth Disease Virus and Models'' *Science* **150:**1303–1305.

114 Brisse, F. and Sundararajan, P. R. (1975) ''A Practical Method of Simulating X-ray Diffraction'' *J. Chem. Educ.* **52:**414–415.

115 Brisson, A. (1978) ''Acetylcholine Receptor Protein Structure'' *Proc. Ninth Int. Congr. Electron Microsc.* (*Toronto*) **2:**180–181.

116 Brown, L. M., Ferrier, R. P., Toms, N., and Woods, P. J. (1970) ''Optical Analysis of Electron Metallographs'' *Proc. Seventh Int. Congr. Electron Microsc.* (*Grenoble*) **1:**363–364.

117 Buckman, A. P. and Woolley, R. A. (1979) ''Spatial Filtering with a Photographic Replica of the Fourier Transform of a Half-Tone Picture'' *J. Phys. E: Sci. Instrum.* **12:**95–97.

118 Buerger, M. J. (1939) "The Photography of Interatomic Distance Vectors and of Crystal Patterns" *Proc. Natl. Acad. Sci. U.S.A.* **25**:383–388.

119 Buerger, M. J. (1941) "Optically Reciprocal Gratings and Their Application to Synthesis of Fourier Series" *Proc. Natl. Acad. Sci. U.S.A.* **27**:117–124.

120 Buerger, M. J. (1950) "The Photography of Atoms in Crystals" *Proc. Natl. Acad. Sci. U.S.A.* **36**:330–335.

121 Buerger, M. J. (1950) "Generalized Microscopy and the Two-Wavelength Microscope" *J. Appl. Phys.* **21**:909–917.

122 Buerger, M. J. (1970) In *Contemporary Crystallography*, McGraw-Hill, New York.

123 Bullard, B., Luke, B., and Winkleman, L. (1973) "The Paramyosin of Insect Flight Muscle" *J. Mol. Biol.* **75**:359–367.

124 Bullivant, S. (1970) "Present Status of Freezing Techniques," in *Some Biological Techniques in Electron Microscopy* (D. F. Parsons, Ed.), Academic Press, New York, pp. 101–146.

125 Bullivant S. (1973) "Freeze-Etching and Freeze-Fracturing," in *Advanced Techniques in Biological Electron Microscopy*, Vol. 1 (J. K. Koehler, Ed.), Springer-Verlag, New York, pp. 67–112.

126 Burge, R. E. (1976) "Contrast and Image Formation of Biological Specimens," in *Principles and Techniques in Electron Microscopy*, Vol. 6 (M. A. Hayat, Ed.), Van Nostrand Reinhold, New York, pp. 85–116.

127 Burge, R. E., Dainty, J. C., and Scott, R. F. (1977) "Optical and Digital Image Processing in High-Resolution Electron Microscopy" *Ultramicroscopy* **2**:169–178.

128 Burge, R. E., Fowler, A. G., and Reaveley, D. A. (1977) "Structure of the Peptidoglycan of Bacterial Cell Walls. I" *J. Mol. Biol.* **117**:927–953.

129 Burge, R. E. and Scott, R. F. (1975) "Binary Filters for High Resolution Electron Microscopy" *Optik* **43**:53–64.

130 Burge, R. E. and Scott, R. F. (1975) "Binary Filters for High Resolution Electron Microscopy. II" *Optik* **44**:159–172.

131 Burroughs, J. N., Doel, T. R., Smale, C. J., and Brown, F. (1978) "A Model for Vesicular Exanthema Virus, the Prototype of the Calicivirus Group" *J. Gen. Virol.* **40**:161–174.

132 Burton, P. R. (1966) "Substructure of Certain Cytoplasmic Microtubules: An Electron Microscopic Study" *Science* **154**:903–905.

133 Burton, P. R. (1970) "Optical Diffraction and Translational Reinforcement of Microtubules Having a Prominent Helical Wall Structure" *J. Cell Biol.* **44**:693–699.

134 Burton, P. R. and Silveira, M. (1971) "Electron Microscopic and Optical Diffraction Studies of Negatively Stained Axial Units of Certain Platyhelminth Sperm" *J. Ultrastruct. Res.* **36**:757–767.

135 Camatini, M., Castellani, L. C., Franchi, E., Lanzavecchia, G., and Paoletti, L. (1976) "Thick Filaments and Paramyosin of Annelid Muscles" *J. Ultrastruct. Res.* **55**:433–447.

136 Carlemalm, E. and Weibull, C. (1975) "The Response to Electrons and Developing Conditions of Two Photographic Films" *J. Ultrastruct. Res.* **53**:298–305.

137 Carrascosa, J. L. and Kellenberger, E. (1978) "Head Maturation Pathway of Bacteriophages T4 and T2. III. Isolation and Characterization of Particles Produced by Mutants in Gene 17" *J. Virol.* **25**:831–844.

138 Carstrom, D. (1957) "The Crystal Structure of α-Chitin" *J. Biophys. Biochem. Cytol.* **3**:669–683.

139 Cartaud, J., Benedetti, E. L., Sobel, A., and Changeux, J. P. (1978) "A Morphological Study of the Cholinergic Receptor Protein from *Torpedo marmorata* in Its Membrane Environment and in its Detergent-Extracted Purified Form" *J. Cell Sci.* **29**:313–337.

140 Caspar, D. L. D. (1964) "Design and Assembly of Organized Biological Structures," in *Molecular Architecture in Cell Physiology* (T. Hayashi and A. G. Szent-Gyorgyi, Eds.), Prentice-Hall, Englewood Cliffs, NJ, pp. 191–207.

141 Caspar, D. L. D. (1966) "An Analogue for Negative Staining" *J. Mol. Biol.* **15:**365–371.

142 Caspar, D. L. D., Cohen, C., and Longley, W. (1969) "Tropomyosin: Crystal Structure, Polymorphism and Molecular Interactions" *J. Mol. Biol.* **41:**87–107.

143 Caspar, D. L. D., Goodenough, D. A., Makowski, L., and Phillips, W. C. (1977) "Gap Junction Structures. I. Correlated Electron Microscopy and X-Ray Diffraction" *J. Cell Biol.* **74:**605–628.

144 Caspar, D. L. D. and Klug, A. (1962) "Physical Principles in the Construction of Regular Viruses" *Cold Spring Harb. Symp. Quant. Biol.* **27:**1–24.

145 Catt, J. W., Hills, G. J., and Roberts, K. (1976) "A Structural Glycoprotein, Containing Hydroxyproline, Isolated from the Cell Wall of *Chlamydomonas reinhardi*" *Planta (Berlin)* **13:**165–171.

146 Chambers, T. C., Francki, R. I. B., and Randles, J. W. (1965) "The Fine Structure of Gladiolus Virus" *Virology* **25:**15–21.

147 Champness, J. N. (1971) "X-ray and Optical Diffraction Studies of Bacterial Flagella" *J. Mol. Biol.* **56:**295–310.

148 Chang, B. B. and Parsons, D. F. (1978) "Low Dose Electron Diffraction of Wet Protein Crystals" *Proc. Ninth Int. Congr. Electron Microsc.* (*Toronto*) **2:**6–7.

149 Chasey, D. (1972) "Subunit Arrangement in Ciliary Microtubules from *Tetrahymena pyriformis*" *Exp. Cell Res.* **74:**140–146.

150 Chasey, D. (1974) "Left-Handed Subunit Helix in Flagellar Microtubules" *Nature* (*London*) **248:**611–612.

151 Chaudhuri, B. (1972) "Determination of Crystal Structure," in *Optical Transforms* (H. Lipson, Ed.), Academic Press, New York, pp. 71–113.

152 Chiu, W. (1978) "Factors in High Resolution Biological Structure Analysis by Conventional Transmission Electron Microscopy" *Scanning Electron Microsc.* **1:**569–580.

153 Chiu, W. and Hosoda, J. (1978) "Crystallization and Preliminary Electron Diffraction Study to 3.7 angstroms of DNA Helix-Destabilizing Protein gp32*I" *J. Mol. Biol.* **122:**103–107.

154 Chiu, W. and Hosoda, J. (1978) "Electron Microscopy of a DNA Helix Destabilizing Protein Crystal (gp32*I)" *Proc. Ninth Int. Congr. Electron Microsc.* (*Toronto*) **2:**178–179.

155 Clark, M. W., Hammons, M., Langer, J. A., and Lake, J. A. (1979) "Helical Arrays of *Escherichia coli* Small Ribosomal Subunits Produced *in Vitro*" *J. Mol. Biol.* **135:**507–512.

156 Cochran, W., Crick, F. H. C., and Vand, V. (1952) "The Structure of Synthetic Polypeptides. I. The Transform of Atoms on a Helix" *Acta Crystallogr.* **5:**581–586.

157 Cohen, N. D., Utter, M. F., Wrigley, N. G., and Barrett, A. N. (1979) "Quaternary Structure of Yeast Pyruvate Carboxylase: Biochemical and Electron Microscope Studies" *Biochemistry* **18:**2197–2203.

158 Cosslett, V. E. (1970) "Beam and Specimen: Radiation Damage and Image Resolution" *Ber. Bunsen-Gesellschaft Phys. Chem.* **74:**1171–1175.

159 Cosslett, V. E. (1978) "Radiation Damage in the High Resolution Electron Microscopy of Biological Materials: A Review" *J. Microsc.* **113:**113–129.

160 Costello, M. J. and Gulik-Krzywicki, T. (1976) "Correlated X-ray Diffraction and Freeze-Fracture Studies on Membrane Model Systems: Perturbation Induced by Freeze-Fracture Preparative Procedures" *Biochim. Biophys. Acta* **455:**412–432.

161 Cowley, J. M. and Bridges, R. E. (1979) "Phase and Amplitude Contrast in Electron Microscopy of Stained Biological Objects" *Ultramicroscopy* **4**:419–427.

162 Cowley, J. M. and Grinton, G. R. (1970) "Phase and Amplitude Contrast from Biological Materials" *Proc. Seventh Int. Congr. Electron Microsc.* (*Grenoble*) **1**:59–60.

163 Craig, R. (1977) "Structure of A-Segments from Frog and Rabbit Skeletal Muscle" *J. Mol. Biol.* **109**:69–81.

164 Craig, R. and Megerman, J. (1977) "Assembly of Smooth Muscle Myosin into Side-Polar Filaments" *J. Cell Biol.* **75**:990–996.

165 Cremers, A. F. M., Fischer, J. C., and Mellema, J. E. (1979) "The Reconstruction of a Helical Structure from Projections Applied to a Phage Tail" *Ultramicroscopy* **4**:91–96.

166 Cremers, A. F. M., Fischer, J. C., Schilstra, M. J., and Mellema, J. E. (1979) "Low-Dose Electron Image Reconstruction of Negatively Stained Contractile Phage Sheath from *Bacillus subtilis* (PBS-Z)" *Ultramicroscopy* **4**:395–412.

167 Cremers, A. F. M. and Krijgsman, P. C. J. (1978) "A Quantitative EM Study of the Packing and Shape of the Protein Subunits Organized in Contractile Sheaths" *Proc. Ninth Intr. Congr. Electron Microsc.* (*Toronto*) **2**:26–27.

168 Crepeau, R. H., Dykes, G., and Edelstein, S. J. (1977) "Structure of the Fibers of Sickle Cell Hemoglobin in the Presence of 2,3-Diphosphoglycerate" *Biochem. Biophys. Res. Commun.* **75**:496–502.

169 Crepeau, R. H., Dykes, G., Garrell, R., and Edelstein, S. J. (1978) "Diameter of Haemoglobin S Fibres in Sickled Cells" *Nature* (*London*) **274**:616–617.

170 Crepeau, R. H., McEwen, B., Dykes, G., and Edelstein, S. J. (1977) "Structural Studies on Porcine Brain Tubulin in Extended Sheets" *J. Mol. Biol.* **116**:301–315.

171 Crepeau, R. H., McEwen, B., and Edelstein, S. J. (1978) "Differences in α and β Polypeptide Chains of Tubulin Resolved by Electron Microscopy with Image Reconstruction" *Proc. Natl. Acad. Sci. U.S.A.* **75**:5006–5010.

172 Crewe, A. V. (1973) "Considerations of Specimen Damage for the Transmission Electron Microscope, Conventional Versus Scanning" *J. Mol. Biol.* **80**:315–325.

173 Crewe, A. V., Isaacson, M., and Johnson, D. (1970) "Electron Beam Damage in Biological Molecules" *Electron Microsc. Soc. Am. Proc.* **28**:264–265.

174 Cronshaw, J., Gilder, J., and Stone, D. (1973) "Fine Structural Studies of P-Proteins in *Cucurhita, Cucumis*, and *Nicotiana*" *J. Ultrastruct. Res.* **45**:192–205.

175 Crowther, R. A. (1971) "Three-Dimensional Reconstruction and the Architecture of Spherical Viruses" *Endeavour* **30**:124–129.

176 Crowther, R. A. (1971) "Procedures for Three-Dimensional Reconstruction of Spherical Viruses by Fourier Synthesis from Electron Micrographs" *Phil. Trans. R. Soc. London, B* **261**:221–230.

177 Crowther, R. A. (1974) "Criteria and Methods for Reliable Three-Dimensional Image Reconstruction" *Electron Microsc. Soc. Am. Proc.* **32**:330–331.

178 Crowther, R. A. (1976) "The Interpretation of Images Reconstructed from Electron Micrographs of Biological Particles," in *Structure-Function Relationships of Proteins* (R. Markham and R. W. Horne, Eds.), North-Holland, Amsterdam, pp. 15–25.

179 Crowther, R. A. and Amos, L. A. (1971) "Harmonic Analysis of Electron Microscope Images with Rotational Symmetry" *J. Mol. Biol.* **60**:123–130.

180 Crowther, R. A. and Amos, L. A. (1972) "Three-Dimensional Image Reconstruction of Some Small Spherical Viruses" *Cold Spring Harbor Symp. Quant. Biol.* **36**:489–494.

181 Crowther, R. A., Amos, L. A., Finch, J. T., DeRosier, D. J., and Klug, A. (1970) "Three-Dimensional Reconstructions of Spherical Viruses by Fourier Synthesis from Electron Micrographs" *Nature* (*London*) **226**:421–425.

182 Crowther, R. A., Amos, L. A., and Finch, J. T. (1975) "Three-Dimensional Image Reconstructions of Bacteriophages R17 and f2" *J. Mol. Biol.* **98:**631–635.

183 Crowther, R. A., DeRosier, D. J., and Klug, A. (1970) "The Reconstruction of a Three-Dimensional Structure from Projections and Its Application to Electron Microscopy" *Proc. R. Soc. London*, **A317:**319–340.

184 Crowther, R. A., Finch, J. T., and Pearse, B. M. F. (1976) "On the Structure of Coated Vesicles" *J. Mol. Biol.* **103:**785–798.

185 Crowther, R. A. and Franklin, R. M. (1972) "The Structure of the Groups of Nine Hexons from Adenovirus" *J. Mol. Biol.* **68:**181–184.

186 Crowther, R. A., Geelen, J. L. M. C., and Mellema, J. E. (1974) "A Three-Dimensional Image Reconstruction of Cowpea Mosaic Virus" *Virology* **57:**20–27.

187 Crowther, R. A. and Klug, A. (1971) "ART and Science or Conditions for Three-Dimensional Reconstruction from Electron Microscope Images" *J. Theor. Biol.* **32:**199–203.

188 Crowther, R. A., and Klug, A. (1974) "Three-Dimensional Image Reconstruction on an Extended Field—A Fast, Stable Algorithm" *Nature* (*London*) **251:**490–492.

189 Crowther, R. A. and Klug, A. (1975) "Structural Analysis of Macromolecular Assemblies by Image Reconstruction from Electron Micrographs" *Annu. Rev. Biochem.* **44:**161–182.

190 Crowther, R. A., Lenk, E. V., Kikuchi, Y., and King, J. (1977) "Molecular Reorganization in the Hexagon to Star Transition of the Baseplate of Bacteriophage T4" *J. Mol. Biol.* **116:**489–523.

191 Crowther, R. A. and Sleytr, U. B. (1977) "An Analysis of the Fine Structure of the Surface Layers from Two Strains of *Clostridia*, Including Correction for Distorted Images" *J. Ultrastruct. Res.* **58:**41–49.

192 Cutrona, L. J. (1965) "Recent Developments in Coherent Optical Technology," in *Optical and Electronoptical Information Processes* (J. T. Tippett, D. A. Berkowitz, L. C. Clapp, C. J. Koester, and A. Vanderburgh, Jr., Eds.), MIT Press, Cambridge, MA, pp. 83–123.

193 Dales, S., Schulze, I. T., and Ratner, S. (1971) "Tubular Organization of Crystalline Argininosuccinase" *Biochim. Biophys. Acta* **229:**771–778.

194 David, M. M., Ar, A., Ben-Shaul, Y., Schejter, A., and Daniel, E. (1977) "Subunit Structure of Hemoglobin from the Clam Shrimp *Cyzicus*" *J. Mol. Biol.* **111:**211–214.

195 David, M. M. and Daniel, E. (1974) "Subunit Structure of Earthworm Erythrocruorin" *J. Mol. Biol.* **87:**89–101.

196 de Harven, E. (1968) "The Centriole and the Mitotic Spindle," in *The Nucleus* (A. J. Dalton and F. Haguenau, Eds.), Academic Press, New York, pp. 197–227.

197 de Harven, E., Leonard, K. R., and Kleinschmidt, A. K. (1971) "Dark Field Procedures in Electron Microscopy of Particulate Biological Material" *Electron Microsc. Soc. Am. Proc.* **29:**426–427.

198 de Lang, H. and Premsela, H. F. (1970) "The Study of Aberrations with the Aid of the Diffractometer" *Proc. Seventh Int. Congr. Electron Microsc.* (*Grenoble*) **2:**3–4.

199 DeRosier, D. J. (1970) "Reconstruction of Three-Dimensional Images from Electron Micrographs" *Electron Microsc. Soc. Am. Proc.* **28:**246–247.

200 DeRosier, D. J. (1970) "Three-Dimensional Image Reconstruction of Helical Structures" *Ber. Bunsen-Gesellschaft Phys. Chem.* **74:**1127–1128.

201 DeRosier, D. J. (1971) "The Reconstruction of Three-Dimensional Images from Electron Micrographs" *Contemp. Phys.* **12:**437–452.

202 DeRosier, D. J. (1971) "Three-Dimensional Image Reconstruction of Helical Structures" *Phil. Trans. R. Soc. London, B.* **261:**209–210.

203 DeRosier, D. J. (1973) "Structure of a Dehydrogenase Enzyme Complex by Electron Microscopy and X-Ray Diffraction" *Am. Crystallogr. Assoc. Trans.* **9**:1–9.

204 DeRosier, D. J. and Klug, A. (1968) "Reconstruction of Three-Dimensional Structures from Electron Micrographs" *Nature (London)* **217**:130–134.

205 DeRosier, D. J. and Klug, A. (1969) "Positions of Ribosomal Subunits" *Science* **163**:1470.

206 DeRosier, D. J. and Klug, A. (1972) "Structure of the Tubular Variants of the Head of Bacteriophage T4 (Polyheads). I. Arrangement of Subunits in Some Classes of Polyheads (Appendix: Indexing and Filtering of the Diffraction Patterns)" *J. Mol. Biol.* **65**:469–488.

207 DeRosier, D. J., Mandelkow, E., Silliman, A., Tilney, L., and Kane, R. (1977) "Structure of Actin-Containing Filaments from Two Types of Non-Muscle Cells" *J. Mol. Biol.* **113**:679–695.

208 DeRosier, D. J. and Moore, P. B. (1970) "Reconstruction of Three-Dimensional Images from Electron Micrographs of Structures with Helical Symmetry" *J. Mol. Biol.* **52**:355–369.

209 DeRosier, D. J. and Oliver, R. M. (1972) "A Low Resolution Electron-Density Map of Lipoyl Transsuccinylase, the Core of the α-Ketoglutarate Dehydrogenase Complex" *Cold Spring Harbor Symp. Quant. Biol.* **36**:199–203.

210 DeRosier, D. J., Oliver, R. M., and Reed, L. J. (1971) "Crystallization and Preliminary Structural Analysis of Dihydrolipoyl Transsuccinylase, the Core of the 2-Oxoglutarate Dehydrogenase Complex" *Proc. Natl. Acad. Sci. U.S.A.* **68**:1135–1137.

211 de Vos, P. J. G. (1948) "The Use of the 'Fly's Eye' Apparatus to Study Crystal Structures Containing Atoms of Different Scattering Powers" *Acta Crystallogr.* **1**:118–123.

212 Dewey, M. M., Levine, R. J. C., and Colflesh, D. E. (1973) "Structure of *Limulus* Striated Muscle: The Contractile Apparatus at Various Sarcomere Lengths" *J. Cell Biol.* **58**:574–593.

213 De Zoeten, G. A. and Gaard, G. (1969) "Possibilities for Inter- and Intracellular Translocation of Some Icosahedral Plant Viruses" *J. Cell Biol.* **40**:814–823.

214 Dickerson, R. E. (1964) "X-Ray Analysis and Protein Structure," in *The Proteins*, Vol. 2, (H. Neurath, Ed.), Academic Press, New York, pp. 603–778.

215 Dietrich, I., Formanek, H., Fox, F., Knapek, E., and Weyl, R. (1979) "Reduction of Radiation Damage in an Electron Microscope with a Superconducting Lens System" *Nature (London)* **277**:380–381.

216 Dobbs, B. C., Pangborn, W. A., and Parsons, D. F. (1975) "Electron Diffraction of Wet, Unstained, Unfixed Rat Hemoglobin" *Electron Microsc. Soc. Am. Proc.* **33**:216–217.

217 Donelli, G., D'Uva, V., and Paoletti, L. (1975) "Ultrastructure of Gliosomes in Ependymal Cells of the Lizard" *J. Ultrastruct. Res.* **50**:253–263.

218 Donelli, G., Griso, G., Paoletti, L., and Rebessi, S. (1976) "Capsomeric Arrangement in the Bacteriophage G Head" *Sixth Eur. Reg. Conf. Electron Microsc. (Jerusalem)* **2**:502–503.

219 Donelli, G., Guglielmi, F., and Paoletti, L. (1972) "Structure and Physico-Chemical Properties of Bacteriophage G. I. Arrangement of Protein Subunits and Contraction Process of Tail Sheath" *J. Mol. Biol.* **71**:113–125.

220 Donelli, G., Guglielmi, F., and Paoletti, L. (1971) "The Structural Arrangement of Protein Subunits in Bacteriophage G" *First Eur. Biophys. Congr. (Baden)* **1**:547–551.

221 Dorignac, D., Maclachlan, M. E. C., and Jouffrey, B. (1979) "Low-Noise Boron Supports for High Resolution Electron Microscopy" *Ultramicroscopy* **4**:85–89.

222 Dowell, W. C. T., Farrant, J. L., and McLean, J. D. (1968) "The Enhancement of the Signal-to-Noise Ratio in Micrographs" *Fourth Eur. Reg. Conf. Electron Microsc. (Rome)* **1**:583–584.

223 Downing, K. H. (1979) "Possibilities of Heavy Atom Discrimination Using Single-Sideband Techniques" *Ultramicroscopy* **4**:13–31.

224 Downing, K. H. and Siegel, B. M. (1974) "Image Enhancement in Electron Microscopy by Single-Side Band Holographic Methods" *Proc. Eighth Int. Congr. Electron Microsc. (Canberra)* **1**:326–327.

225 Doyle, B. B., Hulmes, D. J. S., Miller, A., Parry, D. A. D., Piez, K. A., and Woodhead-Galloway, J. (1974) "A D-Periodic Narrow Filament in Collagen" *Proc. R. Soc. London, B* **186**:67–74.

226 Drahos, V. and DeLong, A. (1960) "A Simple Method for Obtaining Perforated Supporting Membranes for Electron Microscopy" *Nature (London)* **186**:104.

227 Driedonks, R. A. (1976) "The Assembly of Alfalfa Mosaic Virus Protein and Nucleic Acid" *Sixth Eur. Reg. Conf. Electron Microsc. (Jerusalem)* **2**:504–506.

228 Driedonks, R. A., Krijgsman, P. C. J., and Mellema, J. E. (1977) "Alfalfa Mosaic Virus Protein Polymerization" *J. Mol. Biol.* **113**:123–140.

229 Driedonks, R. A., Krijgsman, P. C. J., and Mellema, J. E. (1978) "Coat Protein Polymerization of Alfalfa Mosaic Virus Strain VRU" *J. Mol. Biol.* **124**:713–719.

230 Dubochet, J. (1973) "High Resolution Dark-Field Electron Microscopy" *J. Microsc.* **98**:334–344.

231 Dubochet, J. (1973) "High Resolution Dark-Field Electron Microscopy" in *Principles and Techniques in Electron Microscopy,* Vol. 3 (M. A. Hayat, Ed.), Van Nostrand Reinhold, New York, pp. 113–151.

232 Dubochet, J. (1975) "Carbon Loss During Irradiation of T4 Bacteriophages and *E. coli* Bacteria in Electron Microscopes" *J. Ultrastruct. Res.* **52**:276–288.

233 Dubochet, J. and Engel, A. (1976) "Dark-Field Electron Microscopy of Biomacromolecules" *Sixth Eur. Reg. Conf. Electron Microsc. (Jerusalem)* **2**:134–136.

234 Dubochet, J. and Noll, M. (1978) "Nucleosome Arcs and Helices" *Science* **202**:280–286.

235 Dunkerley, B. D. and Lipson, H. (1955) "A Simple Version of Bragg's X-ray Microscope" *Nature (London)* **176**:81–82.

236 Duthie, A., Humpherys, S., and Probyn, B. A. (1963) "The Preparation of Evaporated Micro-Circuits" *Elec. Eng.* **35**:430–433.

237 Dykes, G., Crepeau, R. H., and Edelstein, S. J. (1978) "Three-Dimensional Reconstruction of the Fibres of Sickle Cell Haemoglobin" *Nature (London)* **272**:506–510.

238 Dykes, G. W., Crepeau, R., and Edelstein, S. J. (1979) "Three-Dimensional Reconstruction of the 14-Filament Fibers of Hemoglobin S" *J. Mol. Biol.* **130**:451–472.

239 Eagles, P. A. M., and Johnson, L. N. (1972) "Electron Microscopy of Phosphorylase *b* Crystals" *J. Mol. Biol.* **64**:693–695.

240 Edgell, M. H., Hutchison, C. A., III, and Sinsheimer, R. L. (1969) "The Process of Infection with Bacteriophage phi-X174. XXVIII. Removal of the Spike Proteins from the Phage Capsid" *J. Mol. Biol.* **42**:547–557.

241 Edintsov, I. M., Ivanitskii, G. R., and Kuniskii, A. S. (1975) "Three-Dimensional Reconstruction of Extended Tail of Bacteriophage H17 of *Bacillus mycoides*" *Dokl. Biophys.* **224**:160–162.

242 Edintsov, I. M., Ivanitskii, G. R., and Kuniskii, A. S. (1976) "Three-Dimensional Reconstruction of Contracted Tail of Bacteriophage H17 of *Bacillus mycoides*" *Dokl. Biophys.* **228**:80–83.

243 Eisenberg, D. and Crothers, D. (1979) In *Physical Chemistry with Applications to the Life Sciences*, Benjamin/Cummings, Menlo Park, CA, pp. 749–846.

244 Elliott, A. (1974) "The Arrangement of Myosin on the Surface of Paramyosin Filaments in the White Adductor Muscle of *Crassostrea angulata*" *Proc. R. Soc. London, B* **186**:53–66.

245 Elliott, A. (1979) "Structure of Molluscan Thick Filaments: A Common Origin for Diverse Appearances (Appendix: Three-Dimensional Reconstruction of a Paramyosin Filament, S. D. Dover and A. Elliott)" *J. Mol. Biol.* **132**:323–341.

246 Elliott, A. and Lowy, J. (1970) "A Model for the Coarse Structure of Paramyosin Filaments" *J. Mol. Biol.* **53**:181–203.

247 Elliott, A., Lowy, J., and Squire, J. M. (1968) "Convolution Camera to Reveal Periodicities in Electron Micrographs" *Nature (London)* **219**:1224–1226.

248 Elliott, A. and Malcolm, B. R. (1959) "Chain Arrangement and Sense of the α-Helix in Poly-L-Alanine Fibres" *Proc. R. Soc. London, A* **249**:30–41.

249 Elliott, A. and Robertson, P. (1955) "Note on a Reflecting Optical Diffraction Spectrometer" *Acta Crystallogr.* **8**:736.

250 Elsner, P. R. (1971) "A Simple, Reliable Method for Preparing Perforated Formvar Films" *Electron Microsc. Soc. Am. Proc.* **29**:460–461.

251 Engel, A. (1978) "The STEM: An Attractive Tool for the Biologist" *Ultramicroscopy* **3**:355–357.

252 Engel, A., Dubochet, J., and Kellenberger, E. (1976) "Some Progress in the Use of a Scanning Transmission Electron Microscope for the Observation of Biomacromolecules" *J. Ultrastruct. Res.* **57**:322–330.

253 Ensor, D. R., Jensen, C. G., Fillery, J. A., and Baker, R. J. K. (1978) "Microdensitometer—Computer Correlation Analysis of Ultrastructural Periodicity" *Proc. Ninth Int. Congr. Electron Microsc. (Toronto)* **2**:32–33.

254 Erickson, H. P. (1973) "The Fourier Transform of an Electron Micrograph: First-Order and Second-Order Theory of Image Formation" *Adv. Opt. Electron Microsc.* **5**:163–199.

255 Erickson, H. P. (1974) "2-D Image Enhancement by Optical and Computer Fourier Techniques: Microtubules and Catalase Crystals" *Proc. Eighth Int. Congr. Electron Microsc. (Canberra)* **1**:310–311.

256 Erickson, H. P. (1974) "Microtubule Surface Lattice and Subunit Structure and Observations on Reassembly" *J. Cell Biol.* **60**:153–167.

257 Erickson, H. P. (1974) "Assembly of Microtubules from Preformed, Ring-Shaped Protofilaments and 6-S Tubulin" *J. Supramol. Struct.* **2**:393–411.

258 Erickson, H. P. (1975) "The Structure and Assembly of Microtubules" *Ann. Trans. N. Y. Acad. Sci.* **253**:60–77.

259 Erickson, H. P. (1978) "The Structure of One and Two Dimensional Polymers of Tubulin and Their Role in Nucleation of Microtubule Assembly" *Proc. Ninth Int. Congr. Electron Microsc. (Toronto)* **3**:483–494.

260 Erickson, H. P. and Klug, A. (1970) "The Fourier Transform of an Electron Micrograph: Effects of Defocusing and Aberrations, and Implications for the Use of Underfocus Contrast Enhancement" *Ber. Bunsen-Gesellschaft Phys. Chem.* **74**:1129–1137.

261 Erickson, H. P. and Klug, A. (1970) "Phase Contrast Electron Microscopy and Compensation of Aberrations by Fourier Image Processing" *Electron Microsc. Soc. Am. Proc.* **28**:248–249.

262 Erickson, H. P. and Klug, A. (1971) "Measurement and Compensation of Defocusing and Aberrations by Fourier Processing of Electron Micrographs" *Phil. Trans. R. Soc. London, B* **261**:105–118.

263 Erickson, H. P., Voter, W. A., and Leonard, K. (1978) "Image Reconstruction in Electron Microscopy: Enhancement of Periodic Structure by Optical Filtering" *Methods Enzymol.* **49**:39–63.

264 Erickson, R. O. (1973) "Tubular Packing of Spheres in Biological Fine Structure" *Science* **181**:705–716.

265 Esparza, J. and Gil, F. (1978) "A Study on the Ultrastructure of Human Rotavirus" *Virology* **91**:141–150.

266 Fahrenbach, W. H. (1967) "The Fine Structure of Fast and Slow Crustacean Muscles" *J. Cell Biol.* **35:**69–79.

267 Falconer, D. G. (1966) "Optical Processing of Bubble Chamber Photographs" *Appl. Opt.* **5:**1365–1369.

268 Farnell, G. C. and Flint, R. B. (1973) "The Response of Photographic Materials to Electrons with Particular Reference to Electron Micrography" *J. Microsc.* **97:**271–291.

269 Farnell, G. C. and Flint, R. B. (1975) "Photographic Aspects of Electron Microscopy," in *Principles and Techniques in Electron Microscopy*, Vol. 5 (M. A. Hayat, Ed.), Van Nostrand Reinhold, New York, pp. 19–61.

270 Fernandez-Moran, H., Ohtsuki, M., and Hough, C. (1970) "High Resolution Electron Microscopy of Cell Membranes and Derivatives" *Proc. Seventh Int. Congr. Electron Microsc.* (*Grenoble*) **3:**9–10.

271 Ferrier, R. P. (1969) "Small Angle Electron Diffraction in the Electron Microscope" *Adv. Opt. Electron Microsc.* **3:**155–218.

272 Ferrier, R. P. and Murray, R. T. (1966) "Low-Angle Electron Diffraction" *J. R. Microsc. Soc.* **85:**323–335.

273 Fertig, J. and Rose, H. (1978) "Computer Simulation of Dark-Field Imaging as a Tool for Image Interpretation" *Proc. Ninth Int. Congr. Electron Microsc.* (*Toronto*) **1:**238–239.

274 Finch, J. T. (1969) "The Pitch of Tobacco Mosaic Virus" *Virology* **38:**182–185.

275 Finch, J. T. (1972) "The Hand of the Helix of Tobacco Mosaic Virus" *J. Mol. Biol.* **66:**291–294.

276 Finch, J. T. (1974) "The Surface Structure of Polyoma" *J. Gen. Virol.* **24:**359–364.

277 Finch, J. T. (1975) "Electron Microscopy of Proteins," in *The Proteins*, Vol. 1, 3rd ed. (H. Neurath and R. L. Hill, Eds.), Academic Press, New York, pp. 413–497.

278 Finch, J. T. and Crawford, L. V. (1975) "Structure of Small DNA-Containing Animal Viruses" *Compr. Virol.* **5:**119–154.

279 Finch, J. T., Crowther, R. A., Hendry, D. A., and Struthers, J. K. (1974) "The Structure of *Nudaurelia capensis* beta-Virus: The First Example of a Capsid with Icosahedral Surface Symmetry T = 4" *J. Gen. Virol.* **24:**191–200.

280 Finch, J. T. and Gibbs, A. J. (1970) "Observation on the Structure of the Nucleocapsids of Some Paramyxoviruses" *J. Gen. Virol.* **6:**141–150.

281 Finch, J. T. and Holmes, K. C. (1967) "Structural Studies of Viruses" *Methods Virol.* **3:**351–474.

282 Finch, J. T. and Klug, A. (1965) "The Structure of Viruses of the Papilloma–Polyoma Type. III. Structure of Rabbit Polyoma Virus. Appendix: Topography of Contrast in Negative-Staining for Electron Microscopy" *J. Mol. Biol.* **13:**1–12.

283 Finch, J. T. and Klug, A. (1966) "Arrangement of Protein Subunits and the Distribution of Nucleic Acid in Turnip Yellow Mosaic Virus. II. Electron Microscopic Studies" *J. Mol. Biol.* **15:**344–364.

284 Finch, J. T. and Klug, A. (1967) "Structure of Broad Bean Mottle Virus. I. Analysis of Electron Micrographs and Comparison with Turnip Yellow Mosaic Virus and Its Top Component" *J. Mol. Biol.* **24:**289–302.

285 Finch, J. T. and Klug, A. (1971) "Three-Dimensional Reconstruction of the Stacked-Disk Aggregate of Tobacco Mosaic Virus Protein from Electron Micrographs" *Phil. Trans. R. Soc. London, B* **261:**211–219.

286 Finch, J. T. and Klug, A. (1972) "The Helical Surface Lattice of Bacterial Flagella," in *The Generation of Subcellular Structures* (R. Markham, J. B. Bancroft, D. R. Davies, D. A. Hopwood, and R. W. Horne, Eds.), North Holland, Amsterdam, pp. 167–177.

287 Finch, J. T. and Klug, A. (1974) "The Structural Relationship Between the Stacked Disk and Helical Polymers of Tobacco Mosaic Virus Protein" *J. Mol. Biol.* **87:**633–640.

288 Finch, J. T., Klug, A., and Nermut, M. V. (1967) "The Structure of the Macromolecular Units on the Cell Walls of *Bacillus polymyxa*" *J. Cell Sci.* **2**:587–590.

289 Finch, J. T., Klug, A., and Stretton, A. O. W. (1964) "The Structure of the 'Polyheads' of T4 Bacteriophage" *J. Mol. Biol.* **10**:570–575.

290 Finch, J. T., Leberman, R., Yu-shang, C., and Klug, A. (1966) "Rotational Symmetry of the Two Turn Disk Aggregate of Tobacco Mosaic Virus Protein" *Nature (London)* **212**:349–350.

291 Finch, J. T., Lutter, L. C., Rhodes, D., Brown, R. S., Rushton, B., Levitt, M., and Klug, A. (1977) "Structure of Nucleosome Core Particles of Chromatin" *Nature (London)* **269**:29–36.

292 Finch, J. T., Perutz, M. F., Bertles, J. F., and Dobler, J. (1973) "Structure of Sickled Erythrocytes and of Sickle-Cell Hemoglobin Fibers" *Proc. Natl. Acad. Sci. U.S.A.* **70**:718–722.

293 Fisher, H. W. and Cooper, T. W. (1967) "Electron Microscope Observations on the Nuclear Pores of Hela Cells" *Exp. Cell Res.* **48**:620–622.

294 Fisher, K. A. and Stoeckenius, W. (1977) "Freeze-Fractured Purple Membrane Particles: Protein Content" *Science* **197**:72–74.

295 Fisher, K. A., Yanagimoto, K., and Stoeckenius, W. (1978) "Oriented Adsorption of Purple Membrane to Cationic Surfaces" *J. Cell Biol.* **77**:611–621.

296 Fiskin, A. M. (1972) "Spatial Frequencies in Electron Images of Negatively Stained Protein" *Electron Microsc. Soc. Am. Proc.* **30**:598–599.

297 Fiskin, A. M. (1977) "Real-Space Filtering of Electron Images from Optical Autocorrelation Patterns" *Ultramicroscopy* **2**:397–404.

298 Fiskin, A. M. and Beer, M. (1968) "Autocorrelation Functions of Noisy Electron Micrographs of Stained Polynucleotide Chains" *Science* **159**:1111–1113.

299 Fiskin, A. M., Cohn, D. V., and Peterson, G. S. (1977) "A Model for the Structure of Bovine Parathormone Derived by Dark Field Electron Microscopy" *J. Biol. Chem.* **252**:8261–8268.

300 Folkhard, W., Leonard, K. R., Malsey, S., Marvin, D. A., Duboche, J., Engel, A., Achtman, M., and Helmuth, R. (1979) "X-ray diffraction and Electron Microscope Studies on the Structure of Bacterial F Pili" *J. Mol. Biol.* **130**:145–160.

301 Fox, F., Knapek, E., and Weyl, R. (1978) "High Resolution Imaging of the Rigid Layer of Bacterial Cell Walls at Liquid Helium Temperatures" *Proc. Ninth Int. Congr. Electron Microsc. (Toronto)* **2**:342–343.

302 Franke, W. W. (1966) "Isolated Nuclear Membranes" *J. Cell Biol.* **31**:619–623.

303 Franke, W. W., Grund, C., Schmid, E., and Mandelkow, E. (1977) "Paracrystalline Arrays of Membrane-to-Membrane Cross Bridges Associated with the Inner Surface of Plasma Membrane" *J. Cell Biol.* **77**:323–328.

304 Franke, W. W. and Scheer, U. (1970) "The Ultrastructure of the Nuclear Envelope of Amphibian Oocytes: A Reinvestigation. I. The Mature Oocyte" *J. Ultrastruct. Res.* **30**:288–316.

305 Frank, J. (1972) "Observation of the Relative Phases of Electron Microscopic Phase Contrast Zones with the Aid of the Optical Diffractometer" *Optik* **35**:608–612.

306 Frank, J. (1973) "Computer Processing of Electron Micrographs," in *Advanced Techniques in Biological Electron Microscopy*, Vol. 1 (J. K. Koehler, Ed.), Springer-Verlag, New York, pp. 215–274.

307 Frank, J. (1974) "Correlation Methods in Electron Image Analysis" *Electron Microsc. Soc. Am. Proc.* **32**:336–337.

308 Frank, J. (1975) "Digital Image Processing in High Resolution Electron Microscopy" *Electron Microsc. Soc. Am. Proc.* **33**:12–13.

309 Frank, J. (1975) "Averaging of Low Exposure Micrographs of Non-Periodic Objects" *Ultramicroscopy* **1**:159–162.

310 Frank, J. (1975) "Determination of Energy Spread and Source Size by Optical Diffraction" *Electron Microsc. Soc. Am. Proc.* **33**:182–183.

311 Frank, J. (1976) "Low-Exposure Electron Microscopy of Non-Periodic Objects" *Sixth Eur. Reg. Conf. Electron Microsc. (Jerusalem)* **1**:273–274.

312 Frank, J. (1978) "Reconstruction of Non-Periodic Objects Using Correlation Methods" *Proc. Ninth Int. Congr. Electron Microsc. (Toronto)* **3**:87–93.

313 Frank, J. (1979) "Image Analysis in Electron Microscopy" *J. Microsc.* **117**:25–38.

314 Frank, J. and Al-Ali, L. (1975) "Signal-to-Noise Ratio of Electron Micrographs Obtained by Cross Correlation" *Nature (London)* **256**:376–379.

315 Frank, J., Bussler, P. H., Langer, R., and Hoppe, W. (1970) "A Computer Program System for Image Reconstruction and Its Application to Electron Micrographs of Biological Objects" *Proc. Seventh Int. Congr. Electron Microsc. (Grenoble)* **1**:17–18.

316 Frank, J., Goldfarb, W., Eisenberg, D., and Baker, T. S. (1978) "Reconstruction of Glutamine Synthetase Using Computer Averaging" *Ultramicroscopy* **3**:283–290.

317 Frank, J., Goldfarb, W., Eisenberg, D., and Baker, T. S. (1979) "Addendum to Reconstruction of Glutamine Synthetase Using Computer Averaging' " *Ultramicroscopy* **4**:247.

318 Frank, J., Goldfarb, W., and Kessel, M. (1978) "Image Reconstruction of Low and High Dose Micrographs of Negatively Stained Glutamine Synthetase" *Proc. Ninth Int. Congr. Electron Microsc. (Toronto)* **2**:8–9.

319 Frank, J., Salih, S. M., and Cosslett, V. E. (1974) "Radiation Damage Assessment by Digital Correlation of Images" *Proc. Eighth Int. Congr. Electron Microsc. (Canberra)* **2**:678–679.

320 Frank, J. and Shimkin, B. (1978) "A New Image Processing Software System for Structural Analysis and Contrast Enhancement" *Proc. Ninth Int. Congr. Electron Microsc. (Toronto)* **1**:210–211.

321 Franzini-Armstrong, C. (1973) "The Structure of a Simple Z-Line" *J. Cell Biol.* **58**:630–642.

322 Fraser, R. D. B. and MacRae, T. P. (1959) "Molecular Organisation in Feather Keratin" *J. Mol. Biol.* **1**:387–397.

323 Fraser, R. D. B. and MacRae, T. P. (1961) "The Molecular Configuration of α-Keratin" *J. Mol. Biol.* **3**:640–647.

324 Fraser, R. D. B. and MacRae, T. P. (1963) "Structural Organization in Feather Keratin" *J. Mol. Biol.* **7**:272–280.

325 Fraser, R. D. B., MacRae, T. P., Suzuki, E., and Davey, C. L. (1976) "Image Processing of Electron Micrographs of Deformed Filaments" *J. Microsc.* **108**:343–348.

326 Fraser, R. D. B., and Millward, G. R. (1970) "Image Averaging by Optical Filtering" *J. Ultrastruct. Res.* **31**:203–211.

327 Frey, T. G. (1978) "The Structures of Glutamine Synthetase and Cytochrome c Oxidase—Studies by Electron Microscopy and Image Analysis" *Proc. Ninth Int. Congr. Electron Microsc. (Toronto)* **3**:107–119.

328 Frey, T. G., Chan, S. H. P., and Schatz, G. (1978) "Structure and Orientation of Cytochrome *c* Oxidase in Crystalline Membranes" *J. Biol. Chem.* **253**:4389–4395.

329 Frey, T. G., Eisenberg, D., and Eiserling, F. A. (1975) "Glutamine Synthetase Forms Three- and Seven-Stranded Helical Cables" *Proc. Natl. Acad. Sci. U.S.A.* **72**:3402–3406.

330 Frey, T. G., Eisenberg, D. S., and Smith, P. R. (1978) "The Variation of Intermolecular

Contacts in Helical Aggregates of Glutamine Synthetase'' *Proc. Ninth Int. Congr. Electron Microsc. (Toronto)* **2**:174–175.

331 Frieder, G. and Herman, G. T. (1971) ''Resolution in Reconstructing Objects from Electron Micrographs'' *J. Theor. Biol.* **33**:189–211.

332 Friedman, M. H. (1970) ''A Reevaluation of the Markham Rotation Technique Using Model Systems'' *J. Ultrastruct. Res.* **32**:226–236.

333 Frigon, R. P. and Timasheff, S. N. (1975) ''Magnesium-Induced Self-Association of Calf Brain Tubulin. I. Stoichiometry'' *Biochemistry* **14**:4559–4566.

334 Frisch, D. (1969) ''A Photographic Reinforcement Analysis of Neurotubules and Cytoplasmic Membranes'' *J. Ultrastruct. Res.* **29**:357–372.

335 Fujiwara, K. and Tilney, L. G. (1975) ''Substructural Analysis of the Microtubule and its Polymorphic Forms'' *Ann. Trans. N. Y. Acad. Sci.* **253**:27–50.

336 Fukami, A. and Adachi, K. (1965) ''A New Method of Preparation of a Self-Perforated Micro Plastic Grid and Its Application'' *J. Electron Microsc.* **14**:112–118.

337 Fuller, S. D., Capaldi, R. A., and Henderson, R. (1979) ''Structure of Cytochrome *c* Oxidase in Deoxycholate-Derived Two-Dimensional Crystals'' *J. Mol. Biol.* **134**:305–327.

338 Furlong, D. (1978) ''Use of the STEM to Obtain Information from One Level of a Complex Biological Object'' *Proc. Ninth Int. Congr. Electron Microsc. (Toronto)* **2**:108–109.

339 Furlong, D. (1978) ''Direct Evidence for 6-Fold Symmetry of the Herpesvirus Hexon Capsomere'' *Proc. Natl. Acad. Sci. U.S.A.* **75**:2764–2766.

340 Furneaux, P. J. S. and Mackay, A. L. (1970) ''Periodic Protein Structure in Insect Egg Shells'' *Proc. Seventh Int. Congr. Electron Microsc. (Grenoble)* **1**:621–622.

341 Furneaux, P. J. S. and Mackay, A. L. (1972) ''Crystalline Protein in the Chorion of Insect Egg Shells'' *J. Ultrastruct. Res.* **38**:343–359.

342 Gachet, J. and Thiery, J. P. (1964) ''Application de la Methode de Tirage Photographique avec Rotations ou Translations à l'Étude de Macromolecules et de Structures Biologiques'' *J. Microscop.* **3**:253–268.

343 Gall, J. G. (1966) ''Microtubule Fine Structure'' *J. Cell Biol.* **31**:639–643.

344 Gall, J. G. (1967) ''The Light Microscope as an Optical Diffractometer'' *J. Cell Sci.* **2**:163–168.

345 Gall, J. G. (1967) ''Octagonal Nuclear Pores'' *J. Cell Biol.* **32**:391–399.

346 Gannon, J. R. and Tilley, R. J. D. (1976) ''A Technique for Preparing Holey Carbon Films for High Resolution Electron Microscopy'' *J. Microsc.* **106**:59–61.

347 Garrell, R. L., Crepeau, R. H., and Edelstein, S. J. (1979) ''Cross-Sectional Views of Hemoglobin S Fibers by Electron Microscopy and Computer Modeling'' *Proc. Natl. Acad. Sci. U.S.A.* **76**:1140–1144.

348 Gerchberg, R. W. (1972) ''Holography Without Fringes in the Electron Microscope'' *Nature (London)* **240**:404–406.

349 Gerchberg, R. W. and Saxton, W. O. (1972) ''A Practical Algorithm for the Determination of Phase from Image and Diffraction Plane Pictures'' *Optik* **35**:237–246.

350 Gibbs, A. J. and Rowe, A. J. (1973) ''Reconstruction of Images from Transforms by an Optical Method'' *Nature (London)* **246**:509–511.

351 Gilbert, P. (1972) ''Iterative Methods for the Three-Dimensional Reconstruction of an Object from Projections'' *J. Theor. Biol.* **36**:105–117.

352 Gilbert, P. F. C. (1972) ''The Reconstruction of a Three-Dimensional Structure from Projections and Its Application to Electron Microscopy. II. Direct Methods'' *Proc. R. Soc. London, B* **182**:89–102.

353 Gilev, V. P. (1979) "A Simple Method of Optical Filtration" *Ultramicroscopy* **4**:323–336.

354 Gillis, J. M. and O'Brien, E. J. (1975) "The Effect of Calcium Ions on the Structure of Reconstituted Muscle Thin Filaments" *J. Mol. Biol.* **99**:445–459.

355 Gilula, N. B., Branton, D., and Satir, P. (1970) "The Septate Junction: A Structural Basis for Intercellular Coupling" *Proc. Natl. Acad. Sci. U.S.A.* **67**:213–220.

356 Glaeser, R. M. (1970) "Radiation Damage, Contrast and Statistics in High Resolution Biological Electron Microscopy" *Electron Microsc. Soc. Am. Proc.* **28**:260–261.

357 Glaeser, R. M. (1971) "Limitations to Significant Information in Biological Electron Microscopy as a Result of Radiation Damage" *J. Ultrastruct. Res.* **36**:466–482.

358 Glaeser, R. M. (1973) "Radiation Damage and High Resolution Biological Electron Microscopy" *Electron Microsc. Soc. Am. Proc.* **31**:226–227.

359 Glaeser, R. M. (1975) "Radiation Damage and Biological Electron Microscopy," in *Physical Aspects of Electron Microscopy and Microbeam Analysis* (B. Siegel and D. R. Beaman, Eds.), Wiley, New York, pp. 205–229.

360 Glaeser, R. M., Budinger, T. F., Aebersold, P. M., and Thomas, G. (1970) "Radiation Damage in Biological Specimens" *Proc. Seventh Int. Congr. Electron Microsc. (Grenoble)* **2**:463–464.

361 Glaeser, R. M., Chiu, W., and Grano, D. (1979) "Structure of the Surface Layer Protein of the Outer Membrane of *Spirillium serpens*" *J. Ultrastruct. Res.* **66**:235–242.

362 Glaeser, R. M., Cosslett, V. E., and Valdre, U. (1971) "Low Temperature Electron Microscopy: Radiation Damage in Crystalline Biological Materials" *J. Microscop.* **12**:133–138.

363 Glaeser, R. M. and Hayward, S. B. (1978) "Measurement and Reduction of Radiation Damage in Frozen Hydrated Crystalline Specimens" *Proc. Ninth Int. Congr. Electron Microsc. (Toronto)* **3**:70–77.

364 Glaeser, R. M. and Hayward, S. B. (1978) "Measurement and Reduction of Damage in Frozen Hydrated Crystalline Specimens" *Proc. Ninth Int. Congr. Electron Microsc. (Toronto)* **2**:10–11.

365 Glaeser, R. M. and Hobbs, L. W. (1975) "Radiation Damage in Stained Catalase at Low Temperature" *J. Microsc.* **103**:209–214.

366 Glaeser, R. M., Kuo, I., and Budinger, T. F. (1971) "Method for Processing of Periodic Images at Reduced Levels of Electron Irradiation" *Electron Microsc. Soc. Am. Proc.* **29**:466–467.

367 Glaeser, R. M. and Taylor, K. A. (1976) "Frozen Hydrated Biological Specimens for Electron Microscopy and Structure Analysis" *Sixth Europ. Reg. Conf. Electron Microsc. (Jerusalem)* **2**:69–72.

368 Glaeser, R. M. and Taylor, K. A. (1978) "Radiation Damage Relative to Transmission Electron Microscopy of Biological Specimens at Low Temperature: A Review" *J. Microsc.* **112**:127–138.

369 Glaeser, R. M. and Thomas, G. (1969) "Application of Electron Diffraction to Biological Electron Microscopy" *Biophys. J.* **9**:1073–1099.

370 Glauert, A. M. (1965) "Factors Influencing the Appearance of Biological Specimens in Negatively Stained Preparations" *Lab. Invest.* **14**:331–341.

371 Glauert, A. M. (1974) "The High Voltage Electron Microscope in Biology" *J. Cell. Biol.* **63**:717–748.

372 Glauert, A. M. (1975) "Fixation, Dehydration and Embedding of Biological Specimens," in *Practical Methods of Electron Microscopy*, Vol. 3 (A. M. Glauert, Ed.), North Holland, Amsterdam, pp. 1–207.

373 Glauert, A. M. & Thornley, M. J. (1973) "Self-Assembly of a Surface Component of a Bacterial Outer Membrane," in *The Generation of Subcellular Structures* (R. Markham,

J. B. Bancroft, D. R. Davies, D. A. Hopwood, and R. W. Horne, Eds.), North Holland, Amsterdam, pp. 297–305.

374 Glenner, G. G., Keiser, H. R., Bladen, H. A., Cuatrecasas, P., Eanes, E. D., Ram, J. S., Kanfer, J. N., and DeLellis, R. A. (1968) "Amyloid. VI. A Comparison of Two Morphogenic Components of Human Amyloid Deposits" *J. Histol. Cytochem.* **16**:633–644.

375 Glusker, J. P. and Trueblood, K. N. (1972) in *Crystal Structure Analysis: A Primer*, Oxford University Press, New York.

376 Goldfarb, W. and Frank, J. (1978) "Digital Image Enhancement for Transmission Electron Microscopy" *Proc. Ninth Int. Congr. Electron Microsc.* (*Toronto*) **2**:22–23.

377 Goldstein, M. A. and Entman, M. L. (1979) "Microtubules in Mammalian Heart Muscle" *J. Cell Biol.* **80**:183–195.

378 Goldstein, M. A., Schroeter, J. P., and Sass, R. L. (1977) "Optical Diffraction of the Z Lattice in Canine Cardiac Muscle" *J. Cell Biol.* **75**:818–836.

379 Goldstein, M. A., Schroeter, J. P., and Sass, R. L. (1979) "The Z Lattice in Canine Cardiac Muscle" *J. Cell Biol.* **83**:187–204.

380 Gonzales-Beltran, C. and Balyuzi, H. H. M. (1973) "The Use of Microfilm Computer Plotting Facilities for Generating Optical Diffraction Masks" *J. Appl. Crystallogr.* **6**:487–488.

381 Gonzales-Beltran, C. and Burge, R. E. (1974) "Subunit Arrangements in Bacterial Flagella" *J. Mol. Biol.* **88**:711–716.

382 Goodman, J. W. (1968) "Spatial Filtering and Optical Information Processing," in *Introduction to Fourier Optics* McGraw-Hill, New York, pp. 141–144.

383 Goodman, R. M., Horne, R. W., and Hobart, J. M. (1975) "Reconstitution of Potato Virus X *in Vitro* II. Characterization of the Reconstituted Product" *Virology* **68**:299–308.

384 Goodman, R. M., McDonald, J. G., Horne, R. W., and Bancroft, J. B. (1976) "Assembly of Flexuous Plant Viruses and Their Proteins" *Phil. Trans. R. Soc. London B* **276**:173–179.

385 Gordon, R. and Bender, R. (1971) "New Three-Dimensional Algebraic Reconstruction Techniques (ART)" *Electron Microsc. Soc. Am. Proc.* **29**:82–83.

386 Gordon, R., Bender, R., and Herman, G. T. (1970) "Algebraic Reconstruction Techniques (ART) for Three-Dimensional Electron Microscopy and X-Ray Photography" *J. Theor. Biol.* **29**:471–481.

387 Gordon, R. and Herman, G. T. (1974) "Three-Dimensional Reconstruction from Projections: A Review of Algorithms" *Int. Rev. Cytol.* **38**:111–151.

388 Gotte, L. (1976) "Recent Observations on the Structure and Composition of Elastin," in *Structure-Function Relationships of Proteins* (R. Markham and R. W. Horne, Eds.), North-Holland, Amsterdam, pp. 39–57.

389 Gotte, L., Volpin, D., Horne, R. W., and Mammi, M. (1976) "Electron Microscopy and Optical Diffraction of Elastin" *Micron* **7**:95–102.

390 Grimstone, A. V. and Klug, A. (1966) "Observations on the Substructure of Flagellar Fibres" *J. Cell Sci.* **1**:351–362.

391 Grinton, G. R. and Cowley, J. M. (1971) "Phase and Amplitude Contrast in Electron Micrographs of Biological Material" *Optik* **34**:221–233.

392 Gross, H., Bas, E., Kubler, O., and Moor, H. (1976) "An Experimental System for Freeze-Fracturing in Ultra High Vacuum at Temperatures to −196°C" *Sixth Eur. Reg. Conf. Electron Microsc.* (*Jerusalem*) **1**:402–404.

393 Gross, H., Bas, E., and Moor, H. (1978) "Freeze-Fracturing in Ultrahigh Vacuum at −196°C" *J. Cell Biol.* **76**:712–728.

394 Gross, H., Kubler, O., Bas, E., and Moor, H. (1978) "Decoration of Specific Sites on Freeze-Fractured Membranes" *J. Cell Biol.* **79**:646–656.

395 Gross, H. and Moor, H. (1978) "Decoration of Specific Sites on Freeze-Fractured Membranes" *Proc. Ninth Int. Congr. Electron Microsc.* (*Toronto*) **2:**140–141.

396 Grubb, D. T. and Groves, G. W. (1971) "Rate of Damage of Polymer Crystals in the Electron Microscope: Dependence on Temperature and Beam Voltage" *Phil. Mag.* **24:**815–828.

397 Grubb, D. T. and Keller, A. (1972) "Beam-Induced Radiation Damage in Polymers and Its Effect on the Image Formed in the Electron Microscope" *Fifth Eur. Reg. Conf. Electron Microsc.* (*Manchester*) **1:**554–560.

398 Guckenberger, R. and Hoppe, W. (1978) "On-Line Electron-Optical Correlation Computing in the CTEM" *Proc. Ninth Int. Congr. Electron Microsc.* (*Toronto*) **1:**88–89.

399 Guerritore, D., Bonacci, M. L., Bruneri, M., Antonini, E., Wyman J., and Rossi-Fanelli, A. (1965) "Studies on Chlorocruorin. III. Electron Microscope Observations on *Spirographis* Chlorocruorin" *J. Mol. Biol.* **13:**234–237.

400 Gurskaya, G. V. (1975) "An Increase in Resolution in the Study of the Structure of Protein Molecules by X-ray Diffraction and Electron Microscopy. Structure of Catalase with a Resolution of 25 Angstroms" *Sov. Phys. Cryst.* **20:**315–319.

401 Gurskaya, G. V., Lobanova, G. M., and Vainshtein, B. K. (1972) "X-ray Diffraction and Electron-Microscope Study of Hexagonal Catalase Crystals. II. X-ray Diffraction Study" *Sov. Phys. Cryst.* **16:**662–669.

402 Hahn, M., Seredynski, J., and Baumeister, W. (1976) "Inactivation of Catalase Monolayers by Irradiation with 100 keV Electrons" *Proc. Natl. Acad. Sci. U.S.A.* **73:**823–827.

403 Hall, C. E. (1966) In *Introduction to Electron Microscopy*, 2nd ed., McGraw-Hill, New York.

404 Handley, D. A. and Olsen, B. R. (1979) "Butvar B-98 as a Thin Support Film" *Ultramicroscopy* **4:**479–480.

405 Hanson, A. W. (1952) "Improvements in Optical Fourier Synthesis" *Nature* (*London*) **170:**580.

406 Hanson, A. W. and Lipson, H. (1952) "A Simplified Fly's Eye Procedure" *Acta Crystallogr.* **5:**145.

407 Hanson, A. W. and Lipson, H. (1952) "Optical Methods in X-Ray Analysis. III. Fourier Synthesis by Optical Interference" *Acta Crystallogr.* **5:**362–366.

408 Hanson, A. W., Lipson, H., and Taylor, C. A. (1953) "The Application of the Principles of Physical Optics to Crystal-Structure Determination" *Proc. R. Soc. London, A* **218:** 371–384.

409 Hanson, A. W., Taylor, C. A., and Lipson, H. (1951) "Fourier Synthesis by Optical Interference" *Nature* (*London*) **168:**460.

410 Hanson, A. W., Taylor, C. A., and Lipson, H. (1952) "Determination of Crystal Structure by Optical Diffraction Methods" *Nature* (*London*) **169:**1086.

411 Hanson, J., O'Brien, E. J., and Bennett, P. M. (1971) "Structure of the Myosin-Containing Filament Assembly (A-Segment) Separated from Frog Skeletal Muscle" *J. Mol. Biol.* **58:**865–871.

412 Hanson, K. R. (1966) "Symmetry of Protein Oligomers Formed by Isologous Association" *J. Mol. Biol.* **22:**405–409.

413 Hanszen, K. J. (1971) "The Optical Transfer Theory of the Electron Microscope: Fundamental Principles and Applications" *Adv. Opt. Electron Microsc.* **4:**1–84.

414 Harburn, G. (1972) "Optical Fourier Synthesis," in *Optical Transforms* (H. Lipson, Ed.), Academic Press, New York, pp. 189–227.

415 Harburn, G., Miller, J. S., and Welberry, T. R. (1974) "Optical-Diffraction Screens Containing Large Numbers of Apertures" *J. Appl. Crystallogr.* **7:**36–38.

416 Harburn, G. and Ranniko, J. K. (1971) "Details for an Optical Gate" *J. Phys. E: Sci. Instrum.* **4**:394–395.

417 Harburn, G. and Ranniko, J. K. (1972) "An Improved Optical Diffractometer" *J. Phys. E: Sci. Instrum.* **5**:757–762.

418 Harburn, G. and Taylor, C. A. (1961) "Three-Dimensional Optical Transforms" *Proc. R. Soc. London, A* **264**:339–354.

419 Harburn, G., Taylor, C. A., and Welberry, T. R. (1975) In *Atlas of Optical Transforms*, Cornell University Press, Ithaca, NY.

420 Harburn, G., Taylor, C. A., and Yeadon, E. C. (1965) "Optical Diffractometer Masks Representing Continuous Amplitude Distributions and Antiphase Regions" *Br. J. Appl. Phys.* **16**:1367–1375.

421 Harburn, G., Walkley, K., and Taylor, C. A. (1965) "Gas-Phase Laser as a Source of Light for an Optical Diffractometer" *Nature (London)* **205**:1095–1096.

422 Harris, J. R. (1969) "Some Negative Contrast Staining Features of a Protein from Erythrocite Ghosts" *J. Mol. Biol.* **46**:329–335.

423 Harris, W. J. (1962) "Holey Films for Electron Microscopy" *Nature (London)* **196**:499–500.

424 Harris, W. W. (1970) "Reducing the Effect of Substrate Noise in Electron Images of Biological Objects," in *Some Biological Techniques in Electron Microscopy* (D. F. Parsons, Ed.), Academic Press, New York, pp. 147–163.

425 Hartman, R. E. and Hartman, R. S. (1971) "Residual Gas Reactions in the Electron Microscope: IV. A Factor in Radiation Damage" *Electron Microsc. Soc. Am. Proc.* **29**:74–75.

426 Hart, R. G. (1968) "Electron Microscopy of Unstained Biological Material: The Polytropic Montage" *Science* **159**:1464–1467.

427 Hart, R. G. and Yoshiyama, J. M. (1970) "Polytropic Montage: Ribosomes Examined in a Tissue Section" *Proc. Natl. Acad. Sci. U.S.A.* **65**:402–408.

428 Hart, R. G. and Yoshiyama, J. M. (1975) "Electron Microscopy with Reduced Beam Damage to the Specimen: A Retractable Image Intensifier" *J. Ultrastruct. Res.* **51**:40–45.

429 Haschemeyer, R. H. (1968) "Electron Microscopy of Enzymes" *Ann. Trans. N. Y. Acad. Sci.* **30**:875–891.

430 Haschemeyer, R. H. (1970) "Electron Microscopy of Enzymes" *Adv. Enzymol.* **33**:71–118.

431 Haschemeyer, R. H. and de Harven, E. (1974) "Electron Microscopy of Enzymes" *Annu. Rev. Biochem.* **43**:279–301.

432 Haschemeyer, R. H. and Haschemeyer, A. E. V. (1973) "Symmetry in Protein Structures," in *Proteins: A Guide to Study by Physical and Chemical Methods*, Wiley, New York, pp. 386–395.

433 Haschemeyer, R. H. & Myers, R. J. (1970) "Negative Staining," in *Principles and Techniques in Electron Microscopy*, Vol. 2 (M. A. Hayat, Ed.), Van Nostrand Reinhold, New York, pp. 99–147.

434 Hatta, T. and Francki, R. I. B. (1977) "Morphology of Fiji Disease Virus" *Virology* 76:797–807.

435 Hawkes, P. W. (1978) "Coherence in Electron Optics" *Adv. Opt. Electron Microsc.* **7**:101–184.

436 Hawkins, H., Mergner, W. J., Henkens, R., Kinney, T. D., and Trump, B. F. (1970) "Attempts of Resolution of the Subunit Structure of Ferritin and Apoferritin by Electron Microscopy" *Electron Microsc. Soc. Am. Proc.* **28**:268–269.

437 Hayat, M. A. (1970) "Fixation, Embedding, Sectioning, Staining and Support Films" in *Principles and Techniques in Electron Microscopy*, Vol. 1 (M. A. Hayat, Ed.), Van Nostrand Reinhold, New York, pp. 3–412.

438 Hayat, M. A. & Zirkin, B. R. (1973) "Critical Point-Drying Method," in *Principles and Techniques in Electron Microscopy*, Vol. 3 (M. A. Hayat, Ed.), Van Nostrand Reinhold, New York, pp. 297–313.

439 Haydon, G. B. (1969) "On the Interpretation of High Resolution Electron Micrographs of Macromolecules" *J. Ultrastruct. Res.* **25**:349–361.

440 Haydon, G. B. (1970) "Light Optical Diffraction Studies of Macromolecular Ultrastructure" *Electron Microsc. Soc. Am. Proc.* **28**:258–259.

441 Haydon, G. B. and Scales, D. J. (1973) "Pitfalls in Optical Filtering Techniques" *Electron Microsc. Soc. Am. Proc.* **31**:276–277.

442 Haydon, G. B. and Taylor, D. A. (1966) "The Optimal Under-Focus Enhancement of Contrast in Electron Microscopy" *J. R. Microsc. Soc.* **85**:305–312.

443 Hayward, S. B. and Glaeser, R. M. (1979) "Radiation Damage of Purple Membrane at Low Temperature" *Ultramicroscopy* **4**:201–210.

444 Hayward, S. B., Grano, D. A., Glaeser, R. M., and Fisher, K. A. (1978) "Molecular Orientation of Bacteriorhodopsin Within the Purple Membrane of Halobacterium halobium" *Proc. Natl. Acad. Sci. U.S.A.* **75**:4320-4324.

445 Hecht, E. and Zajac, A. (1974) in *Optics*, Addison-Wesley, Reading, MA.

446 Heide, H. G. (1965) "Contamination and Irradiation Effects and Their Dependence on the Composition of Residual Gases in the Electron Microscope" *Lab. Invest.* **14**:396–401.

447 Heinemann, K. and Poppa, H. (1972) "Electron Microscope Image Intensification and Specimen Exposure" *Electron Microsc. Soc. Am. Proc.* **30**:610–611.

448 Henderson, D., Eibl, H., and Weber, K. (1979) "Structure and Biochemistry of Mouse Hepatic Gap Junctions" *J. Mol. Biol.* **132**:193–218.

449 Henderson, R., Capaldi, R. A., and Leigh, J. S. (1977) "Arrangement of Cytochrome Oxidase Molecules in Two-Dimensional Vesicle Crystals" *J. Mol. Biol.* **112**:631–648.

450 Henderson, R. and Unwin, P. N. T. (1975) "Three-Dimensional Model of Purple Membrane Obtained by Electron Microscopy" *Nature* (*London*) **257**:28–32.

451 Henderson, W. J. and Griffiths, K. (1972) "Shadow Casting and Replication," in *Principles and Techniques in Electron Microscopy*, Vol. 3 (M. A. Hayat, Ed.), Van Nostrand Reinhold, New York, pp. 149–193.

452 Hendrix, R. W. (1979) "Purification and Properties of groE, a Host Protein Involved in Bacteriophage Assembly" *J. Mol. Biol.* **129**:375–392.

453 Herman, G. T. (1972) "Two Direct Methods for Reconstructing Pictures from Their Projections: A Comparative Study" *Comput. Graph. Image Process.* **1**:123–144.

454 Herman, G. T., Lent, A., and Rowland, S. W. (1973) "ART: Mathematics and Applications. A Report on the Mathematical Foundations and on the Applicability to Real Data of the Algebraic Reconstruction Techniques" *J. Theor. Biol.* **42**:1–32.

455 Herman, G. T. and Rowland, S. (1971) "Resolution in ART: An Experimental Investigation of the Resolving Power of an Algebraic Picture Reconstruction Technique" *J. Theor. Biol.* **33**:213–223.

456 Herman, G. T. and Rowland, S. W. (1973) "Three Methods for Reconstructing Objects from X-Rays: A Comparative Study" *Comput. Graph. Image Process.* **2**:151–178.

457 Herrmann, K. H. and Krahl, D. (1974) "Experiments with Thermoplastic Recording Material for 'Real-Time' Diffractometry" *Proc. Eighth Int. Congr. Electron Microsc.* (*Canberra*) **1**:102–103.

458 Hibi, T., Yada, K., Takahashi, S., and Shibata, K. (1973) "High Resolution Electron Microscopy of Tobacco Mosaic Virus" *J. Electron Microsc.* **22:**243–253.

459 Hill, A. E. and Rigby, P. A. (1969) "The Precision Manufacture and Registration of Masks for Vacuum Evaporation" *J. Phys. E: Sci. Instrum.* **2:**1084–1086.

460 Hillier, J., Mudd, S., Smith, A. G., and Beutner, E. H. (1950) "The 'Fixation' of Electron Microscopic Specimens by the Electron Beam" *J. Bacteriol.* **60:**641–654.

461 Hills, G. J. and Campbell, R. N. (1968) "Morphology of Broccoli Necrotic Yellow Virus" *J. Ultrastruct. Res.* **24:**134–144.

462 Hills, G. J., Gurney-Smith, M., and Roberts, K. (1973) "Structure, Composition and Morphogenesis of the Cell Wall of *Chlamydomonas reinhardi*. II. Electron Microscopy and Optical Diffraction Analysis" *J. Ultrastruct. Res.* **43:**179–192.

463 Hills, G. J., Phillips, J. M., Gay, M. R., and Roberts, K. (1975) "Self-Assembly of a Plant Cell Wall *in Vitro*" *J. Mol. Biol.* **96:**431–441.

464 Hirai, K., Kato, N., Fujiura, A., and Shimakura, S. (1979) "Further Morphological Characterization and Structural Proteins of Infectious Bursal Disease Virus" *J. Virol.* **32:**323–328.

465 Hitchborn, J. H. and Hills, G. J. (1967) "Tubular Structures Associated with Turnip Yellow Mosaic Virus *in Vivo*" *Science* **157:**705–706.

466 Hitchborn, J. H. and Hills, G. J. (1968) "A Study of Tubes Produced in Plants Infected with a Strain of Turnip Yellow Mosaic Virus" *Virology* **35:**50–70.

467 Hoglund, S., Ruttkay-Nedecky, G., and Hjerten, S. (1969) "Some Structural Studies on the Nucleocapsid of the Influenza A2 Virus," in *Symmetry and Function of Biological Systems at the Macromolecular Level* (A. Engstrom and B. Strandberg, Eds.), Wiley, New York, pp. 341–347.

468 Hohn, T., Hohn, B., Engel, A., Wurtz, M., and Smith, P. R. (1979) "Isolation and Characterization of the Host Protein groE Involved in Bacteriophage λ Assembly" *J. Mol. Biol.* **129:**359–373.

469 Hohn, T., Wurtz, M., and Engel, A. (1978) "Sevenfold Rotational Symmetry of a Protein Complex" *J. Ultrastruct. Res.* **65:**90–93.

470 Holmes, K. C. and Blow, D. M. (1965) "The Use of X-ray Diffraction in the Study of Protein and Nucleic Acid Structure" *Methods Biochem. Anal.* **13:**113–239.

471 Hooper, C. W., Seeds, W. E., and Stokes, A. R. (1955) "Photographic Preparation of Masks of Large Molecules for the Lipson Diffractometer" *Nature* (*London*) **175:**679–681.

472 Hoppe, W. (1970) "Principles of Electron Structure Research at Atomic Resolution Using Conventional Electron Microscopes for the Measurement of Amplitudes and Phases" *Acta Crystallogr.* **A26:**414–426.

473 Hoppe, W. (1970) "Principles of Structure Analysis at High Resolution Using Conventional Electron Microscopes and Computers" *Ber. Bunsen-Gesellschaft Phys. Chem.* **74:**1090–1100.

474 Hoppe, W. (1971) "Use of Zone Correction Plates and Other Techniques for Structure Determination of Aperiodic Objects at Atomic Resolution Using a Conventional Electron Microscope" *Phil. Trans. R. Soc. London, B* **261:**71–94.

475 Hoppe, W. (1974) "High Resolution Studies Using Computerized Image Reconstruction Methods" *Proc. Eighth Int. Congr. Electron Microsc.* (*Canberra*) **1:**240–241.

476 Hoppe, W. (1974) "Towards Three-Dimensional 'Electron Microscopy' at Atomic Resolution" *Naturwissenschaften* **61:**239–249.

477 Hoppe, W. (1974) "Contributions to Three-Dimensional Image Reconstruction" *Proc. Eighth Int. Congr. Electron Microsc.* (*Canberra*) **1:**308–309.

478 Hoppe, W., Gassman, J., Hunsmann, N., Schramm, J., and Sturm, M. (1974) "Three-Dimensional Reconstruction of Individual Negatively Stained Yeast Fatty-Acid Synthetase Molecules from Tilt Series in the Electron Microscope" *Hoppe-Seyler's Z. Phys. Chem.* **355:**1483–1487.

479 Hoppe, W. and Grill, B. (1977) "Prospects of Three-Dimensional High Resolution Electron Microscopy of Non-Periodic Structures" *Ultramicroscopy* **2:**153–168.

480 Hoppe, W., Hunsmann, N., Schramm, H. J., Sturm, M., Grill, B., and Gassmann, J. (1976) "Three-Dimensional Electron Microscopy of Individual Objects" *Sixth Eur. Reg. Conf. Electron Microsc.* (*Jerusalem*) **1:**8–13.

481 Hoppe, W., Katerbau, K. H., Langer, R., Mollenstedt, G., Speidel, R., and Thon, F. (1969) "Electron Microscopical Imaging Using Zonal Correction Plates" *Siemens Rev.* **36:**24–32.

482 Hoppe, W. and Kostler, D. (1976) "Experimental Results in High Resolution Electron Microscopy Using the Tilt Image Reconstruction Method" *Sixth Eur. Reg. Conf. Elec. Microsc.* (*Jerusalem*) **1:**99–104.

483 Hoppe, W., Schramm, H. J., Sturm, M., Hunsmann, N., and Gassmann, J. (1976) "Three-Dimensional Electron Microscopy of Individual Biological Objects. Part I. Methods" *Z. Naturforsch., A* **31:**645–655.

484 Hoppe, W., Schramm, H. J., Sturm, M., Hunsmann, N., and Gassmann, J. (1976) "Three-Dimensional Electron Microscopy of Individual Biological Objects. Part II. Test Calculations" *Z. Naturforsch., A* **31:**1370–1379.

485 Hoppe, W., Schramm, H. J., Sturm, M., Hunsmann, N., and Gassmann, J. (1976) "Three-Dimensional Electron Microscopy of Individual Biological Objects. Part III. Experimental Results on Yeast Fatty Acid Synthetase" *Z. Naturforsch., A* **31:**1380–1390.

486 Hoppe, W., Wenzl, H., and Schramm, H. J. (1976) "New 3D Reconstruction Techniques at Minimal Doses Conditions" *Sixth Eur. Reg. Conf. Electron Microsc.* (*Jerusalem*) **2:**58–60.

487 Horgen, H. M., Villagrana, R. E., and Maher, D. M. (1974) "Comments on the Processing of Periodic and Near-Periodic Images by Computer" *Proc. Eighth Int. Congr. Electron Microsc.* (*Canberra*), **1:**316–317.

488 Horne, R. W. (1964) "Electron Microscopy of Viruses" *Sci. Prog.* **52:**525–542.

489 Horne, R. W. (1973) "Contrast and Resolution from Biological Objects Examined in the Electron Microscope with Particular Reference to Negatively Stained Specimens" *J. Microsc.* **98:**286–298.

490 Horne, R. W. (1975) "Recent Advances in the Application of Negative Staining Techniques to the Study of Virus Particles Examined in the Electron Microscope" *Adv. Opt. Electron Microsc.* **6:**227–274.

491 Horne, R. W. (1978) "Special Specimen Preparation Methods for Image Processing in Transmission Electron Microscopy: A Review" *J. Microsc.* **113:**241–256.

492 Horne, R. W. (1978) "Recent Applications of High Resolution Electron Microscopy to Crystalline Arrays of Virus Particles" *Proc. Ninth Int. Congr. Electron Microsc.* (*Toronto*) **3:**470–482.

493 Horne, R. W. (1979) "The Formation of Virus Crystalline and Paracrystalline Arrays for Electron Microscopy and Image Analysis" *Adv. Virol. Res.* **24:**173–221.

494 Horne, R. W. and Davies, D. R. (1970) "High Resolution Electron Microscopy and Optical Diffraction Studies of Chlamydomonas Cell Walls" *Proc. Seventh Int. Congr. Electron Microsc.* (*Grenoble*) **3:**117–118.

495 Horne, R. W., Davies, D. R., Norton, K., and Gurney-Smith, M. (1971) "Electron

Microscope and Optical Diffraction Studies on Isolated Cell Walls from *Chlamydomonas*'' *Nature (London)* **232:**493–495.

496 Horne, R. W., Harnden, J. M., and Hull, R. (1977) ''The *in Vitro* Crystalline Formations of Turnip Rosette Virus. I. Electron Microscopy of Two- and Three-Dimensional Arrays'' *Virology* **82:**150–162.

497 Horne, R. W., Hobart, J. M., and Markham, R. (1976) ''Electron Microscopy of Tobacco Mosaic Virus Prepared with the Aid of Negative Staining–Carbon Film Techniques'' *J. Gen. Virol.* **31:**265–269.

498 Horne, R. W., Hobart, J. M., and Pasquali-Ronchetti, I. (1975) ''Application of the Negative Staining–Carbon Technique to the Study of Virus Particles and Their Components by Electron Microscopy'' *Micron* **5:**233–261.

499 Horne, R. W., Hobart, J. M., and Pasquali-Ronchetti, I. (1975) ''A Negative Staining–Carbon Film Technique for Studying Viruses in the Electron Microscope. III. The Formation of Two-Dimensional and Three-Dimensional Crystalline Arrays of Cowpea Chlorotic Mottle Virus'' *J. Ultrastruct. Res.* **53:**319–330.

500 Horne, R. W. and Markham, R. (1972) ''Application of Optical Diffraction and Image Reconstruction Techniques to Electron Micrographs'' in *Practical Methods in Electron Microscopy*, Vol. 1 (A. M. Glauert, Ed.), North Holland, Amsterdam, pp. 361–379.

501 Horne, R. W. and Pasquali-Ronchetti, I. (1974) ''A Negative Staining–Carbon Film Technique for Studying Viruses in the Electron Microscope'' *J. Ultrastruct. Res.* **47:**361–383.

502 Horne, R. W., Pasquali-Ronchetti, I., and Hobart, J. M. (1975) ''A Negative Staining–Carbon Film Technique for Studying Viruses in the Electron Microscope. II. Application to Adenovirus Type 5'' *J. Ultrastruct. Res.* **51:**233–252.

503 Horne, R. W. and Wildy, P. (1961) ''Symmetry in Virus Architecture'' *Virology* **15:**348–373.

504 Horne, R. W. and Wildy, P. (1963) ''Virus Structure Revealed by Negative Staining'' *Adv. Virol. Res.* **10:**101–170.

505 Horne, R. W. and Wildy, P. (1979) ''An Historical Account of the Development and Applications of the Negative Staining Technique to the Electron Microscopy of Viruses'' *J. Microsc.* **117:**103–122.

506 Howatson, A. F. and Kemp, C. L. (1975) ''The Structure of Tubular Head Forms of Bacteriophage λ: Relation to the Capsid Structure of Petit λ and Normal λ Heads'' *Virology* **67:**80–84.

507 Howitt, D. G. (1975) ''The Accurate Measurement of the Energy Dependence of Radiation Damage Rates in Organic Materials'' *Electron Microsc. Soc. Am. Proc.* **33:**210–211.

508 Howitt, D. G., Glaeser, R. M., and Thomas, G. (1975) ''The Energy Dependence of Radiation Damage in L-Valine'' *Electron Microsc. Soc. Am. Proc.* **33:**246–247.

509 Howitt, D. G., Glaeser, R. M., and Thomas, G. (1976) ''The Energy Dependence of Electron Radiation Damage in L-Valine'' *J. Ultrastruct. Res.* **55:**457–461.

510 Hughes, W. and Taylor, C. A. (1953) ''Apparatus Used in the Development of Optical-Diffraction Methods for the Solution of Problems in X-Ray Analysis'' *J. Sci. Instrum.* **30:**105–110.

511 Hui, S. W. and Parsons, D. F. (1974) ''Electron Microscopy and Electron Diffraction of Hydrated Biological Material'' *Electron Microsc. Soc. Am. Proc.* **32:**296–297.

512 Hui, S. W. and Parsons, D. F. (1974) ''Electron Diffraction of Wet Biological Membranes'' *Science* **184:**77–78.

513 Hui, S. W. and Parsons, D. F. (1978) ''Electron Microscopy and Electron Diffraction Studies on Hydrated Membranes'' in *Advanced Techniques in Biological Electron Microscopy*, Vol. 2 (J. K. Koehler, Ed.), Springer Verlag, Berlin, pp. 213–235.

514 Hui, S. W., Parsons, D. F., and Cowden, M. (1974) "Electron Diffraction of Wet Phospholipid Bilayers" *Proc. Natl. Acad. Sci. U.S.A.* **71**:5068–5072.

515 Hull, R., Hills, G. J., and Markham, R. (1969) "Studies on Alfalfa Mosaic Virus. II. The Structure of the Virus Components" *Virology* **37**:416–428.

516 Hull, R., Hills, G. J., and Plaskitt, A. (1969) "Electron Microscopy on *in Vivo* Aggregation Forms of a Strain of Alfalfa Mosaic Virus" *J. Ultrastruct. Res.* **26**:465–479.

517 Huxley, H. E. (1968) "Structural Difference Between Resting and Rigor Muscle: Evidence from Intensity Changes in the Low-Angle Equatorial X-ray Diagram" *J. Mol. Biol.* **37**:507–520.

518 Huxley, H. E. and Zubay, G. (1960) "Electron Microscope Observations on the Structure of Microsomal Particles from *Escherichia coli*" *J. Mol. Biol.* **2**:10–18.

519 Isaacson, M. (1973) "Inelastic Scattering and Beam Damage of Biological Molecules" *Electron Microsc. Soc. Am. Proc.* **31**:478–479.

520 Isaacson, M., Collins, M. L., and Listvan, M. (1978) "Electron Beam Damage of Biomolecules Assessed by Energy Loss Spectroscopy" *Proc. Ninth Int. Congr. Electron Microsc.* (*Toronto*) **3**:61–69.

521 Isaacson, M., Johnson, D., and Crewe, A. V. (1973) "Electron Beam Excitation and Damage of Biological Molecules; Its Implications for Specimen Damage in Electron Microscopy" *Radiat. Res.* **55**:205–224.

522 Isaacson, M., Langmore, J., Lamvik, M., Lin, P., Golladay, S., Hainfeld, J., Pullman, J., and Furlong, D. (1975) "Mass Loss of Biological Molecules in the Electron Microscope" *Electron Microsc. Soc. Am. Proc.* **33**:676–677.

523 Ishii, T. and Yanagida, M. (1975) "Molecular Organization of the Shell of the T-Even Bacteriophage Head" *J. Mol. Biol.* **97**:655–660.

524 Ivanitskii, G. R., Kuniskii, A. S., and Tsyganov, M. A. (1978) "Determination of Relative Orientation of Structures of the 'Stacked-Disk' Type on Electron Microscope Images" *Dokl. Biophys.* **238**:33–36.

525 Jack, A., Harrison, S. C., Crowther, R. A. (1975) "Structure of Tomato Bushy Stunt Virus. II. Comparison of Results Obtained by Electron Microscopy and X-Ray Diffraction" *J. Mol. Biol.* **97**:163–172.

526 Jacobs, M., Bennett, P. M., and Dickens, M. J. (1975) "Duplex Microtubule Is a New Form of Tubulin Assembly Induced by Polycations" *Nature* (*London*) **257**:707–709.

527 James, W. and Mulvey, T. (1970) "Simplified Optical Diffractometer for High Resolution Electron Microscopy" *Proc. Seventh Int. Congr. Electron Microsc.* (*Grenoble*) **2**:65–66.

528 Jennison, R. C. (1961) in *Fourier Transforms and Convolutions for the Experimentalist*, Pergamon Press, Oxford.

529 Johannssen, W., Schutte, H., Mayer, F., and Mayer, H. (1979) "Quaternary Structure of the Isolated Restriction Endonuclease EndoR.Bgl 1 from *Bacillus globigii* as Revealed by Electron Microscopy" *J. Mol. Biol.* **134**:707–726.

530 Johansen, B. V. (1972) "An Optical Diffractometer of Simplified Design and Short Camera Length for the Analysis of High Resolution Electron Micrographs" *Micron* **3**:256–270.

531 Johansen, B. V. (1973) "Bright Field Electron Microscopy of Biological Specimens. I. Obtaining the Optimum Contribution of Phase Contrast to Image Formation" *Micron* **4**:446–472.

532 Johansen, B. V. (1974) "Bright Field Electron Microscopy of Biological Specimens. II. Preparation of Ultra-Thin Carbon Support Films" *Micron* **5**:209–221.

533 Johansen, B. V. (1975) "Optical Diffractometry" in *Principles and Techniques in Electron Microscopy,*" Vol. 5 (M. A. Hayat, Ed.), Van Nostrand Reinhold, New York, pp. 114–173.

534 Johansen, B. V. (1975) "Bright Field Electron Microscopy of Biological Specimens. III. Expanded Phase Contrast Information Using *in Situ* Double Exposures with Complementary Objective Apertures" *Micron* **6**:153–163.

535 Johansen, B. V. (1975) "Bright Field Electron Microscopy of Biological Specimens. IV. Ultrasonic Exfoliated Graphite as 'Low-Noise' Support Films" *Micron* **6**:165–173.

536 Johansen, B. V. (1976) "Bright Field Electron Microscopy of Biological Specimens. V. A Low Dose Pre-Irradiation Procedure Reducing Beam Damage" *Micron* **7**:145–156.

537 Johansen, B. V. (1976) "Bright Field Electron Microscopy of Biological Specimens. VI. Signal-to-Noise Ratio in Specimens Prepared on Amorphous Carbon and Graphite Crystal Supports" *Micron* **7**:157–170.

538 Johansen, B. V. (1977) "High Resolution Bright Field Electron Microscopy of Biological Specimens" *Ultramicroscopy* **2**:229–239.

539 Johansen, B. V. and Hoglund, S. (1975) "Report on a Symposium on Contrast Problems in Transmission Electron Microscopy" *Ultramicroscopy* **1**:83.

540 Johnson, D. E. (1975) "An Image Reconstruction System Applied to High Voltage Microscopy" *Electron Microsc. Soc. Am. Proc.* **33**:292–293.

541 Johnson, D. E. and Pfeifer, J. (1974) "Image Reconstruction Applied to High Voltage Microscopy" *Election Microsc. Soc. Am. Proc.* **32**:396–397.

542 Johnson, D. J. and Crawford, D. (1973) "Defocusing Phase Contrast Effects in Electron Microscopy" *J. Microsc.* **98**:313–324.

543 Johnson, H. M. (1973) "In-Focus Phase Contrast Electron Microscopy" in *Principles and Techniques in Electron Microscopy*, Vol. 3 (M. A. Hayat, Ed.), Van Nostrand Reinhold, New York, pp. 153–198.

544 Johnson, H. M. and Parsons, D. F. (1970) "In-Focus Phase Contrast Electron Microscopy" *Proc. Seventh Int. Congr. Electron Microsc. (Grenoble)* **1**:71–72.

545 Johnson, H. M. and Parsons, D. F. (1970) "Phase Contrast in Electron Microscopy" *Electron Microsc. Soc. Am. Proc.* **28**:48–49.

546 Johnston, H. S. and Reid, O. (1971) "An Improved Method for Preparing Perforated Carbon Films for Electron Microscopy Using Ultrasonic Vibration" *J. Microsc.* **94**:283–286.

547 Jones, A. V. and Leonard, K. R. (1978) "Scanning Transmission Electron Microscopy of Unstained Biological Sections" *Nature (London)* **271**:659–660.

548 Josephs, R. (1975) "An Analysis of the Mechanism of Crystallization of Glutamic Dehydrogenase" *J. Mol. Biol.* **97**:127–138.

549 Josephs, R. and Borisy, G. (1972) "Self-Assembly of Glutamic Dehydrogenase into Ordered Superstructures: Multichain Tubes Formed by Association of Single Molecules" *J. Mol. Biol.* **65**:127–155.

550 Joy, R. T. (1973) "The Electron Microscopical Observation of Aqueous Biological Specimens" *Adv. Opt. Electron Microsc.* **5**:297–352.

551 Junger, E., Hahn, M. H., and Reinauer, H. (1970) "The Structure of Lysolecithin-Water Phases: Negative Staining and Optical Diffraction Analysis of the Electron Micrographs" *Biochim. Biophys. Acta* **21**:381–388.

552 Kaftanova, A. S., Kiselev, N. A., Novikov, V. K., and Atabekov, J. (1975) "Structure of Products of Protein Reassembly and Reconstruction of Potato Virus X" *Virology* **67**:283–287.

553 Kamiya, R., Asakura, S., Wakabayashi, K., and Namba, K. (1979) "Transition of Bacterial Flagella from Helical to Straight Forms with Different Subunit Arrangements" *J. Mol. Biol.* **131**:725–742.

554 Karasaki, S. (1966) "Size and Ultrastructure of the H-Viruses as Determined with the Use of Specific Antibodies" *J. Ultrastruct. Res.* **16**:109–122.

555 Katsura, I. (1978) "Structure and Inherent Properties of the Bacteriophage λ Head Shell I. Polyheads Produced by Two Defective Mutants in the Major Head Protein. Appendix: Topological Irregularities in Tubes Made of P6 Nets" *J. Mol. Biol.* **121**:71–93.

556 Kaye, J. S. (1970) "The Fine Structure and Arrangement of Microcylinders in the Lumina of Flagellar Fibers in Cricket Spermatids" *J. Cell Biol.* **45**:416–430.

557 Kellenberger, E. (1978) "Possibility to Detect Reproducibly Some 5–10 Angstrom Conformational Differences by Conventional Techniques: Physiologically Defined Lattice Transformations in Bacteriophage T4" *Proc. Ninth Int. Congr. Electron Microsc.* (*Toronto*) **3**:441–449.

558 Kennard, C. H. L. and Shields, K. G. (1979) "Model Drawing Programs: Manufacture of Masks to Produce Optical Transforms" *J. Appl Crystallogr.* **12**:135–136.

559 Keown, S. R. (1974) "The Representation of Electron Diffraction Patterns by Optical Diffraction" *Proc. Eighth Int. Congr. Electron Microsc.* (*Canberra*) **1**:366–367.

560 Kessel, M. (1978) "A Unique Crystalline Wall Layer in the Cyanobacterium *Microcystis marginata*" *J. Ultrastruct. Res.* **62**:203–212.

561 Kim, H., Binder, L. I., and Rosenbaum, J. L. (1979) "The Periodic Association of MAP(2) with Brain Microtubules *in Vitro*" *J. Cell Biol.* **80**:266–276.

562 King, M. V. and Parsons, D. F. (1977) "Recording of Electron-Diffraction Patterns of Radiation-Sensitive Materials in the High-Voltage Electron Microscope with Luminescent Radiographic Screens" *J. Appl. Crustallogr.* **10**:62–63.

563 King, M. V. and Parsons, D. F. (1978) "A Perspicuous Technique for Directly Visualizing Radiation-Damage Artifacts in Biological Electron Microscopy" *J. Microsc.* **113**:301–305.

564 Kingsbury, E. W., Nauman, R. K., Morgan, R. S., and Voelz, H. (1970) "Structure and Function of Ribosomal Helices in *Escherichia coli*" *Proc. Seventh Int. Congr. Electron Microsc.* (*Grenoble*) **3**:69–70.

565 Kirk, R. G. and Ginzburg, M. (1972) "Ultrastructure of Two Species of *Halobacterium*" *J. Ultrastruct. Res.* **41**:80–94.

566 Kiselev, L. L., Favorova, O. O., Parin, A. V., Stel'mashchuk, V. Ya., and Kiselev, N. A. (1971) "Crystallization (Polymerization) of a Complex of Tryptophanyl-tRNA-Synthetase with Tryptophan" *Dokl. Biophys.* **199**:82–84.

567 Kiselev, N. A. (1978) "Reconstruction of the Structure of Enzymes from Their Images" *Proc. Ninth Int. Congr. Electron Microsc.* (*Toronto*) **3**:94–106.

568 Kiselev, N. A., DeRosier, D. J., and Atabekov, J. G. (1969) "A Double-Helical Structure Found on the Reaggregation of the Protein of Barley Stripe Mosaic Virus" *J. Mol. Biol.* **39**:673–674.

569 Kiselev, N. A., DeRosier, D. J., and Klug, A. (1968) "Structure of the Tubes of Catalase: Analysis of Electron Micrographs by Optical Filtering" *J. Mol. Biol.* **35**:561–566.

570 Kiselev, N. A. and Klug, A. (1969) "The Structure of Viruses of the Papilloma-Polyoma Type V. Tubular Variants Built of Pentamers (Appendix: Helical Plotting Program, T. H. Gossling)" *J. Mol. Biol.* **40**:155–171.

571 Kiselev, N. A., Lerner, F. Ya., and Livanova, N. B. (1971) "Electron Microscopy of Muscle Phosphorylase *b*" *J. Mol. Biol.* **62**:537–549.

572 Kiselev, N. A., Lerner, F. Ya., and Livanova, N. B. (1971) "Electron Microscopy of Muscle Phosphorylase *b*" *Mol. Biol.* **5**:510–523.

573 Kiselev, N. A., Lerner, F. Ya., and Livanova, N. B. (1974) "Electron Microscopy of Muscle Phosphorylase *a*" *J. Mol. Biol.* **86**:587–599.

574 Kiselev, N. A., Orlova, E. V., and Stel'mashchuk, V. Ya. (1979) "Investigation of Fine Structure of Small Ribosome Subparticles by Optical Filtration of Their Electron-Microscopic Images" *Dokl. Biophys.* **246**:118–121.

575 Kiselev, N. A., Stel'mashchuk, V. Y., Lutsch, H., and Noll, F. (1978) "Structure of Small Subparticles of Liver Ribosomes" *J. Mol. Biol.* **126**:109–115.

576 Kiselev, N. A., Stel'mashchuk, V. Y., Tsuprun, V. L., Lerman, M. I., and Abakumova, O. Y. (1974) "The Structure of Liver Ribosomes" *Acta Biol. Med. Germ.* **33**:795–807.

577 Kiselev, N. A., Stel'mashchuk, V. Y., Tsuprun, V. L., Ludewig, M., and Hanson, H. (1977) "Electron Microscopy of Leucine Aminopeptidase" *J. Mol. Biol.* **115**:33–43.

578 Kiselev, N. A. and Vainshtein, B. K. (1974) "The Investigation of Tubular Crystals of Enzymes by Optical Filtering and 3-D Reconstruction" *Proc. Eighth Int. Congr. Electron Microsc.* (*Canberra*) **1**:730–731.

579 Kistler, J., Aebi, U., and Kellenberger, E. (1976) "Image Processing Applied to Freeze Dried-Shadowed Specimen" *Sixth Eur. Reg. Conf. Electron Microsc.* (*Jerusalem*) **2**:156–158.

580 Kistler, J., Aebi, U., and Kellenberger, E. (1977) "Freeze-Drying and Shadowing a Two-Dimensional Periodic Specimen" *J. Ultrastruct. Res.* **59**:76–86.

581 Kistler, J., Aebi, U., Onorato, L., ten Heggeler, B., and Showe, M. K. (1978) "Structural Changes During the Transformation of Bacteriophage T4 Polyheads: Characterization of the Initial and Final States by Freeze-Drying and Shadowing Fab-Fragment-Labeled Preparations" *J. Mol. Biol.* **126**:571–589.

582 Kistler, J., Aebi, U., ten Heggeler, B., Onorato, L., and Showe, M. (1978) "Structural Rearrangements Studied by Topographical Mapping of Antigenic Sites by Electron Microscopy" *Proc. Ninth Int. Congr. Electron Microsc.* (*Toronto*) **2**:160–161.

583 Kistler, J. and Kellenberger, E. (1977) "Collapse Phenomena in Freeze-Drying" *J. Ultrastruct. Res.* **59**:70–75.

584 Kitamura, M., Miyazono, J., Mori, R., Oda, H., and Nonomura, Y. (1978) "Optical Diffraction Analysis of Electron Microscopic Images of Tubular Structures Found in Cells Infected with Herpes Simplex Virus Type 2" *J. Gen. Virol.* **41**:167–170.

585 Klug, A. (1967) "The Design of Self-Assembling Systems of Equal Units," in *Formation and Fate of Organelles* (K. B. Warren, Ed.), Academic Press, New York, pp. 1–18.

586 Klug, A. (1969) "The Structure of the Tubular Variants of the Papilloma Viruses," in *Symmetry and Function of Biological Systems at the Macromolecular Level* (A. Engstrom and B. Strandberg, Eds.), Wiley, New York, pp. 313–327.

587 Klug, A. (1969) "Point Groups and the Design of Aggregates," in *Symmetry and Function of Biological Systems at the Macromolecular Level* (A. Engstrom and B. Strandberg, Eds.), Wiley, New York, pp. 425–436.

588 Klug, A. (1971) "III. Applications of Image Analysis Techniques in Electron Microscopy: Optical Diffraction and Filtering and Three-Dimensional Reconstructions from Electron Micrographs" *Phil. Trans. R. Soc. London, B* **261**:173–179.

589 Klug, A. and Berger, J. E. (1964) "An Optical Method for the Analysis of Periodicities in Electron Micrographs, and Some Observations on the Mechanism of Negative Staining" *J. Mol. Biol.* **10**:565–569.

590 Klug, A., Crick, F. H. C., and Wyckoff, H. W. (1958) "Diffraction by Helical Structures" *Acta Crystallogr.* **11**:199–213.

591 Klug, A. and Crowther, R. A. (1972) "Three-Dimensional Image Reconstruction from the Viewpoint of Information Theory" *Nature* (*London*) **238**:435–440.

592 Klug, A. and DeRosier, D. J. (1966) "Optical Filtering of Electron Micrographs: Reconstruction of One Sided Images" *Nature* (*London*) **212**:29–32.

593 Klug, A. and Durham, A. C. H. (1972) "The Disk of TMV Protein and Its Relation to the Helical and Other Modes of Aggregation" *Cold Spring Harbor Symp. Quant. Biol.* **36**:449–468.

594 Klug, A. and Finch, J. T. (1968) "Structure of Viruses of the Papilloma-Polyoma Type. IV. Analysis of Tilting Experiments in the Electron Microscope" *J. Mol. Biol.* **31**:1–12.

595 Klug, A., Lutter, L. C., Rhodes, D., Brown, R. S., Rushton, B., and Finch, J. T. (1977) "X-Ray Crystallographic and Enzymic Analysis of Nucleosome Cores" *FEBS*, 11th Meeting (B. F. C. Clark,, et al., Eds.), Copenhagen **43**:233–244.

596 Knauer, V., Schramm, H. J., and Hoppe, W. (1978) "A Minimal Dose Technique in 3D-Electron Microscopy" *Proc. Ninth Int. Congr. Electron Microsc.* (*Toronto*) **2**:4–5.

597 Knutton, S. and Robertson, J. D. (1976) "Regular Substructures in Membranes: The Luminal Plasma Membrane of the Cow Urinary Bladder" *J. Cell Sci.* **22**:355–370.

598 Kobayashi, K. and Sakaoku, K. (1965) "Irradiation Changes in Organic Polymers at Various Accelerating Voltages" *Lab. Invest.* **14**:359–376.

599 Koehler, J. K. (1972) "The Freeze-Etching Technique," in *Principles and Techniques in Electron Microscopy*, Vol. 2 (M. A. Hayat, Ed.), pp. 51–98.

600 Kondoh, H. and Yanagida, M. (1975) "Structure of Straight Flagellar Filaments from a Mutant of *Escherichia coli*" *J. Mol. Biol.* **96**:641–652.

601 Kosourov, G. I., Lifshits, I. E., and Kiselev, N. A. (1972) "An Optical Diffractometer" *Sov. Phys. Cryst.* **16**:702–708.

602 Krakow, W., Downing, K. H., and Siegel, B. M. (1973) "A Technique for the Rapid Determination of the Contrast Transfer Characteristics of an Electron Microscope" *Electron Microsc. Soc. Am. Proc.* **31**:278–279.

603 Krakow, W. and Siegel, B. (1972) "Characteristics of an Electrostatic Phase Plate" *Electron Microsc. Soc. Am. Proc.* **30**:618–619.

604 Krimm, S. and Anderson, T. F. (1967) "Structure of Normal and Contracted Tail Sheaths of T4 Bacteriophage" *J. Mol. Biol.* **27**:197–202.

605 Kubler, O., Gross, H., and Moor, H. (1978) "Complementary Structures of Membrane Fracture Faces Obtained by Ultrahigh Vacuum Freeze-Fracturing at −196°C and Digital Image Processing" *Ultramicroscopy* **3**:161–168.

606 Kubler, O., Hahn, M., and Baumeister, W. (1978) "Topography of a Bacterial Membrane: The HPI-Layer of *M. radiodurans*" *Proc. Ninth Int. Congr. Electron Microsc.* (*Toronto*) **2**:28–29.

607 Kubler, O., Hahn, M., and Seredynski, J. (1976) "Image Processing as a Tool for Radiation Damage Assessment and Minimization" *Sixth Eur. Reg. Conf. Electron Microsc.* (*Jerusalem*) **1**:306–308.

608 Kubler, O. and Gross, H. (1978) "UHV Freeze-Fracturing and Image Processing Applied to the Purple Membrane" *Proc. Ninth Int. Congr. Electron Microsc.* (*Toronto*) **2**:142–143.

609 Kuo, I. A. M. and Glaeser, R. M. (1975) "Development of Methodology for Low Exposure, High Resolution Electron Microscopy of Biological Specimens" *Electron Microsc. Soc. Am. Proc.* **33**:662–663.

610 Kuo, I. A. M. and Glaeser, R. M. (1975) "Development of Methodology for Low Exposure, High Resolution Electron Microscopy of Biological Specimens" *Ultramicroscopy* **1**:53–66.

611 Labaw, L. W. (1971) "The External Shape of Human γ-G1 Immunoglobulin from Crystal Sections" *Electron Microsc. Soc. Am. Proc.* **29**:428–429.

612 Labaw, L. W. (1974) "Molecular Packing in Two Fab′ Crystals" *Electron Microsc. Soc. Am. Proc.* **32**:68–69.

613 Labaw, L. W. and Davies, D. R. (1971) "An Electron Microscopic Study of Human γ-G1 Immunoglobulin Crystals: Preliminary Results" *J. Biol. Chem.* **246**:3760–3762.

614 Labaw, L. W. and Davies, D. R. (1972) "The Molecular Outline of Human γ-G1 Immunoglobulin from an EM Study of Crystals" *J. Ultrastruct. Res.* **40**:349–365.

615 Labaw, L. W. and Olson, R. A. (1970) "Further Electron Microscopic Observations of Bacteriochlorophyll Protein Crystals" *J. Ultrastruct. Res.* **31**:456–464.

616 Labaw, L. W., Padlan, E. A., Segal, D. M., and Davies, D. R. (1975) "An EM Study of

Phosphorylcholine-Binding Fab′ Immunoglobulin Fragment Crystals" *J. Ultrastruct. Res.* **51**:326–339.

617 Labaw, L. W. and Rossmann, M. G. (1969) "Electron Microscopic Observations of L-Lactate Dehydrogenase Crystals" *J. Ultrastruct. Res.* **27**:105–117.

618 Laemmli, U. K., Amos, L. A., and Klug, A. (1976) "Correlation Between Structural Transformation and Cleavage of the Major Head Protein of T4 Bacteriophage" Cell **7**:191–203.

619 Laemmli, U. K., Paulson, J. R., and Hitchins, V. (1974) "Maturation of the Head of Bacteriophage T4 V. A Possible DNA Packaging Mechanism: *In Vitro* Cleavage of the Head Proteins and the Structure of the Core of the Polyhead" *J. Supramol. Struct.* **2**:276–301.

620 Lake, J. A. (1971) "Reconstruction of Three-Dimensional Structures from Electron Micrographs: The Equivalence of Two Methods" *Electron Microsc. Soc. Am. Proc.* **29**:90–91.

621 Lake, J. A. (1972) "Reconstruction of Three-Dimensional Structures from Sectioned Helices by Deconvolution of Partial Data" *J. Mol. Biol.* **66**:255–269.

622 Lake, J. A. (1972) "Biological Studies," in *Optical Transforms* (H. Lipson, Ed.), Academic Press, New York, pp. 153–188.

623 Lake, J. A. (1974) "Ribosome Structural Organization" *Electron Microsc. Soc. Am. Proc.* **32**:332–333.

624 Lake, J. A. and Leonard, K. R. (1973) "Computer Fourier Analysis of Electron Micrographs from *Caulobacter crescentus* Bacteriophage ϕ-CbK" *Electron Microsc. Soc. Am. Proc.* **31**:272–273.

625 Lake, J. A. and Leonard, K. R. (1974) "Structure and Protein Distribution for the Capsid of *Caulobacter crescentus* Bacteriophage ϕ-CbK" *J. Mol. Biol.* **86**:499–518.

626 Lake, J. A., Leonard, K. R. (1974) "Bacteriophage Structure: Determination of Head-Tail Symmetry Mismatch for *Caulobacter crescentus* Phage ϕ-CbK" *Science* **183**:744–747.

627 Lake, J. A. and Slayter, H. S. (1970) "Three-Dimensional Fourier Analysis of the Ribonucleoprotein Particle (Ribosome) Helix of *Entamoeba invadens*" *Electron Microsc. Soc. Am. Proc.* **28**:266–267.

628 Lake, J. A. and Slayter, H. S. (1970) "Three-Dimensional Structure of the Chromatoid Body of *Entamoeba invadens*" *Nature* (*London*) **227**:1032–1037.

629 Lake, J. A. and Slayter, H. S. (1972) "Three-Dimensional Structure of the Chromatoid Body Helix of *Entamoeba invadens*" *J. Mol. Biol.* **66**:271–282.

630 Lakshminarayanan, A. V. and Lent, A. (1979) "Methods of Least Squares and SIRT in Reconstruction" *J. Theor. Biol.* **76**:267–295.

631 Lamvik, M. K. and Groves, T. (1976) "Minimization of Dose as a Criterion for the Selection of Imaging Modes in Electron Microscopy of Amorphous Specimens" *Ultramicroscopy* **2**:69–75.

632 Langer, R., Poppe, Ch., Schramm, H. J., and Hoppe, W. (1975) "Electron Microscopy of Thin Protein Crystal Sections" *J. Mol. Biol.* **93**:159–165.

633 Langridge, R., Marvin, D. A., Seeds, W. E., Wilson, H. R., Hooper, C. W., Wilkins, M. H. F., and Hamilton, L. D. (1960) "The Molecular Configuration of Deoxyribonucleic Acid. II. Molecular Models and Their Fourier Transforms (Appendix: Calculation of the Fourier Transform of a Helical Molecule, R. Langridge, M. P. Barnett, and A. F. Mann)" *J. Mol. Biol.* **2**:38–64.

634 Ledbetter, M. C. and Porter, K. R. (1964) "Morphology of Microtubules of Plant Cells" *Science* **144**:872–874.

635 Lenz, F. A. (1974) "Retrieval of Information from Electron Micrographs" *Proc. Eighth Int. Congr. Electron Microsc.* (*Canberra*) **1**:726–727.

636 Leonard, K. R., Kleinschmidt, A. K., Agabian-Keshishian, N., Shapiro, L., and Maizel, J. V. Jr. (1972) "Structural Studies on the Capsid of *Caulobacter crescentus* Bacteriophage ϕ-CbK" *J. Mol. Biol.* **71**:201–216.

637 Leonard, K. R., Kleinschmidt, A. K., and Lake, J. A. (1973) "*Caulobacter crescentus* Bacteriophage ϕ-CbK: Structure and *in Vitro* Self-Assembly of the Tail" *J. Mol. Biol.* **81**:349–365.

638 Lessing, R. (1976) "Optical Fourier's Transformation of Two-Dimensional Picture Signals" *Sixth Eur. Reg. Conf. Electron Microsc.* (*Jerusalem*) **1**:309–311.

639 Lewis, P. R. and Knight, D. P. (1977) "Staining Methods for Sectioned Material" in *Practical Methods in Electron Microscopy*, Vol. 5 (A. M. Glauert, Ed.), North Holland, Amsterdam, pp. 1–311.

640 Linck, R. and Olson, G. (1976) "Structural Chemistry of Sperm Flagellar Central Pair and Outer Doublet Microtubules," in *Contractile Systems in Non-Muscle Tissues* (S. V. Perry et al., Eds.), Elsevier/North Holland Biomedical Press, Amsterdam, pp. 229–240.

641 Linck, R. W. and Amos, L. A. (1974) "The Hands of Helical Lattices in Flagellar Doublet Microtubules" *J. Cell Sci.* **14**:551–559.

642 Lipson, H. and Taylor, C. A. (1951) "Optical Methods in X-ray Analysis. II. Fourier Transforms and Crystal-Structure Determination" *Acta Crystallogr.* **4**:458–462.

643 Lipson, H. and Taylor, C. A. (1958) in *Fourier Transforms and X-ray Diffraction*, G. Bell & Sons, London.

644 Lipson, H. and Taylor, C. A. (1966) "X-ray Crystal-Structure Determination as a Branch of Physical Optics" *Prog. Opt.* **5**:287–350.

645 Lipson, S. G. (1972) "Optical Transforms in Teaching," in *Optical Transforms* (H. Lipson, Ed.), Academic Press, New York, pp. 349–400.

646 Lipson, S. G. and Lipson, H. (1969) in *Optical Physics*, Cambridge University Press, Cambridge.

647 Longley, W. (1967) "The Crystal Structure of Bovine Liver Catalase: A Combined Study by X-ray Diffraction and Electron Microscopy" *J. Mol. Biol.* **30**:323–327.

648 Longley, W. (1975) "The Double Helix of Tropomyosin" *Nature* (*London*) **253**:126–127.

649 Longley, W. (1977) "A New Crystalline Form of Tropomyosin" *J. Mol. Biol.* **115**:381–387.

650 Lucy, J. A. and Glauert, A. M. (1964) "Structure and Assembly of Macromolecular Lipid Complexes Composed of Globular Micelles" *J. Mol. Biol.* **8**:727–748.

651 Luftig, R. B., Kilham, S. S., Hay, A. J., Zweerink, H. J., and Joklik, K. (1972) "An Ultrastructural Study of Virions and Cores of Reovirus Type 3" *Virology* **48**:170–181.

652 Luft, J. H. (1973) "Embedding Media—Old and New" in *Advanced Techniques in Biological Electron Microscopy* (J. K. Koehler, Ed.), Springer-Verlag, New York, pp. 1–34.

653 Lund, B. M., Gee, J. M., King, N. R., Horne, R. W., and Harnden, J. M. (1978) "The Structure of the Exosporium of a Pigmented *Clostridium*" *J. Gen. Microbiol.* **105**:165–174.

654 Luther, P. and Squire, J. (1978) "Three-Dimensional Structure of the Vertebrate Muscle M-Region" *J. Mol. Biol.* **125**:313–324.

655 Mackall, J. C., Lane, M. D., Leonard, K. R., Pendergast, M., and Kleinschmidt, A. K. (1978) "Subunit Size and Paracrystal Structure of Avian Liver Acetyl-CoA Carboxylase" *J. Mol. Biol.* **123**:595–606.

656 MacLeod, R., Hills, G. J., and Markham, R. (1963) "Formation of True Three-Dimensional Crystals of thc Tobacco Mosaic Virus Protein" *Nature* (*London*) **200**:932–934.

657 Maeda, V. (1979) "X-ray Diffraction Patterns from Molecular Arrangements with 38 nm Periodicities Around Muscle Thin Filaments" *Nature (London)* **277:**670–672.

658 Malech, H. L. and Alpert, J. P. (1979) "Negative Staining of Protein Macromolecules: A Simple Rapid Method" *J. Ultrastruct. Res.* **69:**191–195.

659 Malsberger, R. G. and Cerini, C. P. (1965) "Morphology of Infectious Pancreatic Necrosis Virus" *Ann. Trans. N. Y. Acad. Sci.* **126:**315–319.

660 Mandelkow, E., Thomas, J., and Cohen, C. (1977) "Microtubule Structure at Low Resolution by X-ray Diffraction" *Proc. Natl. Acad. Sci. U.S.A.* **74:**3370–3374.

661 Mandelkow, E. M. and Mandelkow, E. (1978) "Image Reconstruction of Tubulin Hoops (Abstract)" *Proc. Ninth Int. Congr. Electron Microsc. (Toronto)* **2:**695.

662 Mandelkow, E. M. and Mandelkow, E. (1979) "Junctions Between Microtubule Walls" *J. Mol. Biol.* **129:**135–148.

663 Mandelkow, E. M., Mandelkow, E., and Schultheiss, R. (1979) "Correlation Between Structural Polarity and Polar Assembly of Brain Tubulin" *J. Mol. Biol.* **135:**293–299.

664 Mandelkow, E. M., Mandelkow, E., Unwin, P. N. T., and Cohen, C. (1977) "Tubulin Hoops" *Nature (London)* **265:**655–657.

665 Maniloff, J. (1970) "Structure of Ribosome Helices of *Mycoplasma gallisepticum*" *Proc. Seventh Int. Congr. Electron Microsc. (Grenoble)* **3:**71–72.

666 Maniloff, J. (1971) "Analysis of the Helical Ribosome Structures of *Mycoplasma gallisepticum*" *Proc. Natl. Acad. Sci. U.S.A.* **68:**43–47.

667 Maniloff, J., Vanderkooi, G., Hayashi, H., and Capaldi, R. A. (1973) "Optical Analysis of Electron Micrographs of Cytochrome Oxidase Membranes" *Biochim. Biophys. Acta* **298:**180–183.

668 Margaritis, L. H., Elgsaeter, A., and Branton, D. (1977) "Rotary Replication for Freeze-Etching" *J. Cell Biol.* **72:**47–56.

669 Margaritis, L. H., Miller, K. R., and Branton, D. (1976) "Application of Rotary Shadowing to Freeze Etching" *Sixth Eur. Reg. Conf. Electron Microsc. (Jerusalem)* **2:**128–130.

670 Markham, R. (1968) "The Optical Diffractometer" *Methods Virol.* **4:**503–529.

671 Markham, R., Frey, S., and Hills, G. J. (1963) "Methods for the Enhancement of Image Detail and Accentuation of Structure in Electron Microscopy" *Virology* **20:**88–102.

672 Markham, R., Hitchborn, J. H., Hills, G. J., and Frey, S. (1964) "The Anatomy of the Tobacco Mosaic Virus" *Virology* **22:**342–359.

673 Marston, S. B., Rodger, C. D., and Tregear, R. T. (1976) "Changes in Muscle Crossbridges When β,γ–Imido-ATP Binds to Myosin" *J. Mol. Biol.* **104:**263–276.

674 Massover, W. H. (1971) "Intramitochondrial Yolk-Crystals of Frog Oocytes. II. Expulsion of Intramitochondrial Yolk-Crystals to Form Single-Membrane-Bound Hexagonal Crystalloids" *J. Ultrastruct. Res.* **36:**603–620.

675 Massover, W. H. (1975) "Borates as Agents for Negative Staining" *Electron Microsc. Soc. Am. Proc.* **33:**630–631.

676 Matricardi, V. R., Hausner, G. G., Jr., and Parsons, D. F. (1971) "Microscopy of Hydrated Biological Specimens with JEM 200 Stage (Mark II)" *Electron Microsc. Soc. Am. Proc.* **29:**468–469.

677 Matricardi, V. R., Moretz, R. C., and Parsons, D. F. (1972) "Electron Diffraction of Wet Proteins: Catalase" *Science* **177:**268–270.

678 Mattern, C. F. T., Takemoto, K. K., and DeLeva, A. M. (1967) "Electron Microscopic Observations on Multiple Polyoma Virus-Related Particles" *Virology* **32:**378–392.

679 Matthews, B. W. and Bernhard, S. A. (1973) "Structure and Symmetry of Oligomeric Enzymes" *Annu. Rev. Biophys. Bioeng.* **2:**257–317.

680 Matthews, B. W., Fenna, R. E., and Remington, S. J. (1977) "An Evaluation of Electron Micrographs of Bacteriochlorophyll **a**–Protein Crystals in Terms of the Structure Determined by X-Ray Crystallography" *J. Ultrastruct. Res.* **58**:316–330.

681 Maul, G. G. (1970) "Ultrastructure of Pore Complexes of Annulate Lamellae" *J. Cell Biol.* **46**:604–610.

682 Maul, G. G. (1971) "On the Octagonality of the Nuclear Pore Complex" *J. Cell Biol.* **51**:558–563.

683 Maul, G. G. (1971) "Structure and Formation of Pores in Fenestrated Capillaries" *J. Ultrastruct. Res.* **36**:768–782.

684 Mazza, A. and Felluga, B. (1973) "Electron Microscope Observations on λ Bacteriophage Ultrastructure" *J. Ultrastruct. Res.* **45**:259–278.

685 McDonald, J. G. and Bancroft, J. B. (1977) "Assembly Studies on Potato Virus Y and Its Coat Protein" *J. Gen. Virol.* **35**:251–263.

686 McDonald, J. G., Beveridge, T. J., and Bancroft, J. B. (1976) "Self-Assembly of Protein from a Flexuous Virus" *Virology* **69**:327–331.

687 McDonald, K. L., Edwards, M. K., and McIntosh, J. R. (1979) "Cross-Sectional Structure of the Central Mitotic Spindle of *Diatoma vulgare:* Evidence for Specific Interactions Between Antiparallel Microtubules" *J. Cell Biol.* **83**:443–461.

688 McDonnel, A. and Staehelin, L. A. (1980) "Adhesion Between Liposomes Mediated by the Chlorophyll *a/b* Light-Harvesting Complex Isolated from Chloroplast Membranes" *J. Cell Biol.* **84**:40–56.

689 McIntosh, J. R. (1973) "The Axostyle of *Saccinobaculus* II. Motion of the Microtubule Bundle and a Structural Comparison of Straight and Bent Axostyles" *J. Cell Biol.* **56**:324–339.

690 McIntosh, J. R. (1974) "Bridges Between Microtubules" *J. Cell Biol.* **61**:166–187.

691 McLachlan, D. Jr. (1958) "Crystal Structure and Information Theory" *Proc. Natl. Acad. Sci. U.S.A.* **44**:948–956.

692 McPherson, A. Jr. (1976) "The Analysis of Biological Structure with X-ray Diffraction Techniques," in *Principles and Techniques in Electron Microscopy*, Vol. 6 (M. A. Hayat, Ed.), Van Nostrand Reinhold, New York, pp. 117–240.

693 McPherson, A. Jr. and Rich, A. (1973) "Preliminary Study of *B. subtilis* α-Amylase Crystals by Electron Microscopy and Optical Diffraction" *J. Ultrastruct. Res.* **44**:75–84.

694 McPherson, A. Jr. and Rich, A. (1973) "X-ray Crystallographic Study of the Quaternary Structure of Canavalin" *J. Biochem.* **74**:155–160.

695 Meek, G. A. (1970) In *Practical Electron Microscopy for Biologists*, Wiley-Interscience, London.

696 Mellema, J. E. (1975) "Model for the Capsid Structure of Alfalfa Mosaic Virus" *J. Mol. Biol.* **94**:643–648.

697 Mellema, J. E. and Amos, L. A. (1972) "Three-Dimensional Image Reconstruction of Turnip Yellow Mosaic Virus" *J. Mol. Biol.* **72**:819–822.

698 Mellema, J. E. and Klug, A. (1972) "Quaternary Structure of Gastropod Haemocyanin" *Nature (London)* **239**:146–150.

699 Mellema, J. E. and Schepman, A. M. H. (1976) "Digital Image Analysis in Structural Biology" *Sixth Eur. Reg. Conf. Electron Microsc. (Jerusalem)* **1**:4–7.

700 Mellema, J. E., van Bruggen, E. F. J., and Gruber, M. (1967) "Uranyl Oxalate as a Negative Stain for Electron Microscopy of Proteins" *Biochim. Biophys. Acta* **140**:180–182.

701 Mellema, J. E. and van den Berg, H. J. N. (1974) "The Quaternary Structure of Alfalfa Mosaic Virus" *J. Supramol. Struct.* **2**:17–31.

702 Mendelsohn, M. L., Mayall, B. H., Prewitt, J. M. S., Bostrom, R. C., and Holcomb, W. G. (1968) "Digital Transformation and Computer Analysis of Microscopic Images" *Adv. Opt. Electron Microsc.* **2**:77–150.

703 Meyers, T. R. and Hirai, K. (1980) "Morphology of a Reo-Like Virus Isolated from Juvenile American Oysters (*Crassostrea virginica*)" *J. Gen. Virol.* **46**:249–253.

704 Michel, H., Oesterhelt, D., and Henderson, R. (1980) "Orthorhombic Two-Dimensional Crystal Form of Purple Membrane" *Proc. Natl. Acad. Sci. U.S.A.* **77**:338–342.

705 Mikhailov, A. M. (1971) "Determining the Helical-Structure Parameters of the Outer Sheath of Phages DD6 and T2 by the Optical-Diffraction Method" *Sov. Phys. Cryst.* **15**:701–702.

706 Mikhailov, A. M., Andriashvili, I. A., Petrovskii, G. V., and Kaftanova, A. S. (1978) "Structure of Bacteriophage FI-5 in Intact State" *Dokl. Biophys.* **239**:54–57.

707 Mikhailov, A. M. and Belyaeva, N. N. (1971) "Determination of the Helical Parameters of Biological Objects" *Mol. Biol.* **5**:253–255.

708 Mikhailov, A. M., Belyaeva, N. N., Zograf, O. N., and Vainshtein, B. K. (1975) "Electron Microscope Determination of the Three-Dimensional Structure of the Elongated Tail of the Defective Bacteriophage ϕ *Bacillus subtilis*" *Sov. Phys. Cryst.* **20**:41–44.

709 Mikhailov, A. M., Belyaeva, N. N., Zograf, O. N., and Tikhonenko, A. S. (1978) "Structure of Tail of Temperate Phage No. 1 of *Bacillus megaterium*" *Dokl. Biophys.* **242**:151–154.

710 Mikhailov, A. M., Shomodi, P., Petrovskii, G. V., Grigor'ev, V. B., Kaftanova, A. S., and Vainshtein, B. K. (1976) "Electron-Microscopic Investigation of the Three-Dimensional Structure of the Tail of the *Butyricum* mycobacteriophage" *Dokl. Biophys.* **231**:164–167.

711 Mikhailov, A. M. and Vainshtein, B. K. (1971) "Electron Microscope Determination of the Three-Dimensional Structure of the Extended Tail of the T6 Bacteriophage" *Sov. Phys. Cryst.* **16**:428–436.

712 Mikhailov, A. M. and Vainshtein, B. K. (1972) "Three-Dimensional Structure of Tails of *Escherichia coli* B Phages T2 and DD6" *Dokl. Biophys.* **203**:36–39.

713 Miller, A. (1968) "A Short Periodicity in the Thick Filaments of the Anterior Byssus Retractor Muscle of *Mytilus edulis*" *J. Mol. Biol.* **32**:687–688.

714 Miller, K. R. (1979) "Structure of a Bacterial Photosynthetic Membrane" *Proc. Natl. Acad. Sci. U.S.A.* **76**:6415–6419.

715 Millman, B. M. and Bennett, P. M. (1976) "Structure of the Cross-Striated Adductor Muscle of the Scallop" *J. Mol. Biol.* **103**:439–467.

716 Millonig, G. and Marinozzi, V. (1968) "Fixation and Embedding in Electron Microscopy" *Adv. Opt. Electron Microsc.* **2**:251–341.

717 Millward, G. R. (1970) "The Substructure of alpha-Keratin Microfibrils" *J. Ultrastruct. Res.* **31**:349–355.

718 Millman, G., Uzman, B. G., Mitchell, A., and Langridge, R. (1966) "Pentagonal Aggregation of Virus Particles" *Science* **152**:1381–1383.

719 Misell, D. L. (1973) "The Resolution and Contrast in Biological Sections Determined by Inelastic and Elastic Scattering" *Electron Microsc. Soc. Am. Proc.* **31**:224–225.

720 Misell, D. L. (1978) "Image Analysis, Enhancement and Interpretation" *Practical Methods in Electron Microscopy*, Vol. 7 (A. M. Glauert, Ed.), North Holland, Amsterdam, pp. 1–305.

721 Misell, D. L. (1978) "The Phase Problem in Electron Microscopy" *Adv. Opt. Electron Microsc.* **7**:185–279.

722 Mitsova, I. Z. (1976) "Investigation of the Structure of Nitrogenase by Electron Microscopy and Optical Diffraction" *Mol. Biol.* **10**:456–461.

723 Mitzukami, I. and Gall, J. (1966) "Centriole Replication. II. Sperm Formation in the Fern, *Marsilea*, and the Cycad, *Zamia*" *J. Cell Biol.* **29**:97–111.

724 Mollenstedt, G., Speidel, R., Hoppe, W., Langer, R., Katerbau, K. H., and Thon, F. (1968) "Electron Microscopical Imaging Using Zonal Correction Plates" *Fourth Eur. Reg. Conf. Electron Microsc.* (*Rome*) **1**:125–126.

725 Moody, M. F. (1967) "Structure of the Sheath of Bacteriophage T4. I. Structure of the Contracted Sheath and Polysheath (Appendix: The Interpretation of Optical Diffraction Patterns of Electron Micrographs of Negatively Stained Helical Structures) *J. Mol. Biol.* **25**:167–200.

726 Moody, M. F. (1967) "Structure of the Sheath of Bacteriophage T4. II. Rearrangement of the Sheath Subunits During Contraction" *J. Mol. Biol.* **25**:201–208.

727 Moody, M. F. (1968) "Eliminating Spherical Aberration and Preserving Phase Contrast Information in Electron Microscopy" *Nature* (*London*) **218**:263–265.

728 Moody, M. F. (1971) "Optical Diffraction of Electron Micrographs of Argininosuccinase Tubules" *Biochim. Biophys. Acta* **229**:779–784.

729 Moody, M. F. (1971) "Application of Optical Diffraction to Helical Structures in the Bacteriophage Tail" *Phil. Trans. R. Soc. London, B* **261**:181–195.

730 Moody, M. F. (1971) "Structure of the T2 Bacteriophage Tail-Core and Its Relation to the Assembly and Contraction of the Sheath" *First Eur. Biophys. Congr.* (*Baden*) **1**:543–546.

731 Moody, M. F. (1973) "Sheath of Bacteriophage T4. III. Contraction Mechanism Deduced from Partially Contracted Sheaths" *J. Mol. Biol.* **80**:613–635.

732 Moore, P. B., Huxley, H. E., and DeRosier, D. J. (1970) "Three-Dimensional Reconstruction of F-Actin, Thin Filaments and Decorated Thin Filaments" *J. Mol. Biol.* **50**:279–295.

733 Moos, C., Offer, G., Starr, R., and Bennett, P. (1975) "Interaction of C-Protein with Myosin, Myosin Rod and Light Meromyosin" *J. Mol. Biol.* **97**:1–9.

734 Moretz, R. C. and Parsons, D. F. (1969) "Modification of the Electron Microscope for Investigation of Fully Hydrated Biological Specimens" *Electron Microsc. Soc. Am. Proc.* **27**:418–419.

735 Morgan, R. S. (1968) "Structure of Ribosomes of Chromatid Bodies: Three-Dimensional Fourier Synthesis at Low Resolution" *Science* **162**:670–671.

736 Morgan, R. S. and Uzman, B. G. (1966) "Nature of the Packing of Ribosomes Within Chromatid Bodies" *Science* **152**:214–216.

737 Mrena, E. and Dostal, J. (1964) "Interpretation of Symmetrical Structures Evolved by the Rotation Technique in Electron Microscopy of Viruses" *Third Eur. Reg. Conf. Electron Microsc.* (*Prague*) **1**:397–398.

738 Mueller, M., Koller, T., and Moor, H. (1970) "Preparation and Use of Aluminum Films for High Resolution Electron Microscopy of Macromolecules" *Proc. Seventh Int. Congr. Electron Microsc.* (*Grenoble*) **1**:633–634.

739 Mukhopadhyay, U. and Taylor, C. A. (1971) "The Application of Optical Transform Methods to the Interpretation of X-Ray Fibre Photographs" *J. Appl. Crystallogr.* **4**:20–30.

740 Mulvey, T. (1973) "Instrumental Aspects of Image Analysis in the Electron Microscope" *J. Microsc.* **98**:232–250.

741 Munn, E. A. (1970) "Fine Structure of Basal Bodies (Kinetosomes) and Associated Components of Tetrahymena" *Tissue and Cell* **2**:499–512.

742 Murakami, S., Fukushima, K., and Fukami, A. (1978) "Dynamic Observation of Hydrated Biological Structures Using Film-Sealed Environmental Cell Method" *Proc. Ninth Int. Congr. Electron Microsc.* (*Toronto*) **2**:54–55.

743 Murant, A. F. (1965) "The Morphology of Cucumber Mosaic Virus" *Virology* **26**:538–544.

744 Murata, Y. (1978) "Studies of Radiation Damage Mechanism—By Optical Diffraction Analysis and High Resolution Image" *Proc. Ninth Int. Congr. Electron Microsc.* (*Toronto*) **3**:49–60.

745 Murata, Y., Fryer, J. R., Baird, T., and Murata, H. (1977) "Radiation Damage of an Organic Crystal" *Acta Crystallogr.* **A33**:198–200.

746 Murray, R. T. (1973) "The Possibilities for Electron Diffraction in Biology" *J. Microsc.* **98**:345–351.

747 Murray, R. T. and Ferrier, R. P. (1968) "Biological Applications of Electron Diffraction" *J. Ultrastruct. Res.* **21**:361–377.

748 Nagai, R. and Hayama, T. (1979) "Ultrastructure of the Endoplasmic Factor Responsible for Cytoplasmic Streaming in *Chara* Internodal Cells" *J. Cell Sci.* **36**:121–136.

749 Nagano, T. and Ohtsuki, I. (1971) "Reinvestigation of the Fine Structure of Reinke's Crystal in the Human Testicular Interstitial Cell" *J. Cell Biol.* **51**:148–161.

750 Nagy, I. Z., Salanki, J., and Garamvolgyi, N. (1971) "The Contractile Apparatus of the Adductor Muscles in *Anodonta cygnea* L. (Mollusca, Pelecypoda)" *J. Ultrastruct. Res.* **37**:1–16.

751 Nathan, R. (1970) "Computer Enhancement of Electron Micrographs" *Electron Microsc. Soc. Am. Proc.* **28**:28–29.

752 Nathan, R. (1971) "Image Processing for Electron Microscopy: I. Enhancement Procedures" *Adv. Opt. Electron Microsc.* **4**:85–125.

753 Nathan, R. (1974) "Biological Atomic Resolution via Synthetic Aperture" *Proc. Eighth Int. Congr. Electron Microsc.* (*Canberra*) **1**:306–307.

754 Nermut, M. V. (1975) "Fine Structure of Adenovirus Type 5. I. Virus Capsid" *Virology* **65**:480–495.

755 Nermut, M. V. and Williams, L. (1976) "Freeze-Fracturing of Monolayers of Cells, Membranes and Viruses" *Sixth Eur. Reg. Conf. Electron Microsc.* (*Jerusalem*) **2**:131–133.

756 Neugebauer, D. C., Oesterhelt, D., and Zingsheim, H. P. (1978) "The Two Faces of the Purple Membrane. II. Differences in Surface Charge Properties Revealed by Ferritin Binding" *J. Mol. Biol.* **125**:123–135.

757 Neugebauer, D. C. and Zingsheim, H. P. (1978) "The Two Faces of the Purple Membrane: Structural Differences Revealed by Metal Decoration" *J. Mol. Biol.* **123**:235–246.

758 Neville, A. C., Parry, D. A. D., and Woodhead-Galloway, J. (1976) "The Chitin Crystallite in Arthropod Cuticle" *J. Cell Sci.* **21**:73–82.

759 Neville, D. M., Jr. and Davies, D. R. (1966) "The Interaction of Acridine Dyes with DNA: An X-ray Diffraction and Optical Investigation" *J. Mol. Biol.* **17**:57–74.

760 Nonomura, Y. (1974) "Fine Structure of the Thick Filament in Molluscan Catch Muscle" *J. Mol. Biol.* **88**:445–455.

761 Nonomura, Y. and Kohama, K. (1974) "Determination of the Absolute Hand of the Helix in the Nucleocapsid of Haemagglutinating Virus (Japan)" *J. Mol. Biol.* **86**:621–626.

762 Norman, R. S. (1966) "Rotation Technique in Radially Symmetric Electron Micrographs: Mathematical Analysis" *Science* **152**:1238–1239.

763 O'Brien, E. J. and Bennett, P. M. (1972) "Structure of Straight Flagella from a Mutant *Salmonella*" *J. Mol. Biol.* **70**:133–152.

764 O'Brien, E. J. and Couch, J. (1976) "Optical and Fourier Analysis of Actin + Tropomyosin Paracrystals" *Sixth Eur. Reg. Conf. Electron Microsc.* (*Jerusalem*) **2**:153–155.

765 O'Brien, E. J., Gillis, J. M., and Couch, J. (1975) "Symmetry and Molecular Arrangement in Paracrystals of Reconstituted Muscle Thin Filaments" *J. Mol. Biol.* **99**:461–475.

766 Oda, T. and Takanami, M. (1972) "Observations on the Structure of the Termination Factor rho and its Attachment to DNA" *J. Mol. Biol.* **71**:799–802.

767 Ohlendorf, D. H., Collins, M. L., and Banaszak, L. J. (1975) "Analysis of Optical Diffraction Patterns from Electron Micrographs of Lattices" *J. Mol. Biol.* **99**:143–151.

768 Ohlendorf, D. H., Collins, M. L., Puronen, E. O., Banaszak, L. J., and Harrison, S. C. (1975) "Crystalline Lipoprotein-Phosphoprotein Complex in Oocytes from *Xenopus laevis:* Determination of Lattice Parameters by X-Ray Crystallography and Electron Microscopy" *J. Mol. Biol.* **99**:153–165.

769 Ohlendorf, D. H., Wrenn, R. F., and Banaszak, L. J. (1978) "Three-Dimensional Structure of the Lipovitellin-phosvitin Complex from Amphibian Oocytes" *Nature* (*London*) **272**:28–32.

770 Ohtsuki, M., Edelstein, C., Sogard, M., and Scanu, A. M. (1977) "Electron Microscopy of Negatively Stained and Freeze-Etched High Density Lipoprotein-3 from Human Serum" *Proc. Natl. Acad. Sci. U.S.A.* **74**:5001–5005.

771 Ohtsuki, M., Isaacson, M. S., and Crewe, A. V. (1979) "Dark Field Imaging of Biological Macromolecules with the Scanning Transmission Electron Microscope" *Proc. Natl. Acad. Sci. U.S.A.* **76**:1228–1232.

772 Ohtsuki, M., White, S. L., Zeitler, E., Wellems, T. E., Fuller, S. D., Zwick, M., Makinen, M. W., and Sigler, P. B. (1977) "Electron Microscopy of Fibers and Discs of Hemoglobin S having Sixfold Symmetry" *Proc. Natl. Acad. Sci. U.S.A.* **74**:5538–5542.

773 Ohtsuki, M. and Zeitler, E. (1975) "Minimal Beam Exposure with a Field Emission Source" *Ultramicroscopy* **1**:163–165.

774 Oliver, R. M. (1973) "Negative Stain Electron Microscopy of Protein Macromolecules" *Methods Enzymol.* **27**:616–672.

775 Olsen, B. R., Jimenez, S. A., Kivirikko, K. I., and Prockop, D. J. (1970) "Electron Microscopy of Protocollagen Proline Hydroxylase from Chick Embryos" *J. Biol. Chem.* **245**:2649–2655.

776 Olsen, B. R., Torgner, I. A., Christensen, T. B., and Kvamme, E. (1973) "Ultrastructure of Pig Renal Glutaminase. Evidence for Conformational Changes During Polymer Formation" *J. Mol. Biol.* **74**:239–251.

777 Olson, G. E., Lifsics, M., Fawcett, D. W., and Hamilton, D. W. (1977) "Structural Specializations in the Flagellar Plasma Membrane of Opossum Spermatozoa" *J. Ultrastruct. Res.* **59**:207–221.

778 Olson, G. E. and Linck, R. W. (1977) "Observations of the Structural Components of Flagellar Axonemes and Central Pair Microtubules from Rat Sperm" *J. Ultrastruct. Res.* **61**:21–43.

779 Orlov, S. S. (1975) "Theory of Three-Dimensional Reconstruction. I. Conditions for a Complete Set of Projections" *Sov. Phys. Cryst.* **20**:312–314.

780 Orlov, S. S. (1976) "Theory of Three-Dimensional Reconstruction. II. The Recovery Operator" *Sov. Phys. Cryst.* **20**:429–433.

781 Ottensmeyer, F. P. (1969) "Macromolecular Finestructure by Dark Field Electron Microscopy" *Biophys. J.* **9**:1144–1149.

782 Ottensmeyer, F. P., Andrew, J. W., Bazett-Jones, D. P., Chan, A. S. K., and Hewitt, J. (1977) "Signal-to-Noise Enhancement in Dark Field Electron Micrographs of Vasopressin: Filtering of Arrays of Images in Reciprocal Space" *J. Microsc.* **109**:259–268.

783 Ottensmeyer, F. P., Bazett-Jones, D. P., Hewitt, J., and Price, G. (1978) "Structure Analysis of Small Proteins by Electron Microscopy: Valinomycin, Bacitracin and Low Molecular Weight Cell Growth Stimulators (Appendix: Calculation of the Probability of

Observing Valinomycin Structures at Random on a Noisy Support)'' *Ultramicroscopy* **3**:303–313.

784 Ottensmeyer, F. P., Bazett-Jones, D. P., and Korn, A. P. (1978) ''High Resolution Structure Determination of Biological Macromolecules'' *Proc. Ninth Int. Congr. Electron Microsc.* (*Toronto*) **3**:147–159.

785 Ottensmeyer, F. P., Bazett-Jones, D. P., Rust, H. P., Weiss, K., Zemlin, F., and Engel, A. (1978) ''Radiation Exposure and Recognition of Electron Microscopic Images of Protamine at High Resolution'' *Ultramicroscopy* **3**:191–202.

786 Ottensmeyer, F. P. and Pear, M. (1975) ''Contrast in Unstained Sections: A Comparison of Bright and Dark Field Electron Microscopy'' *J. Ultrastruct. Res.* **51**:253–259.

787 Ottensmeyer, F. P., Schmidt, E. E., Jack, T., and Powell, J. (1972) ''Molecular Architecture: The Optical Treatment of Dark Field Electron Micrographs of Atoms'' *J. Ultrastruct. Res.* **40**:546–555.

788 Ottensmeyer, F. P., Schmidt, E. E., and Olbrecht, A. J. (1973) ''Image of a Sulfur Atom'' *Science* **179**:175–176.

789 Ottensmeyer, F. P., Whiting, R. F., Schmidt, E. E., and Clemens, R. S. (1975) ''Electron Microtephroscopy of Proteins: A Close Look at the Ashes of Myokinase and Protamine'' *J. Ultrastruct. Res.* **52**:193–201.

790 Palmer, E. L. and Martin, M. L. (1977) ''The Fine Structure of the Capsid of Reovirus Type 3'' *Virology* **76**:109–113.

791 Papoulis, A. (1962) In *The Fourier Integral and Its Applications*, McGraw-Hill, New York.

792 Papoulis, A. (1968) In *Systems and Transforms with Applications in Optics*, McGraw-Hill, New York.

793 Paradies, H. H. (1979) ''Crystallization of Coupling Factor 1 (CF1) from Spinach Chloroplast'' *Biochem. Biophys. Res. Commun.* **91**:685–692.

794 Parry, D. A. D. and Squire, J. M. (1973) ''Structural Role of Tropomyosin in Muscle Regulation: Analysis of the X-Ray Diffraction Patterns from Relaxed and Contracting Muscles'' *J. Mol. Biol.* **75**:33–55.

795 Parsons, D. F. (1970) ''Problems in High Resolution Electron Microscopy of Biological Materials in their Natural State,'' in *Some Biological Techniques in Electron Microscopy* (D. F. Parsons, Ed.), Academic Press, New York, pp. 1–68.

796 Parsons, D. F. (1972) ''Beam Efficiency, Inelastic Scatter, and Radiation Damage in the High-Voltage Microscope'' *J. Appl. Phys.* **43**:2885–2890.

797 Parsons, D. F. (1972) ''Visualizing the Structure of Macromolecules and Cell Structures in the Natural State'' *Electron Microsc. Soc. Am. Proc.* **30**:176–177.

798 Parsons, D. F. (1973) ''Radiation Damage in Biological Materials'' *Electron Microsc. Soc. Am. Proc.* **31**:482–483.

799 Parsons, D. F. (1973) ''Environmental Wet Cells for Biological Medium Voltage and High Voltage Microscopy'' *Electron Microsc. Soc. Am. Proc.* **31**:14–15.

800 Parsons, D. F. (1974) ''Electron Microscopy and Electron Diffraction of Wet Biological Materials'' *Proc. Eighth Int. Congr. Electron Microsc.* (*Canberra*) **2**:32–33.

801 Parsons, D. F. (1975) ''What Improvement in Electron Microscopy and Electron Diffraction Results from Using Wet Biological Materials?'' *Electron Microsc. Soc. Am. Proc.* **33**:296–297.

802 Parsons, D. F. (1976) ''Biological Application of Electron Microscope Environmental Chambers'' *Sixth Eur. Reg. Conf. Electron Microsc.* (*Jerusalem*) **2**:79–84.

803 Pasquali-Ronchetti, I., Fornieri, C., Mori, G., and Baccarani-Contri, M. (1978) ''Contribution to the Study of Elastin Ultrastructure'' *Proc. Ninth Int. Congr. Electron Microsc.* (*Toronto*) **2**:184–185.

804 Pate, J. L., Johnson, J. L., and Ordal, E. J. (1967) "The Fine Structure of *Chondrococcus columnaris*. II. Structure and Formation of Rhapidosomes" *J. Cell Biol.* **35**:15–35.

805 Paulson, J. R. and Laemmli, U. K. (1977) "Morphogenetic Core of the Bacteriophage T4 Head. Structure of the Core in Polyheads" *J. Mol. Biol.* **111**:459–485.

806 Peachey, L. D., Damsky, C. H., and Veen, A. (1974) "Computer Assisted Three-Dimensional Reconstruction from High Voltage Electron Micrographs of Serial Slices of Biological Material" *Proc. Eighth Int. Congr. Electron Microsc.* (*Canberra*) **1**:330–331.

807 Pease, D. C. (1975) "Micronets for Electron Microscopy" *Micron* **6**:85–92.

808 Pedersen, H. (1970) "Observations on the Axial Filament Complex of the Human Spermatozoon" *J. Ultrastruct. Res.* **33**:451–462.

809 Pepe, F. A. and Dowben, P. (1977) "The Myosin Filament. V. Intermediate Voltage Electron Microscopy and Optical Diffraction Studies of the Substructure" *J. Mol. Biol.* **113**:199–218.

810 Pepe, F. A. and Drucker, B. (1972) "The Myosin Filament. IV. Observation of the Internal Structural Arrangement" *J. Cell Biol.* **52**:255–260.

811 Peracchia, C. (1973) "Low Resistance Junctions in Crayfish. I. Two Arrays of Globules in Junctional Membranes" *J. Cell Biol.* **57**:54–65.

812 Peracchia, C. (1973) "Low Resistance Junctions in Crayfish. II. Structural Details and Further Evidence for Intercellular Channels by Freeze-Fracture and Negative Staining" *J. Cell Biol.* **57**:66–76.

813 Pereira, H. G. and Wrigley, N. G. (1974) "*In Vitro* Reconstitution, Hexon Bonding and Handedness of Incomplete Adenovirus Capsid" *J. Mol. Biol.* **85**:617–631.

814 Perkins, F. O. (1970) "Formation of Centriole-Like Structures During Meiosis and Mitosis in *Labyrinthula* Sp. (Rhizopodea, Labyrinthulida): An Electron Microscope Study" *J. Cell Sci.* **6**:629–653.

815 Perkins, W. J., Barrett, A. N., and Wrigley, N. G. (1977) "Computer Modelling of Protein With Negative Stain and Comparison to X-Ray Analogues" *Micron* **8**:225–227.

816 Perkins, W. J. and Hammond, B. J. (1975) "Computer-Aided Thought in Biomedical Research" *Nature* (*London*) **256**:171–175.

817 Perkins, W. J., Piper, E. A., and Thornton, J. (1975) "Computer Techniques for Conformational Studies of Biological Molecules" *Comput. Biol. Med.* **6**:23–31.

818 Perkins, W. J., Polihroniadis, P., Piper, E. A., and Smart, P. (1976) "A Three-Dimensional Building Procedure for Computer Modelling of Microbiological Structures" *Med. Biol. Eng.* **14**:274–281.

819 Pfister, H. and Burkardt, H. J. (1978) "Protein Composition and Structure of the Sheath and Core of a Defective Bacteriophage from Rhizobium" *J. Ultrastruct. Res.* **64**:159–172.

820 Phillips, D. M. (1966) "Substructure of Flagellar Tubules" *J. Cell Biol.* **31**:635–638.

821 Piez, K. A. and Miller, A. (1974) "The Structure of Collagen Fibrils" *J. Supramol. Struct.* **2**:121–137.

822 Pincus, H. J. (1978) "Optical Diffraction Analysis in Microscopy" *Adv. Opt. Electron. Microsc.* **7**:17–71.

823 Pinnock, P. R. and Taylor, C. A. (1955) "The Determination of the Signs of Structure Factors by Optical Methods" *Acta Crystallogr.* **8**:687–691.

824 Poglazov, B. F., Mazzarelli, M., Kosourov, G. I., and Mesyanzhinov, V. V. (1968) "A Study of Structure of the Tail Sheath of Phage DD-VI by Means of Electron Microscopy and Optical Diffraction Method" *Fourth Eur. Reg. Conf. Electron Microsc.* (*Rome*) **2**:149–150.

825 Poglazov, B. F., Mesyanzhinov, V. V., and Kosourov, G. I. (1967) "A Study of the Self-Assembly of Protein of the Bacteriophage T2 Head" *J. Mol. Biol.* **29**:389–394.

826 Poglazov, B. F., Mesyanzhinov, V. V., Kosourov, G. I., and Bogomolova, T. A. (1971) "Self-Assembly of the Head Protein of T2 Bacteriophage" *Mol. Biol.* **5**:667–672.

827 Pyne, C. K. (1970) "High Resolution Electron Microscopic Studies on the Complex Tubules of the Gymnostome Ciliate *Chilodonella uncinata*" *Proc. Seventh Int. Congr. Electron Microsc.* (*Grenoble*) **3**:401–402.

828 Ramachandran, G. N. and Lakshminarayanan, A. V. (1971) "Three-Dimensional Reconstruction from Radiographs and Electron Micrographs: Application of Convolutions Instead of Fourier Transforms" *Proc. Natl. Acad. Sci. U.S.A.* **68**:2236–2240.

829 Ramamurti, K., Crewe, A. V., and Isaacson, M. S. (1975) "Low Temperature Beam Damage to Biological Specimens in the Electron Microscope" *Electron Microsc. Soc. Am. Proc.* **33**:608–609.

830 Ramamurti, K., Crewe, A. V., and Isaacson, M. S. (1975) "Low Temperature Mass Loss of Thin Films of L-Phenylalanine and L-Tryptophan upon Electron Irradiation—A Preliminary Report" *Ultramicroscopy*, **1**:156–158.

831 Rayns, D. G. (1972) "Myofilaments and Cross Bridges as Demonstrated by Freeze-Fracturing and Etching" *J. Ultrastruct. Res.* **40**:103–121.

832 Rebhun, L. I. (1972) "Freeze-Substitution and Freeze-Drying," in *Principles and Techniques in Electron Microscopy*, Vol. 2 (M. A. Hayat, Ed.), Van Nostrand Reinhold, New York, pp. 1–49.

833 Reedy, M. K. (1968) "Ultrastructure of Insect Flight Muscle. I. Screw Sense and Structural Grouping in the Rigor Cross-Bridge Lattice" *J. Mol. Biol.* **31**:155–176.

834 Reichelt, R., Konig, T., and Wangermann, G. (1977) "Preparation of Microgrids as Specimen Supports for High Resolution Electron Microscopy" *Micron* **8**:29–31.

835 Reid, N. (1975) "Ultramicrotomy," in *Practical Methods in Electron Microscopy*, Vol. 3 (A. M. Glauert, Ed.), North Holland, Amsterdam, pp. 213–353.

836 Reimer, L. (1965) "Irradiation Changes in Organic and Inorganic Objects" *Lab. Invest.* **14**:344–358.

837 Reimer, L. (1973) "Review of the Radiation Damage Problem of Organic Specimens in Electron Microscopy" *Electron Microsc. Soc. Am. Proc.* **31**:476–477.

838 Reimer, L., Roessner, A., Themann, H., and Bassewitz, D. B. V. (1973) "Optical Diffraction on Tilting Series of Paracrystalline Intranuclear Inclusions of Dog Liver Parenchymal Cells" *J. Ultrastruct. Res.* **45**:356–365.

839 Rice, R. V., Moses, J. A., McManus, G. M., Brady, A. C., and Blasik, L. M. (1970) "The Organization of Contractile Filaments in a Mammalian Smooth Muscle" *J. Cell Biol.* **47**:183–196.

840 Rich, A. and Crick, F. H. C. (1961) "The Molecular Structure of Collagen" *J. Mol. Biol.* **3**:483–506.

841 Richards, K. E. and Williams, R. C. (1972) "Electron Microscopy of Aspartate Transcarbamylase and Its Catalytic Subunit" *Biochemistry* **11**:3393–3395.

842 Riemersma, J. C. (1970) "Chemical Effects of Fixation on Biological Specimens," in *Some Biological Techniques in Electron Microscopy* (D. F. Parsons, Ed.), Academic Press, New York, pp. 69–99.

843 Ringo, D. L. (1967) "The Arrangement of Subunits in Flagellar Fibers" *J. Ultrastruct. Res.* **17**:266–277.

844 Roberts, K. (1974) "Crystalline Glycoprotein Cell Walls of Algae: Their Structure, Composition and Assembly" *Phil. Trans. R. Soc. London, B* **268**:129–146.

845 Roberts, K., Gurney-Smith, M., and Hills, G. J. (1972) "Structure, Composition and Morphogenesis of the Cell Wall of *Chlamydomonas reinhardi*. I. Ultrastructure and Preliminary Chemical Analysis" *J. Ultrastruct. Res.* **40**:599–613.

846 Roberts, K. and Hills, G. J. (1976) "The Crystalline Glycoprotein Cell Wall of the Green Alga *Chlorogonium elongatum:* A Structural Analysis" *J. Cell Sci.* **21**:59–71.

847 Roberts, K., Phillips, J. M., and Hills, G. J. (1975) "Structure, Composition and Morphogenesis of the Cell Wall of *Chlamydomonas reinhardi.* VI. The Flagellar Collar" *Micron* **5**:341–357.

848 Robértson, J. D., Knutton, S., Limbrick, A. R., Jakoi, E. R., and Zampighi, G. (1976) "Regular Structures in Unit Membranes. III. Further Observations on the Particulate Component of the Suckling Rat Ileum Endocytic Membrane Complex" *J. Cell Biol.* **70**:112–122.

849 Ross, A. (1968) "The Substructure of Centriole Subfibers" *J. Ultrastruct. Res.* **23**:537–539.

850 Ross, M. J., Klymkowsky, M. W., Agard, D. A., and Stroud, R. M. (1977) "Structural Studies of a Membrane-Bound Acetylcholine Receptor from *Torpedo californica*" *J. Mol. Biol.* **116**:635–659.

851 Rowe, A. J. and Weitzman, P. D. J. (1969) "Allosteric Changes in Citrate Synthase Observed by Electron Microscopy" *J. Mol. Biol.* **43**:345–349.

852 Rowe, R. W. D. (1973) "The Ultrastructure of Z Discs from White, Intermediate, and Red Fibers of Mammalian Striated Muscles" *J. Cell Biol.* **57**:261–277.

853 Rust, H. P. (1974) "Real Time Fourier Analysis by Electronic Means" *Proc. Eighth Int. Congr. Electron Microsc.* (*Canberra*) **1**:92–93.

854 Saffir, A. J., Stroke, G. W., and Halioua, M. (1972) "Extension of the Resolution of the SEM by Holographic Image Enhancement" *Electron Microsc. Soc. Am. Proc.* **30**:374–375.

855 Salazar, L. F., Hutcheson, A. M., Tollin, P., and Wilson, H. R. (1978) "Optical Diffraction Studies of Particles of Potato Virus T" *J. Gen. Virol.* **39**:333–342.

856 Salema, R. and Brandao, I. (1978) "Development of Microtubules in Chloroplasts of Two Halophytes Forced to Follow Crassulacean Acid Metabolism" *J. Ultrastruct. Res.* **62**:132–136.

857 Sale, W. S. and Satir, P. (1976) "Splayed *Tetrahymena* Cilia: A System for Analyzing Sliding and Axonemal Spoke Arrangements" *J. Cell Biol.* **71**:589–605.

858 Salih, S. M. and Cosslett, V. E. (1974) "Some Factors Influencing Radiation Damage in Organic Substances" *Proc. Eighth Int. Congr. Electron Microsc.* (*Canberra*) **2**:670–671.

859 Salih, S. M. and Cosslett, V. E. (1974) "Reduction in Electron Irradiation Damage to Organic Compounds by Conducting Coatings" *Phil. Mag.* **30**:225–228.

860 Salih, S. M. and Cosslett, V. E. (1975) "Radiation Damage in Electron Microscopy of Organic Materials: Effect of Low Temperatures" *J. Microsc.* **105**:269–276.

861 Samsonidze, T. G., Kiselev, N. A., Seitova, T. A., Rusinova, N. G., and Doman, N. G. (1978) "Electron Microscopy of Ribulosebisphosphate Carboxylase" *Dokl. Biophys.* **240**:95–99.

862 Sarma, V. R., Davies, D. R., Labaw, L. W., Silverton, E. W., and Terry, W. D. (1972) "Crystal Structure of an Immunoglobulin Molecule by X-Ray Diffraction and Electron Microscopy" *Cold Spring Harbor Symp. Quant. Biol.* **36**:413–419.

863 Savdir, S. (1963) "A Simple Method of Producing Formvar Films for Supporting Electron Microscope Sections on Grids" *Sci. Tools* **10**:12–13.

864 Saxton, W. O. (1974) "A New Computer Language for Electron Image Processing" Comput. Graph. Image Process. **3**:266–276.

865 Saxton, W. O. (1974) "Computer Techniques for Image Processing in Electron Microscopy" *Proc. Eighth Int. Congr. Electron Microsc.* (*Canberra*) **1**:314–315.

866 Saxton, W. O. (1978) "Computer Techniques for Image Processing in Electron Micros-

copy," in *Advances in Electronics and Electron Physics*, Suppl. 10 (L. Marton and C. Marton, Eds.), Academic Press, New York.

867 Saxton, W. O. and Frank, J. (1977) "Motif Detection in Quantum Noise-Limited Electron Micrographs by Cross-Correlation" *Ultramicroscopy* **2**:219–227.

868 Saxton, W. O., Pitt, T. J., and Horner, M. (1979) "Digital Image Processing: The Semper System" *Optik* **4**:343–354.

869 Sayre, D., Kirz, J., Feder, R., Kim, D. M., and Spiller, E. (1977) "Transmission Microscopy of Unmodified Biological Materials: Comparative Radiation Doses with Electrons and Ultrasoft X-ray Photons" *Ultramicroscopy* **2**:337–349.

870 Scheer, U. and Franke, W. W. (1969) "Negative Staining and Adenosine Triphosphatase Activity of Annulate Lamellae of Newt Oocytes" *J. Cell Biol.* **42**:519–533.

871 Schepman, A. M. H., Schutter, W. G., van Bruggen, E. F. J., Steenis, P. J., and Muller, F. (1976) "Electron Microscope Studies on Microcrystals of Parahydroxybenzoate Hydroxylase from *Pseudomonas fluorescens*" *FEBS Lett.* **65**:84–86.

872 Schepman, A. M. H., van der Voort, J. A. P., Kramer, J., and Mellema, J. E. (1978) "Getting Started with a Scanning Transmission Electron Microscope Coupled to a Small Computer" *Ultramicroscopy* **3**:265–269.

873 Schepman, A. M. H., Wichertjes, T., and van Bruggen, E. F. J. (1972) "Visibility of Subunits in Crystals of Globular Proteins: Electron Microscopy and Optical Diffraction of Edestein and Excelsin" *Biochim. Biophys. Acta* **271**:279–285.

874 Schiske, P., Reuber, E., Tesche, B., Wecke, J., and Giesbrecht, P. (1978) "Reconstruction of the Crystal-Like Particle Arrangement Within a Bacterial Membrane Interlayer" *Proc. Ninth Int. Congr. Electron Microsc. (Toronto)* **2**:352–353.

875 Scraba, D. G., Raska, I., Kellenberger, E., and Moor, H. (1973) "Electron Microscopy of Polyheads of Bacteriophage T4 Prepared by Freeze-Etching" *J. Ultrastruct. Res.* **44**:27–40.

876 Septier, A. (1966) "The Struggle to Overcome Spherical Aberration in Electron Optics" *Adv. Opt. Electron Microsc.* **1**:204–274.

877 Serafini-Fracassini, A., Field, J. M., Spina, M., Garbisa, S., and Stuart, R. J. (1978) "The Morphological Organization and Ultrastructure of Elastin in the Arterial Wall of Trout (*Salmo gairdneri*) and Salmon (*Salmo salar*)" *J. Ultrastruct. Res.* **65**:1–12.

878 Serwer, P. (1977) "Flattening and Shrinkage of Bacteriophage T7 After Preparation for Electron Microscopy by Negative Staining" *J. Ultrastruct. Res.* **58**:235–243.

879 Severs, N. J. and Hicks, R. M. (1977) "Frozen-Surface Replicas of Rat Bladder Luminal Membrane" *J. Microsc.* **111**:125–136.

880 Severs, N. J. and Warren, R. C. (1978) "Analysis of Membrane Structure in the Transitional Epithelium of Rat Urinary Bladder. 1. The Luminal Membrane" *J. Ultrastruct. Res.* **64**:124–140.

881 Sherwood, D. (1976) In *Crystals, X-rays and Proteins*, Wiley, New York.

882 Shirahama, T. and Cohen, A. S. (1967) "High-Resolution Electron Microscopic Analysis of the Amyloid Fibril" *J. Cell Biol.* **33**:679–708.

883 Shirakihara, Y. and Wakabayashi, T. (1979) "Three-Dimensional Image Reconstruction of Straight Flagella from a Mutant *Salmonella typhimurium*" *J. Mol. Biol.* **131**:485–507.

884 Sieber, P. (1974) "High Resolution Electron Microscopy with Heated Apertures and Reconstruction of Single-Sideband Micrographs" *Proc. Eighth Int. Congr. Electron Microsc. (Cranberra)* **1**:274–275.

885 Siegel, A., Hills, G. J., and Markham, R. (1966) "*In Vitro* and *in Vivo* Aggregation of the Defective PM2 Tobacco Mosaic Virus Protein" *J. Mol. Biol.* **19**:140–144.

886 Siegel, G. (1970) "The Influence of Low Temperature on the Radiation Damage of Organic Compounds and Biological Objects by Electron Irradiation" *Proc. Seventh Int. Cong. Electron Microsc. (Grenoble)* **2**:221–222.

887 Silveira, M. (1969) "Ultrastructural Studies on a 'Nine Plus One' Flagellum. 1" *J. Ultrastruct. Res.* **26**:274–288.

888 Simpson, R. W. and Hauser, R. E. (1968) "Basic Structure of Group A Arbovirus Strains Middleburg, Sindbis, and Semiliki Forest Examined by Negative Staining" *Virology* **34**:358–361.

889 Sjöstrand, F. S. (1956) "An Improved Method to Prepare Formvar Nets for Mounting Thin Sections for Electron Microscopy" *First Eur. Reg. Conf. Electron Microsc. (Stockholm)* 120–121.

890 Sjöstrand, F. S. (1967) In *Electron Microscopy of Cells and Tissues*, Vol. 1, *Instrumentation and Techniques*, Academic Press, New York.

891 Sjöstrom, M. and Squire, J. M. (1977) "Fine Structure of the A-Band in Cryosections: The Structure of the A-Band of Human Skeletal Muscle Fibres from Ultrathin Cryosections Negatively Stained" *J. Mol. Biol.* **109**:49–68.

892 Sjöstrom, M. and Squire, J. M. (1977) "Cryo-Ultramicrotomy and Myofibrillar Fine Structure: A Review" *J. Microsc.* **111**:239–278.

893 Slayter, E. M. (1969) "Electron Microscopy of Globular Proteins," in *Physical Principles and Techniques of Protein Chemistry*, Part A (S. J. Leach, Ed.), Academic Press, New York, pp. 1–58.

894 Slayter, E. M. (1970) In *Optical Methods in Biology*, Wiley, New York.

895 Sleytr, U. B. and Messner, P. (1978) "Freeze-Fracturing in Normal Vacuum Reveals Ringlike Yeast Plasmalemma Structures" *J. Cell Biol.* **79**:276–280.

896 Sleytr, U. B. and Robards, A. W. (1977) "Plastic Deformation During Freeze-Cleavage: A Review" *J. Microsc.* **110**:1–25.

897 Small, J. V. (1977) "Studies on Isolated Smooth Muscle Cells: The Contractile Apparatus" *J. Cell Sci.* **24**:327–349.

898 Small, J. V. and Sobieszek, A. (1977) "Studies on the Function and Composition of the 10-nm (100 angstrom) Filaments of Vertebrate Smooth Muscle" *J. Cell Sci.* **23**:243–268.

899 Small, J. V. and Squire, J. M. (1972) "Structural Basis of Contraction in Vertebrate Smooth Muscle" *J. Mol. Biol.* **67**:117–149.

900 Smith, P. R. (1978) "An Integrated Set of Computer Programs for Processing Electron Micrographs of Biological Structures" *Ultramicroscopy* **3**:153–160.

901 Smith, P. R. and Aebi, U. (1974) "Computer-Generated Fourier Transforms of Helical Particles" *J. Phys. A: Math. Nucl. Gen.* **7**:1627–1633.

902 Smith, P. R. and Aebi, U. (1974) "A Study of the Three-Dimensional Structure of the Tail of the *E. coli* Phage T4 from Electron Microscope Data" *Proc. Eighth Int. Congr. Electron Microsc. (Canberra)* **2**:646.

903 Smith, P. R., Aebi, U., Josephs, R., and Kessel, M. (1976) "Studies of the Structure of the T4 Bacteriophage Tail Sheath. I. The Recovery of Three-Dimensional Structural Information from the Extended Sheath (Appendix: The Determination of the Helical Screw Angle of a Helical Particle from Its Diffraction Pattern)" *J. Mol. Biol.* **106**:243–275.

904 Smith, P. R. and Kistler, J. (1977) "Surface Reliefs Computed from Micrographs of Heavy Metal-Shadowed Specimens" *J. Ultrastruct. Res.* **61**:124–133.

905 Smith, P. R., Peters, T. M., and Bates, R. H. T. (1973) "Image Reconstruction from Finite Numbers of Projections" *J. Phys. A: Gen. Phys.* **6**:361–382.

906 Sobieszek, A. (1972) "Cross-Bridges on Self-Assembled Smooth Muscle Myosin Filaments" *J. Mol. Biol.* **70**:741–744.

907 Sobieszek, A. (1973) "The Fine Structure of the Contractile Apparatus of the Anterior Byssus Retractor Muscle of *Mytilus edulis*" *J. Ultrastruct. Res.* **43**:313–343.

908 Sossinka, J. and Hess, B. (1974) "Electron Microscopy of Pyruvate Kinase Crystals from *Saccharomyces carlsbergensis*" *J. Mol. Biol.* **83**:285–287.

909 Sperling, R. and Amos, L. A. (1976) "Arrangement of Subunits in Assembled Histone H4 Fibers" *Sixth Eur. Reg. Conf. Electron Microsc.* (*Jerusalem*) **2**:520–521.

910 Sperling, R. and Amos, L. A. (1977) "Arrangement of Subunits in Assembled Histone H4 Fibers" *Proc. Natl. Acad. Sci. U.S.A.* **74**:3772–3776.

911 Sperling, R., Amos, L. A., and Klug, A. (1975) "A Study of the Pairing Interaction Between Protein Subunits in the Tobacco Mosaic Virus Family by Image Reconstruction from Electron Micrographs" *J. Mol. Biol.* **92**:541–558.

912 Spiess, E. (1978) "The Influence of Concentrating Methods on Electron Microscopical Imaging of Negatively Stained 50S Ribosomal Subunits of *Escherichia coli*" *FEBS Lett.* **91**:289–292.

913 Spitsberg, V. and Haworth, R. (1977) "The Crystallization of Beef Heart Mitochondrial Adenosine Triphosphatase" *Biochim. Biophys Acta* **492**:237–240.

914 Spudich, A. and Spudich, J. A. (1979) "Actin Triton-Treated Cortical Preparations of Unfertilized and Fertilized Sea Urchin Eggs" *J. Cell Biol.* **82**:212–226.

915 Spudich, J. A. and Amos, L. A. (1979) "Structure of Actin Filament Bundles from Microvilli of Sea Urchin Eggs" *J. Mol. Biol.* **129**:319–331.

916 Spudich, J. A., Huxley, H. E., and Finch, J. T. (1972) "Regulation of Skeletal Muscle Contraction. II. Structural Studies of the Interaction of the Tropomyosin-Troponin Complex with Actin" *J. Mol. Biol.* **72**:619–632.

917 Staehelin, L. A., Chlapowski, F. J., and Bonneville, M. A. (1972) "Lumenal Plasma Membrane of the Urinary Bladder. I. Three-Dimensional Reconstruction from Freeze-Etch Images" *J. Cell Biol.* **53**:73–91.

918 Stannard, L. M. and Schoub, B. D. (1977) "Observations on the Morphology of Two Rotaviruses" *J. Gen. Virol.* **37**:435–439.

919 Starling, D. and Burns, R. G. (1975) "Ultrastructure of Tubulin Paracrystals from Sea Urchin Eggs, with Determination of Spacings by Electron and Optical Diffraction" *J. Ultrastruct. Res.* **51**:261–268.

920 Steitz, T. A., Richmond, T. J., Wise, D., and Engelman, D. (1974) "The lac Repressor Protein; Molecular Shape, Subunit Structure, and Proposed Model for Operator Interaction Based on Structural Studies of Microcrystals" *Proc. Natl. Acad. Sci. U.S.A.* **71**:593–597.

921 Stenn, K. and Bahr, G. F. (1970) "Specimen Damage Caused by the Beam of the Transmission Electron Microscope, A Correlative Reconsideration" *J. Ultrastruct. Res.* **31**:526–550.

922 Sternlieb, I. and Berger, J. E. (1969) "Optical Diffraction Studies of Crystalline Structures in Electron Micrographs. II. Crystalline Inclusions in Mitochondria of Human Hepatocytes" *J. Cell Biol.* **43**:448–455.

923 Sternlieb, I., Berger, J. E., Biempica, L., Quintana, N., and Hodge, T. (1971) "Cytoplasmic Crystals in Human Hepatocytes" *Lab. Invest.* **25**:503–508.

924 Steven, A. C., Aebi, U., and Showe, M. K. (1976) "Folding and Capsomere Morphology of the P23 Surface Shell of Bacteriophage T4 Polyheads from Mutants in Five Different Head Genes (Appendix by A. C. Steven)" *J. Mol. Biol.* **102**:373–407.

925 Steven, A. C. and Carrascosa, J. L. (1979) "Proteolytic Cleavage and Structural Transformation: Their Relationship in Bacteriophage T4 Capsid Maturation'' *J. Supramol. Struct.* **10:**1–11.

926 Steven, A. C., Couture, E., Aebi, U., and Showe, M. K. (1976) ''Structure of T4 Polyheads. II. Pathway of Polyhead Transformations as a Model for T4 Capsid Maturation'' *J. Mol. Biol.* **106:**187–221.

927 Steven, A. C., Smith, P. R., and Horne, R. W. (1978) ''Capsid Fine Structure of Cowpea Chlorotic Mottle Virus: From a Computer Analysis of Negatively Stained Virus Arrays'' *J. Ultrastruct. Res.* **64:**63–73.

928 Steven, A. C., ten Heggeler, B., Muller, R., Kistler, J., and Rosenbusch, J. P. (1977) ''Ultrastructure of a Periodic Protein Layer in the Outer Membrane of *Escherichia coli*'' *J. Cell Biol.* **72:**292–301.

929 Stewart, M. (1975) ''The Location of the Troponin Binding Site on Tropomyosin'' *Proc. R. Soc. London, B* **190:**257–266.

930 Stirm, S. and Freund-Molbert, E. (1971) ''*Escherichia coli* Capsule Bacteriophages. II. Morphology'' *J. Virol.* **8:**330–342.

931 Stokes, A. R. (1946) ''The Construction and Use of a 'Fly's Eye' for Assisting X-Ray Structure Analysis'' *Phys. Soc. London Proc.* **58:**306–313.

932 Stokes, A. R. (1955) ''The Production of Fibre Diagrams with the Optical Diffraction Spectrometer'' *Acta Crystallogr.* **8:**27–29.

933 Stolinski, C. (1977) ''Freeze-Fracture Replication in Biological Research: Development, Current Practice and Future Prospects'' *Micron* **8:**87–111.

934 Stoltz, D. B. (1971) ''The Structure of Icosahedral Cytoplasmic Deoxyriboviruses'' *J. Ultrastruct. Res.* **37:**219–239.

935 Stout, G. H. and Jensen, L. H. (1968) In *X-Ray Structure Determination: A Practical Guide*, Collier-Macmillan, London.

936 Stroke, G. W. (1971) ''Sharpening Images by Holography'' *New Sci.* **51:**671–674.

937 Stroke, G. W. and Halioua, M. (1972) ''Attainment of Diffraction-Limited Imaging in High-Resolution Electron Microscopy by 'a Posteriori' Holographic Image Sharpening. I.'' *Optik* **35:**50–65.

938 Stuart, R. D. (1961) In *An Introduction to Fourier Analysis*, Methuen, London.

939 Talianker, M. and Brandon, D. G. (1978) ''Contrast from DNA Occluded in Epitaxially Grown Single-Crystal Gold Films'' *Ultramicroscopy* **3:**255–260.

940 Tamarin, A. and Keller, P. J. (1972) ''An Ultrastructural Study of the Byssal Thread Forming System in Mytilus'' *J. Ultrastruct. Res.* **40:**401–416.

941 Tamm, L. K., Crepeau, R. H., and Edelstein, S. J. (1979) ''Three-Dimensional Reconstruction of Tubulin in Zinc-Induced Sheets'' *J. Mol. Biol.* **130:**473–492.

942 Tanaka, M., Higashi-Fujime, S., and Uyeda, R. (1974) ''Ultra-Fine Grids for Specimen Supporting Media in High Resolution Electron Microscopy'' *Proc. Eighth Int. Congr. Electron Microsc.* (*Canberra*) **2:**180–181.

943 Tandler, B. and Moriber, L. G. (1966) ''Microtubular Structures Associated with the Acrosome During Spermiogenesis in the Water-Strider, *Gerris remigis* (Say)'' *J. Ultrastruct. Res.* **14:**391–404.

944 Taylor, A., Carpenter, F. H., and Wlodawer, A. (1979) ''Leucine Aminopeptidase (Bovine Lens): An Electron Microscopic Study'' *J. Ultrastruct. Res.* **68:**92–100.

945 Taylor, C. A. (1952) ''An Illustration of the Optical Basis of Wilson's X-ray Method for Detecting Centres of Symmetry'' *Acta Crystallogr.* **5:**141–142.

946 Taylor, C. A. (1967) ''Optical Diffraction Techniques for Interpreting X-ray Diffraction Photographs of Fiber Structures'' *J. Polym. Sci. C* **20:**19–36.

947 Taylor, C. A. (1969) "Optical Methods as an Aid in Structure Determination" *Pure Appl. Chem.* **18:**533–550.

948 Taylor, C. A. (1972) "Polymer and Fibre Diffraction," in *Optical Transforms* (H. Lipson, Ed.), Academic Press, New York, pp. 115–151.

949 Taylor, C. A., Hinde, R. M., and Lipson, H. (1951) "Optical Methods in X-ray Analysis. I. The Study of Imperfect Structures" *Acta Crystallogr.* **4:**261–266.

950 Taylor, C. A. and Lipson, H. (1951) "Optical Methods in Crystal-Structure Determination" *Nature (London)* **167:**809–810.

951 Taylor, C. A. and Lipson, H. (1964) In *Optical Transforms: Their Preparation and Application to X-Ray Diffraction Problems*, Cornell University Press, Ithaca, NY.

952 Taylor, C. A. and Ranniko, J. K. (1974) "Problems in the Use of Selective Optical Spatial Filtering to Obtain Enhanced Information from Electron Micrographs" *J. Microsc.* **100:**307–314.

953 Taylor, C. A. and Thompson, B. J. (1957) "Some Improvements in the Operation of the Optical Diffractometer" *J. Sci. Instrum.* **34:**439–447.

954 Taylor, K. A. (1975) "Electron Microscopy of Frozen, Hydrated Biological Specimens" *Electron Microsc. Soc. Am. Proc.* **33:**300–301.

955 Taylor, K. A. (1978) "Structure Determination of Frozen, Hydrated, Crystalline Biological Specimens" *J. Microsc.* **112:**115–125.

956 Taylor, K. A. and Glaeser, R. M. (1973) "A Method for Maintaining Specimen Hydration by Sandwiching Between Thin Films" *Electron Microsc. Soc. Am. Proc.* **31:**342–343.

957 Taylor, K. A. and Glaeser, R. M. (1973) "Hydrophilic Support Films of Controlled Thickness and Composition" *Rev. Sci. Instrum.* **44:**1546–1547.

958 Taylor, K. A. and Glaeser, R. M. (1974) "Electron Diffraction of Frozen, Hydrated Protein Crystals" *Science* **186:**1036–1037.

959 Taylor, K. A. and Glaeser, R. M. (1974) "Hydrated, Crystalline Protein Structure Observed at High Resolution in Frozen, Thin Specimens" *Proc. Eighth Int. Congr. Electron Microsc. (Canberra)* **2:**64–65.

960 Taylor, K. A. and Glaeser, R. M. (1975) "Modified Airlock Door for the Introduction of Frozen Specimens into the JEM 100B Electron Microscope" *Rev. Sci. Instrum.* **46:**985–986.

961 Taylor, K. A. and Glaeser, R. M. (1976) "Electron Microscopy of Frozen Hydrated Biological Specimens" *J. Ultrastruct. Res.* **55:**448–456.

962 Telford, J. N., Lad, P. M., and Hammes, G. G. (1975) "Electron Microscope Study of Native and Crosslinked Rabbit Muscle Phosphofructokinase" *Proc. Natl. Acad. Sci. U.S.A.* **72:**3054–3056.

963 Tesche, B., Tischendorf, G. W., and Stoffler, G. (1978) "Morphology of *Escherichia coli* Ribosomes as Obtained by High Resolution Shadow Cast" *Proc. Ninth Int. Congr. Electron Microsc. (Toronto)* **2:**240–241.

964 Thach, R. E. and Thach, S. S. (1970) "Damage to Biological Samples Caused by the Electron Beam" *Proc. Seventh Int. Congr. Electron Microsc. (Grenoble)* **2:**465–466.

965 Thach, R. E. and Thach, S. S. (1971) "Damage to Biological Samples Caused by the Electron Beam During Electron Microscopy" *Biophys. J.* **11:**204–210.

966 Thompson, B. J. (1972) "Optical Data Processing," in *Optical Transforms* (H. Lipson, Ed.), Academic Press, New York, pp. 267–298.

967 Thompson, B. J. (1972) "Coherence Requirements," in *Optical Transforms* (H. Lipson, Ed.), Academic Press, New York, pp. 27–69.

968 Thon, F. (1971) "Phase Contrast Electron Microscopy," in *Electron Microscopy in Materials Science* (U. Valdre, Ed.), Academic Press, New York, pp. 570–625.

969 Thon, F. (1974) "In-Line and a Posteriori Improvement of High Resolution Electron Microscopical Images" *Proc. Eighth Int. Congr. Electron Microsc.* (*Canberra*) **1**:238–239.

970 Thon, F. (1975) "Processing of High Resolution Images by Analogue Methods" *Electron Microsc. Soc. Am. Proc.* **33**:16–17.

971 Thon, F. and Siegel, B. M. (1970) "Zonal Filtering in Optical Reconstruction of High Resolution Phase Contrast Images" *Proc. Seventh Int. Congr. Electron Microsc.* (*Grenoble*) **1**:13–14.

972 Thon, F. and Siegel, B. M. (1970) "Experiments with Optical Image Reconstruction of High Resolution Electron Micrographs" *Ber. Bunsen-Gesellschaft Phys.* **74**:1116–1120.

973 Thon, F. and Willasch, D. (1971) "High Resolution Electron Microscopy Using Phase Plates" *Electron Microsc. Soc. Am. Proc.* **29**:38–39.

974 Thon, F. and Willasch, D. (1971) "A Calibration Method Usable in the Production of Phase Plates for High Resolution Electron Microscopy" *Electron Microsc. Soc. Am. Proc.* **29**:46–47.

975 Thornell, L. E. (1974) "The Fine Structure of Purkinje Fiber Glycogen. A Comparative Study of Negatively Stained and Cytochemically Stained Particles" *J. Ultrastruct. Res.* **49**:157–166.

976 Tichelaar, W., Oostergetel, G. T., Haker, J., and van Bruggen, E. F. J. (1978) "Scanning Transmission Electron Microscopy of Biological Macromolecules" *Proc. Ninth Int. Congr. Electron Microsc.* (*Toronto*) **2**:30–31.

977 Tochigi, H., Nakatsuka, H., Fukami, A., and Kanaya, K. (1970) "The Improvement of the Image Contrast by Using the Phase Plate in the Transmission Electron Microscope" *Proc. Seventh Int. Congr. Electron Microsc.* (*Grenoble*) **1**:73–74.

978 To, C. M. (1971) "Quaternary Structure of Glutamate Decarboxylase of *Escherichia coli* as Revealed by Electron Microscopy" *J. Mol. Biol.* **59**:215–217.

979 Tokuyasu, K. T. (1974) "Spoke Heads in Sperm Tail of *Drosophila melanogaster*" *J. Cell Biol.* **63**:334–337.

980 Tollin, P., Bancroft, J. B., Richardson, J. F., Payne, N. C., and Beveridge, T. J. (1979) "Diffraction Studies of Papaya Mosaic Virus" *Virology* **98**:108–115.

981 Tooney, N. M. and Cohen, C. (1977) "Crystalline States of a Modified Fibrinogen" *J. Mol. Biol.* **110**:363–385.

982 Troyer, D. and Miller, C. (1974) "Unusual Tubular Structures in the Urinary Bladder of the Crayfish *Astacus fluviatilis*" *J. Ultrastruct. Res.* **46**:296–300.

983 Tschopp, J. and Smith, P. R. (1977) "Extra-Long Bacteriophage T4 Tails Produced Under *in Vitro* Conditions" *J. Mol. Biol.* **114**:281–286.

984 Tsukada, H., Koyama, S., Gotoh, M., and Tadano, H. (1971) "Fine Structure of Crystalloid Nucleoids of Compact Type in Hepatocyte Microbodies of Guinea Pigs, Cats, and Rabbits" *J. Ultrastruct. Res.* **36**:159–175.

985 Tsuprun, V. L., Kiselev, N. A., and Vainshtein, B. K. (1978) "Three-Dimensional Reconstruction of the Enzyme Leucine Aminopeptidase" *Sov. Phys. Cryst.* **23**:417–420.

986 Tsuprun, V. L., Samsonidze, T. G., and Kiselev, N. A. (1978) "Indexing Spiral Objects on Photomicrographs by Comparing the Phases of Symmetrical Reflections in an Optical Diffractometer" *Sov. Phys. Cryst.* **23**:37–41.

987 Tsuprun, V. L., Stel'maschuk, V. Ya., Kiselev, N. A., Gel'fand, V. I., and Rozenblat, V. A. (1976) "Structure of Microtubules" *Mol. Biol.* **10**:360–367.

988 Turner, J. N., Hausner, G. G., Jr., and Parsons, D. F. (1975) "An Optimized Faraday Cage Design for Electron Beam Current Measurements" *J. Phys. E. Sci. Instrum.* **8**:954–957.

989 Typke, D., Hoppe, W., Sessler, W., and Burger, M. (1976) "Conception of a 3-D Imaging Electron Microscope" *Sixth Eur. Reg. Conf. Electron Microsc.* (*Jerusalem*) **1**:334–335.

990 Unwin, P. N. T. (1970) "Interference Microscopy with a Six Lens Electron Microscope" *Proc. Seventh Int. Congr. Elec. Microsc.* (*Grenoble*) **1**:65–66.

991 Unwin, P. N. T. (1970) An Electrostatic Phase Plate for the Electron Microscope" *Ber. Bunsen-Gesellschaft Phys. Chem.* **74**:1137–1141.

992 Unwin, P. N. T. (1971) "Phase Contrast and Interference Microscopy with the Electron Microscope" *Phil. Trans. R. Soc. London, B* **261**:95–104.

993 Unwin, P. N. T. (1971) "Electron Microscope Imaging of Biological Structures Using Bright Phase Contrast" *Micron* **2**:406–410.

994 Unwin, P. N. T. (1972) "Electron Microscopy of Biological Specimens by Means of an Electrostatic Phase Plate" *Proc. R. Soc. London, A* **329**:327–359.

995 Unwin, P. N. T. (1973) "Phase Contrast Electron Microscopy of Biological Materials" *J. Microsc.* **98**:299–312.

996 Unwin, P. N. T. (1974) "Electron Microscopy of the Stacked Disk Aggregate of Tobacco Mosaic Virus Protein. II. The Influence of Electron Irradiation on the Stain Distribution" *J. Mol. Biol.* **87**:657–670.

997 Unwin, P. N. T. (1974) "A New Electron Microscope Imaging Method for Enhancing Detail in Thin Biological Specimens" *Z. Naturforsch. A,* **29**:158–163.

998 Unwin, P. N. T. (1975) "Beef Liver Catalase Structure: Interpretation of Electron Micrographs" *J. Mol. Biol.* **98**:235–242.

999 Unwin, P. N. T. (1977) "Three-Dimensional Model of Membrane-Bound Ribosomes Obtained by Electron Microscopy" *Nature* (*London*) **269**:118–122.

1000 Unwin, P. N. T. (1979) "Attachment of Ribosome Crystals to Intracellular Membranes" *J. Mol. Biol.* **132**:69–84.

1001 Unwin, P. N. T. and Henderson, R. (1975) "High Resolution Structure Determination of Unstained Biological Molecules" *Electron Microsc. Soc. Am. Proc.* **33**:298–299.

1002 Unwin, P. N. T. and Henderson, R. (1975) "Molecular Structure Determination by Electron Microscopy of Unstained Crystalline Specimens" *J. Mol. Biol.* **94**:425–440.

1003 Unwin, P. N. T. and Klug, A. (1974) "Electron Microscopy of the Stacked Disc Aggregate of Tobacco Mosaic Virus Protein. I. Three-Dimensional Image Reconstruction" *J. Mol. Biol.* **87**:641–656.

1004 Unwin, P. N. T. and Taddei, C. (1977) "Packing of Ribosomes in Crystals from the Lizard *Lacerta sicula*" *J. Mol. Biol.* **114**:491–506.

1005 Uyemura, D. G., Brown, S. S., and Spudich, J. A. (1978) "Biochemical and Structural Characterization of Actin from *Dictyostelium discoideum*" *J. Biol. Chem.* **253**:9088–9096.

1006 Vainshtein, B. K. (1971) "Finding the Structure of Objects from Projections" *Sov. Phys. Cryst.* **15**:781–787.

1007 Vainshtein, B. K. (1971) "Synthesis of Projecting Functions" *Sov. Phys. Dokl.* **16**:66–69.

1008 Vainshtein, B. K. (1973) "Three-Dimensional Electron Microscopy of Biological Macromolecules" *Sov. Phys.-Usp.* **16**:185–206.

1009 Vainshtein, B. K. (1978) "Electron Microscopical Analysis of the Three-Dimensional Structure of Biological Macromolecules" *Adv. Opt. Electron Microsc.* **7**:281–377.

1010 Vainshtein, B. K., Barynin, V. V., and Gurskaya, G. V. (1968) "Electron Microscope and Diffraction Methods for the Investigation of the Structure of Protein Crystals and Molecules" *Fourth Europ. Reg. Conf. Electron Microsc.* (*Rome*) **1**:611–612.

1011 Vainshtein, B. K., Barynin, V. V., Gurskaya, G. V., and Nikitin, V. Ya. (1968) "The Crystal Structure of Catalase" *Sov. Phys. Cryst.* **12**:750–757.

1012 Vainshtein, B. K. Barynin, V. V., and Gurskaya, G. V. (1969) "The Hexagonal Crystalline Structure of Catalase and Its Molecular Structure" *Sov. Phys. Dokl.* **13**:838–841.

1013 Vainshtein, B. K., Chistyakov, I. G., and Inozemtseva, A. D. (1972) "Modelling Possibilities for the Study of the Structure of Mesophases" *Sov. Phys. Cryst.* **17**:424–428.

1014 Vainshtein, B. K., Kiselev, N. A., Kaftanova, A. S., Orlova, E. V., Bogdanov, V. P., Morozkin, A. D., and Degtyar, R. G. (1973) "Tubular Crystals of Glucose Oxidase of *Penicillium vitale* and its Quaternary Structure" *Dokl. Biophys.* **213**:133–136.

1015 Vainshtein, B. K., and Kosourov, G. I. (1967) "A Laser as a Source for Optical Fourier Transforms" *Sov. Phys. Cryst.* **11**:778–780.

1016 Vainshtein, B. K. and Mikhailov, A. M. (1972) "Some Properties of the Synthesis of Projecting Functions" *Sov. Phys. Cryst.* **17**:217–222.

1017 Vainshtein, B. K. and Mikhailov, A. M. (1978) "Spatial Structure of the Tails of Some Bacterial Viruses" *Proc. Ninth Int. Congr. Electron Microsc.* (*Toronto*) **2**:384–385.

1018 Vainshtein, B. K., Mikhailov, A. M., Andriashvili, I. A., Kaftanova, A. S., and Petrovskii, G. V. (1977) "Electron Microscopic Investigation of Three-Dimensional Structure of Tail of Bacteriophage phi-5" *Dokl. Biophys.* **234**:61–64.

1019 Vainshtein, B. K., Mikhailov, A. M., and Kaftanova, A. S. (1977) "Structure of the Contracted Tail of Bacteriophage DD6" *Sov. Phys. Cryst.* **22**:163–166.

1020 Vainshtein, B. K., Mikhailov, A. M., Tikhonenko, A. S., and Belyaeva, N. N. (1976) "Three-Dimensional Structure of the Tail of Bacteriophage AR9 of *Bacillus subtilis*" *Dokl. Biophys.* **228**:103–106.

1021 Vainshtein, B. K. and Orlov, S. S. (1972) "Theory of the Recovery of Functions from Their Projections" *Sov. Phys. Cryst.* **17**:213–216.

1022 Vainshtein, B. K., Sherman, M. B., and Barynin, V. V. (1976) "Electron-Microscope Study of a New Crystal Form of Catalase Using the Fourier-Transform Method" *Sov. Phys. Cryst.* **21**:287–291.

1023 Valdre, U. and Horne, R. W. (1975) "A Combined Freeze Chamber and Low Temperature Stage for an Electron Microscope" *J. Microsc.* **103**:305–317.

1024 Valentine, R. C. (1965) "Characteristic Emulsions for Electron Microscopy" *Lab. Invest.* **14**:596–602.

1025 Valentine, R. C. (1966) "The Response of Photographic Emulsions to Electrons" *Adv. Opt. Electron Microsc.* **1**:180–202.

1026 Valentine, R. C. (1968) "Shapes and Symmetries of Biological Molecules" *Fourth Eur. Reg. Conf. Electron Microsc.* (*Rome*) **2**:3–10.

1027 Valentine, R. C. (1969) "Subunit Arrangements in Enzyme Molecules as Shown by Electron Microscopy," in *Symmetry and Function of Biological Systems at the Macromolecular Level* (A. Engstrom and B. Strandberg, Eds.), Wiley, New York, pp. 165–179.

1028 Valentine, R. C. and Horne, R. W. (1962) "An Assessment of Negative Staining Techniques for Revealing Ultrastructure," in *The Interpretation of Ultrastructure*, Symp. Int'l. Soc. Cell Biol. Vol. 1 (R. J. C. Harris, Ed.), Academic Press, New York, pp. 263–278.

1029 Valentine, R. C., Wrigley, N. G., Scrutton, M. C., Irias, J. J., and Utter, M. F. (1966) "Pyruvate Carboxylase. VIII. The Subunit Structure as Examined by Electron Microscopy" *Biochemistry* **5**: 3111–3116.

1030 van Bruggen, E. F. J. (1976) "Relationship Between Structure and Function of *Panulirus interruptus* and *Helix pomatia* Haemocyanin," in *Structure-Function Relationships of Proteins* (R. Markham and R. W. Horne, Eds.), North-Holland, Amsterdam, pp. 27–36.

1031 Vanderkooi, J. M., Ierokomas, A., Nakamura, H., and Martonosi, A. (1977) "Fluorescence Energy Transfer Between Calcium Transport ATPase Molecules in Artificial Membranes" *Biochemistry* **16:**1262–1267.

1032 Varma, A., Gibbs, A. J., Woods, R. D., and Finch, J. T. (1968) "Some Observations on the Structure of the Filamentous Particles of Several Plant Viruses" *J. Gen. Virol.* **2:**107–114.

1033 Vasquez, C. and Tournier, P. (1964) "New Interpretation of the Reovirus Structure" *Virology* **24:**128–130.

1034 Veraga, J., Longley, W., and Robertson, J. D. (1969) "A Hexagonal Arrangement of Subunits in Membrane of Mouse Urinary Bladder" *J. Mol. Biol.* **46:**593–596.

1035 Volpin, D. and Pasquali-Ronchetti, I. (1977) "The Ultrastructure of High-Temperature Coacervates from Elastin" *J. Ultrastruct. Res.* **61:**295–302.

1036 Volpin, D., Pasquali-Ronchetti, I., Urry, D. W., and Gotte, L. (1976) "Banded Fibers in High Temperature Coacervates of Elastin Peptides" *J. Biol. Chem.* **251:**6871–6873.

1037 Volpin, D., Urry, D. W., Cox, B. A., and Gotte, L. (1976) "Optical Diffraction of Tropoelastin and α-Elastin Coacervates" *Biochim. Biophys. Acta* **439:**253–258.

1038 Voter, W. A. and Erickson, H. P. (1974) "An Experimental Study of Deviations from the Linear Transfer Theory" *Electron Microsc. Soc. Am. Proc.* **32:**400–401.

1039 Wakabayashi, T., Huxley, H. E., Amos, L. A., and Klug, A. (1975) "Three-Dimensional Image Reconstruction of Actin-Tropomyosin Complex and Actin-Tropomyosin–Troponin T–Troponin I Complex" *J. Mol. Biol.* **93:**477–497.

1040 Wakabayashi, T., Kubota, M., Yoshida, M., and Kagawa, Y. (1977) "Structure of ATPase (Coupling Factor TF_1 from a Thermophilic Bacterium" *J. Mol. Biol.* **117:**515–519.

1041 Walther-Mauruschat, A. and Meyer, F. (1978) "Isolation and Characterization of Polysheaths, Phage Tail-Like Defective Bacteriophages of *Alcaligenes eutrophus* H16" *J. Gen. Virol.* **41:**239–254.

1042 Ward, P. R. and Mitchell, R. F. (1972) "A Facility for Electron Microscopy of Specimens in Controlled Environments" *J. Phys. E: Sci. Instrum.* **5:**160–162.

1043 Warner, F. D. (1969) "The Fine Structure of the Protonephridia in the Rotifer *Asplanchna*" *J. Ultrastruct. Res.* **29:**499–524.

1044 Warner, F. D. (1970) "New Observations on Flagellar Fine Structure: The Relationship Between Matrix Structure and the Microtubule Component of the Axoneme" *J. Cell Biol.* **47:**159–182.

1045 Warner, F. D. (1976) "Cross-Bridge Mechanisms in Ciliary Motility: The Sliding-Bending Conversion," in *Cell Motility* (R. Goldman, T. Pollard, and J. Rosenbaum, Eds.), Cold Spring Harbor Laboratory, Cold Spring Harbor, NY, pp. 891–914.

1046 Warner, F. D. and Mitchell, D. R. (1978) "Structural Conformation of Ciliary Dynein Arms and the Generation of Sliding Forces in *Tetrahymena* Cilia" *J. Cell Biol.* **76:**261–277.

1047 Warner, F. D., Mitchell, D. R., and Perkins, C. R. (1977) "Structural Conformation of the Ciliary ATPase Dynein" *J. Mol. Biol.* **114:**367–384.

1048 Warner, F. D. & Satir, P. (1974) "The Structural Basis of Ciliary Bend Formation: Radial Spoke Positional Changes Accompanying Microtubule Sliding" *J. Cell Biol.* **63:**35–63.

1049 Warren, R. C. and Hicks, R. M. (1970) "Structure of the Subunits in the Thick Luminal Membrane of Rat Urinary Bladder" *Nature (London)* **227:**280–281.

1050 Warren, R. C. and Hicks, R. M. (1971) "A Simple Method of Linear Integration for Resolving Structure in Periodic Lattices: Application to an Animal Cell Membrane and a Crystalline Inclusion" *J. Ultrastruct. Res.* **36**:861–874.

1051 Warren, R. C. and Hicks, R. M. (1973) "Correlation Between the Substructure of the Luminal Plasma Membrane of the Bladder in Man and Other Mammals" *Micron* **4**:257–267.

1052 Warren, R. C. and Hicks, R. M. (1978) "Chemical Dissection and Negative Staining of the Bladder Luminal Membrane" *J. Ultrastruct. Res.* **64**:327–340.

1053 Watanabe, M., Sasaki, S., and Anasawa, N. (1970) "Image Analysis of Electron Micrographs by Computer" *Electron Microsc. Soc. Am. Proc.* **28**:36–37.

1054 Weisel, J. W., Warren, S. G., and Cohen, C. (1978) "Crystals of Modified Fibrinogen: Size, Shape and Packing of Molecules" *J. Mol. Biol.* **126**:159–183.

1055 Weiss, R., Frey, T. G., Burton, Z., Eiserling, F. A., and Eisenberg, D. (1976) "Models for Helical Cables of Glutamine Synthetase and Refinement of the Models to Electron Microscope Data by Least Squares" *Sixth Eur. Reg. Conf. Electron Microsc.* (*Jerusalem*) **1**:20–23.

1056 Weiss, R. L. (1977) "Site-Specific Membrane Particle Arrays in Magnesium-Depleted *Escherichia coli*" *J. Cell Biol.* **73**:505–519.

1057 Welberry, T. R. (1975) "Short-Range Order in Crystals" *Nature* (*London*) **253**:527–528.

1058 Welberry, T. R. and Galbraith, R. (1973) "A Two-Dimensional Model of Crystal-Growth Disorder" *J. Appl. Crystallogr.* **6**:87–96.

1059 Welberry, T. R. and Galbraith, R. (1975) "The Effect of Non-Linearity on a Two-Dimensional Model of Crystal-Growth Disorder" *J. Appl. Crystallogr.* **8**:636–644.

1060 Welberry, T. R. and Taylor, C. A. (1972) "A Method of Producing Optical Diffraction Masks Representing Static Disorder in Crystals Using Computer-Generated Patterns" *J. Appl. Crystallogr.* **5**:133–134.

1061 Wellems, T. E. and Josephs, R. (1979) "Crystallization of Deoxyhemoglobin S by Fiber Alignment and Fusion" *J. Mol. Biol.* **135**:651–674.

1062 Welles, K. (1975) "Methods for Implementing Computer Image Processing to Obtain Heavy-Light Atom Discrimination" *Electron Microsc. Soc. Am. Proc.* **33**:198–199.

1063 Welles, K., Krakow, W., and Siegel, B. M. (1974) "Image Processing by Computer, Heavy-Light Atom Discrimination, Non-Linear Effects, Image Difference Technique" *Proc. Eighth Int. Congr. Electron Microsc.* (*Canberra*) **1**:320–321.

1064 Welton, T. A. (1970) "Computer Synthesis and Filtration of Electron Micrographs of Simple Molecular Objects" *Electron Microsc. Soc. Am. Proc.* **28**:32–33.

1065 Welton, T. A. (1971) "Computational Correction of Aberrations in Electron Microscopy" *Electron Microsc. Soc. Am. Proc.* **29**:94–95.

1066 Welton, T. A. (1972) "Image Theory and Image Processing" *Electron Microsc. Soc. Am. Proc.* **30**:592–593.

1067 Welton, T. A. (1974) "Practical Picture Processing" *Electron Microsc. Soc. Am. Proc.* **32**:338–339.

1068 Welton, T. A. (1975) "Practical Resolution Enhancement in Bright Field Electron Microscopy by Computer Processing" *Electron Microsc. Soc. Am. Proc.* **33**:196–197.

1069 Welton, T. A., Ball, F. L., and Harris, W. W. (1973) "Wiener Processing of Phase Contrast Electron Micrographs" *Electron Microsc. Soc. Am. Proc.* **31**:270–271.

1070 Welton, T. A. and Harris, W. W. (1974) "Object Reconstruction in High Coherence Electron Microscopy with an Adaptive Wiener Filter" *Proc. Eighth Int. Congr. Electron Microsc.* (*Canberra*) **1**:318–319.

1071 Whittaker, E. J. W. (1954) "The Diffraction of X-rays by a Cylindrical Lattice. I" *Acta Crystallogr.* **7**:827–832.

1072 Whittaker, E. J. W. (1954) "The Diffraction of X-rays by a Cylindrical Lattice. II" *Acta Crystallogr.* **8**:261–265.

1073 Whittaker, E. J. W. (1955) "The Diffraction of X-rays by a Cylindrical Lattice. III" *Acta Crystallogr.* **8**:265–271.

1074 Wichertjes, T., Schepman, A. M. H., Mellema, J. E., and van Bruggen, E. F. J. (1970) "Electron Microscopy of Glutamate Dehydrogenase and Hemocyanin Crystals: Structural Analysis by Optical Diffraction" *Proc. Seventh Int. Congr. Electron Microsc. (Grenoble)* **1**:625–626.

1075 Wieland, O. and Siess, E. (1970) "Interconversion of Phospho- and Dephospho Forms of Pig Heart Pyruvate Dehydrogenase" *Proc. Natl. Acad. Sci. U.S.A.* **65**:947–954.

1076 Wilkins, M. H. F., Stokes, A. R., Seeds, W. E., and Oster, G. (1950) "Tobacco Mosaic Virus Crystals and Three-Dimensional Microscopic Vision" *Nature (London)* **166**:127–129.

1077 Williams, R. C. and Fisher, H. W. (1970) "Electron Microscopy of Tobacco Mosaic Virus Under Conditions of Minimal Beam Exposure" *J. Mol. Biol.* **52**:121–123.

1078 Williams, R. C. and Fisher, H. W. (1970) "Electron Microscopy with Minimal Beam Damage" *Electron Microsc. Soc. Am. Proc.* **28**:304–305.

1079 Williams, R. C. and Glaeser, R. M. (1972) "Ultrathin Carbon Support Films for Electron Microscopy" *Science* **175**:1000–1001.

1080 Williams, R. C., Glaeser, R. M., and Richards, K. E. (1971) "Ultrathin Support Films for Visualization of Biomolecules" *Electron Microsc. Soc. Am. Proc.* **29**:482–483.

1081 Willis, B. T. M. (1957) "An Optical Method of Studying the Diffraction from Imperfect Crystals. I. Modulated Structures" *Proc. R. Soc. London, A* **239**:184–191.

1082 Willison, J. H. M. and Davey, M. R. (1976) "Fraction 1 Protein Crystals in Chloroplasts of Isolated Tobacco Leaf Protoplasts: A Thin-Section and Freeze-Etch Morphological Study" *J. Ultrastruct. Res.* **55**:303–311.

1083 Wilson, H. R. (1966) In *Diffraction of X-rays by Proteins, Nucleic Acids and Viruses*, Edward Arnold, London.

1084 Wilson, H. R., Al-Mukhtar, J., Tollin, P., and Hutcheson, A. M. (1978) "Observations on the Structure of Particles of White Clover Mosaic Virus" *J. Gen. Virol.* **39**:361–364.

1085 Wingfield, P., Arad, T., Leonard, K., and Weiss, H. (1979) "Membrane Crystals of Ubiquinone: Cytochrome *c* Reductase from *Neurospora* Mitochondria" *Nature (London)* **280**:696–697.

1086 Wischnitzer, S. (1970) In *Introduction to Electron Microscopy*, Pergamon Press, New York.

1087 Wischnitzer, S. (1973) "The Electron Microscope," in *Principles and Techniques in Electron Microscopy*, Vol. 3 (M. A. Hayat, Ed.), Van Nostrand Reinhold, New York, pp. 1–51.

1088 Witman, G. B., Plummer, J., and Sander, G. (1978) "*Chlamydomonas* Flagellar Mutants Lacking Radial Spokes and Central Tubules: Structure, Composition and Function of Specific Axonemal Components" *J. Cell Biol.* **76**:729–747.

1089 Woodhead-Galloway, J. and Knight, D. P. (1974) "Some Observations on the Fine Structure of Elastoidin" *Proc. R. Soc. London, B* **195**:355–364.

1090 Woods, J. W., Woodward, M. P., and Roth, T. F. (1978) "Common Features of Coated Vesicles from Dissimilar Tissues: Composition and Structure" *J. Cell Sci.* **30**:87–97.

1091 Wurtz, M., Imber, R., Tosi, M., and Hohn, T. (1978) "Surface Structure of the Bacteriophage λ Head: Location of the gpD" *Proc. Ninth Int. Congr. Electron Microsc. (Toronto)* **2**:386–387.

1092 Wurtz, M., Kistler, J., and Hohn, T. (1976) "Surface Structure of *in Vitro* Assembled Bacteriophage λ Polyheads" *J. Mol. Biol.* **101**:39–56.

1093 Wyckoff, H. W., Bear, R. S., Morgan, R. S., and Carlstrom, D. (1957) "Optical Diffractometer for Facilitation of X-ray Diffraction Studies of Macromolecular Structures" *J. Opt. Soc. Am.* **47**:1061–1069.

1094 Yamaguchi, T., Hayashi, M., Wakabayashi, K., and Higashi-Fujime, S. (1972) "X-ray and Optical Diffraction Studies on the Outer Fibres of Sea-Urchin Sperm Tail Flagella" *Biochim. Biophys. Acta* **257**:30–36.

1095 Yamamoto, K., Yanagida, M., Kawamura, M., Maruyama, K., and Noda, H. (1975) "A Study on the Structure of the Paracrystals of F-Actin" *J. Mol. Biol.* **91**:463–469.

1096 Yanagida, M. (1977) 'Molecular Organization of the Shell of T-Even Bacteriophage Head. II. Arrangement of Subunits in the Head Shells of Giant Phages" *J. Mol. Biol.* **109**:515–537.

1097 Yanagida, M., Boy de la Tour, E., Alff-Steinberger, C., and Kellenberger, E. (1970) "Studies on the Morphopoiesis of the Head of Bacteriophage T-Even. VIII. Multilayered Polyheads" *J. Mol. Biol.* **50**:35–58.

1098 Yanagida, M., DeRosier, D. J., and Klug, A. (1972) "Structure of the Tubular Variants of the Head of Bacteriophage T4 (Polyheads). II. Structural Transition from a Hexamer to a 6+1 Morphological Unit" *J. Mol. Biol.* **65**:489–499.

1099 Yu, L. C., Lymn, R. W., and Podolsky, R. J. (1977) "Characterization of a Non-Indexible Equatorial X-ray Reflection from Frog Sartorius Muscle" *J. Mol. Biol.* **115**:455–464.

1100 Zaitsev, V. N., Vainshtein, B. K., and Kosourov, G. I. (1969) "Optical Fourier Transforms and Holography" *Sov. Phys. Cryst.* **13**:507–510.

1101 Zampighi, G. and Robertson, J. D. (1973) "Fine Structure of the Synaptic Discs Separated from the Goldfish Medulla Oblongata" *J. Cell Biol.* **56**:92–105.

1102 Zampighi, G. and Unwin, P. N. T. (1979) "Two Forms of Isolated Gap Junctions" *J. Mol. Biol.* **135**:451–464.

1103 Zeitler, E. (1973) "Reconstruction of Objects from Their Projections Using Orthogonal Functions" *Electron Microsc. Soc. Am. Proc.* **31**:268–269.

1104 Zeitler, E. (1978) "Electron Matter Interaction" *Proc. Ninth Int. Congr. Electron Microsc. (Toronto)* **3**:29–39.

1105 Zemlin, F. (1978) "Image Synthesis from Electron Micrographs Taken at Different Defocus" *Ultramicroscopy* **3**:261–263.

1106 Zimmer, B., Paradies, H. H., and Werz, G. (1977) "Electron Microscopic Studies on Microcrystals of D-Ribulose-1,5-Biphosphate Carboxylase from *Dasycladus clavaeformis* ROTH (Ag.)" *Biochem. Biophys. Res. Commun.* **74**:1496–1500.

1107 Zwick, M. and Zeitler, E. (1973) "Image Reconstruction from Projections" *Optik* **38**:550–565.

The following is a condensed list of pre-1980 image processing articles inadvertently omitted from the text and tables:

Acta Virol. **8**:373 (1964).

Chemica Scripta **14**:245 (1979).

Doklady Biophys. **247**:157 (1979).

Elec. Microsc. Soc. Am. Proc. **27**:260,322 (1969).

Europ. J. Cell Biol. **17**:1 (1978); **19**:120 (1979).

Fifth Europ. Reg. Conf. Elec. Microsc. (*Manchester*) :232,574,576,582,586,593,598,600,602, 605,606,608,612,622,624,626,630,632,636,654 (1972).

J. Supramol. Struct. **11:**237 (1979).

Macromol. **10:**927 (1977).

Micron **9:**199 (1978).

"Microtubules" (Roberts, K. and Hyams, J. S., Eds.) Academic Press, N.Y., pp. 1–64 (1979).

Mol. Biol. **13:**626,922 (1979).

Optik **51:**171,213,235 (1978); **52:**235 (1979).

Proc. Fifth Int. Conf. High Volt. Elec. Microsc. (*Kyoto*) :171,175,289 (1977).

Proc. Ninth Int. Cong. Elec. Microsc. (*Toronto*) **1:**98 (1978).

Tech. Protein Enzyme Biochem. **B111:**1 (1978).

"The Biology of Large RNA Viruses" (Barry, R. D. and Mahy, B. W. J., eds.) Academic Press, N.Y., pp. 109–114 (1970).

The following is a partial list of image processing articles published in 1980, but not included in the text or tables:

Arch. Biochem. Biophys. **201:**445; **202:**431.

Exp. Cell Res. **126:**490; **127:**1.

IEEE Proc. E. **127:**161,225.

J. Bact. **142:**302.

J. Cell Biol. **84:**40,618; **86:**190,244,514; **87:**521.

J. Cell Sci. **41:**245; **42:**401; **43:**1.

J. Microsc. **118:**247

J. Mol. Biol. **136:**19; **137:**375,437; **139:**123,277; **140:**35; **141:**409; **143:**329.

J. Opt. Soc. Am. **70:**755.

J. Theor. Biol. **82:**15.

J. Ultra. Res. **70:**21,204; **71:**14,25,49; **72:**20,223.

Nature (*London*) **283:**545,680; **286:**628; **287:**509; **288:**296,410.

Optik **54:**425.

Proc. Nat. Acad. Sci. (*USA*) **77:**944,952,2260,4051,4721.

Ultramicrosc. **5:**3,27,281,479.

INDEX

Adenovirus, 141, 144, 163, 169
  see also RNA:DNA heteroduplexes
Agarose gel filtration chromatography, 178
Aliasing artifacts, 197, 218
Aperiodic specimen, 207-208
Aperture dependence:
  resolution in thick specimens, 117, 119-123
  resolution in thin specimens and, 102-103, 104, 105
Artifacts, aliasing, 197, 218
Assymetric particles, 205
Astigmatism, 109-113, 114
Au-island substrate, 97-98, 99, 101, 102
  focusing and, 105, 106, 107
Avidin, biotin and, 180

Bacterial flagella, bibliography, 226
Bacteriophage virus, bibliography, 231-232
Balbiani ring, nonribosomal transcription and, 42
Base compositions, DNA melting temperature and, 145-146
Beam damage, 205, 224
Biological applications, of image processing, 201-202, 226-235
  see also specific biological structures
Biological specimens, image processing of, *see* Image processing, of biological specimens
Biotin:
  avidin and, 180
  RNA and, 179-180
Branch migration, R-loop and, 158

Cell membrane wall, bibliography, 226
Centrifugation, density gradient, 178
Chaotropic agents, 144, 149, 180
Chloropast, 3, 4
"Chlorophyll-protein complex II," 25-26
Chromatic aberrations, aperture dependence and, 117
Chromatin structure, 32-33
  bibliography, 227
  higher order:
    chromosome organization, 55-60
    fiber organization, 47-55
  nucleosome organization, 33-38
    core particle, 34
    dinoflagellates, 36-37
    histone H1, 35, 36
    linker region, 35
    Miller procedure, 33, 34, 38
    100 A diameter, 35-36
    replication, 47
    Simian virus 40 minichromosome, 34, 36
    sperm, 37
  replication, 46-47
  transcription, 38
    nonribosomal, 41-45
    ribosomal genes, 38-42
    replication and, 47
    ultrasound data and, 45-46
Chromatography gel filtration, 178
Chromosome organization, higher order chromatin structure and, 55-60
Coiled fibers, chromosome organization and, 56
Colicin El, 156
Computer:
  digital processing and, 197, 198
  distortions and, 207
  filter masks and, 215
  optical diffraction and, 192, 193
  optical filtering and, 219
  three-dimensional reconstruction and, 203
Confidence intervals, RNA:DNA heteroduplexes and, 168-169
Contrast:
  aperture dependence and, 110, 112, 113
  enhancement, 223

thick specimens and, 115-116, 119
Contrast transfer function, 196, 207, 209, 211, 218, 223
Core particle, 34
Correlation methods, 208, 218
Critical point drying, 200, 222

Damage, specimen, 133-136
Dark-field imaging, 223
Deep etching, of photosynthetic membrane, 8, 11-14, 15, 16, 17
Dehydration damage, 200, 205, 206-207, 222
Denaturation, 76-77
of polynucleotides, *see* RNA:DNA heteroduplexes
Densitometer, 218
digital image processing and, 197, 198
filtering, 197
Density gradient centrifugation, 178
Deoxyribonucleic acid, *see* DNA
Depth of field, resolution of thick specimens and, 123-124
Diethylpyrocarbonate, as hybridization buffer, 152
Diffraction, 192, 199, 221
electron, 204, 209, 213, 217, 224
optical, 192-194, 201, 210-214, 220, 222
X-ray, 199, 201, 204, 209, 213, 217, 225
Diffraction limit, 102
Diffractometer
optical, 193, 194, 197, 211-212, 213
optical filtration and, 195, 214
Digital image processing, 196-198, 216-220
Digital transform, 29
Dinoflagellates, chromatin structure of, 36-37
Distortion, 202, 210, 217, 218, 219, 222, 223
DNA:
enrichment of, 179-180
genetic maps, 172
hybridization, 152-153
optimal conditions for, 148-149
rate, 146
length standards, 158
melting temperature of, 144-148
base composition and, 145-146
duplex length and, 147-148
electrolytes concentration and, 146-147
formamide concentration and, 145-146
partial homology and, 148
physical maps, 172
see also RNA:DNA heteroduplexes
DNA binding protein, 180-181

Drop-spreading technique, 71

Edge width, resolution and, 97-98
Elastically scattered electrons, 94, 224
Electrolytes, DNA melting temperature and, 146-147
Electron diffraction, 198, 204, 207, 209, 211, 213, 217, 224
Electrons:
energy loss of, 92, 93
scattering, 94, 115, 124, 125
elastically, 94, 224
inelastically, 94, 223
Enzymes, bibliography, 227-229
Equilibrium density gradient centrifugation, 178
Eucaryotic messenger RNAs, 141, 162, 175
see also RNA:DNA heteroduplexes

Ferritin, RNA labeling and, 181
Fiber organization, higher order chromatin structure and, 47-55
Fibrous molecules, bibliography, 229
Filtering, 192
artifacts, 196, 215
optical, 194-196, 214-216, 219, 220
Filter mask, 192, 195, 196, 215
digital processing and, 196-197, 218, 219-220
Fixation, proteins visualized on ribonucleic acid and, 83
Floating image, 197, 208, 218
Focusing, resolution in thin specimens and, 103, 105-109
"Folded fiber mold," chromosome organization and, 56
Formaldehyde denaturation, 75
Formamide:
DNA melting termperature and, 145-146
hybridization and, 148-149, 151
RNA:DNA heteroduplexes and, 142
Formamaide-spreading technique, 69-71
Fourier image processing, 192, 196-198, 199, 216, 220
Freeze-dry specimen, 222
Freeze-etching, *see* Photosynthetic membrane, freeze-etching of
Freeze-fracture studies, of photosynthetic membrane, 4-5
Fresnel fringe:
astigmatism correction and, 110
voltage dependence, 109, 110
Frozen hydrated specimen, 206, 222

Gene enrichment, 178-180
Glyoxal treatment, 73, 75, 179, 180

Handedness, helical specimens and, 203
Helical organization, higher order fiber organization and, 49
Helical particles, 216
  indexing, 212, 213
Helical specimen, 210
  image processing of, 202-203
  three-dimensional reconstruction of, 208, 216, 217
Helical virus, bibliography, 232-233
Heteroduplexes, *see* RNA:DNA heteroduplexes
Heteroduplex mapping, 75-76
Higher order fiber, *see* Chromatin structure
High-formamide method, 74
High-voltage electron microscopy (HVEM), resolution in, 90-91
  in biology, 91-95, 96, 130-133
  resolution measurements, 95-99
    Au-island substrate, 97-98, 99, 101
    edge width, 97-98
    method, 97
    thickness, 99, 100
  resolution in thick specimens, 113, 115
    aperture dependence, 117, 119-123
    contrast, 115-116, 119
    depth of field, 123-124
    orientation in the beam, 125-127
    thickness dependence, 127-133
  resolution in thin specimens, 100-110
    aperture dependence, 102-103, 104, 105
    astigmatism, 109-113, 114
    focusing, 103, 105-109
  specimen damage, 133-136
Histone H1, 35, 36
  bibliography, 227
  higher order fiber and, 49
HVEM, *see* High-voltage electron microscopy, resolution in
Hybridization, *see* RNA:DNA heteroduplexes
Hydrated specimens, 206, 209, 222

Image analysis, 27
Image averaging, 196
Image processing, of biological specimens, 190-191
  biological applications, 201-202, 226-235
  diffraction, 199, 221
  imaging conditions, 200-201, 223-224
  specimen preparation, 200, 222
  theory and methods of, 192
    digital, 196-198, 216-220
    optical diffraction, 192-194, 210-214
    optical filtering, 194-196, 214-216
    real space and other reconstruction methods, 198-199, 220-221
Imaging conditions, 200-201, 223-224
Immunoglobulin bibliography, 229
Indexing, diffraction pattern, 193-194, 196, 197, 200, 202, 212-213, 215, 217, 219
Inelastically scattered electrons, 94, 223
Interaction, electron, 92, 93
Interference mechanism, focusing and, 106, 108
Interphase nulcei, higher order structure and, 49
Ionic strength, of chromatin fragments, 49
Isometric virus, bibliography, 233-234

Kleinschmidt technique, 142, 156

Laser, optical diffraction and, 193-211
Lattices, particle, 16, 18, 19, 20, 21, 22, 23
Length measurements, ribonucleic acids and, 77-78
Linker region, of nucleosome, 35
Low-dose images, 200, 205, 207, 223

Meiotic chromosomes, 60
Melting temperature, of DNA, 144-148
Mercurated RNA, 179
Metaphase chromosomes, 55
  higher order structures and, 49
Methylmercuric hydroxide, 180
Microdensitometer, 198, 219, tracings, 102
Microtubile, bibliography, 229-230
Miller procedure, 33, 34, 38, 45
Moiré images, 219
  optical filtering and, 194, 215
Multiple scattering, 194-195, 213, 216, 217, 223
Muscle, bibliography, 230-231

Negative staining, 200, 201, 206, 209, 222, 224
  higher order structure and, 49
Noise:
  optical filtering and, 194, 195, 214, 216
  reduction, 192, 201, 220
Nonribosomal transcription, 41-45

Balbiani ring and, 42
packing ratio, 42
silk fibroin and, 42
Nuclease digestion, of ribosomal genes, 41
Nucleosome, *see* Chromatin structure

Optical diffraction, 201, 210-214, 220, 222
Optical diffractometer, 193, 194, 197, 211-212, 213
Optical filtering, 194-196, 214-216, 220
advantages, 219
Optical transform, 29
Orientation in beam, resolution in thick specimens and, 125-127

Packing patterns, higher order structure and, 49-50
Packing ratio:
of DNA, 49, 52
nonribosomal transcription and, 42
of ribosomal genes, 38
Particle lattices, 16, 18, 19, 20, 21, 22, 23
p chromatin, 41
Penetration, 93, 95
Phase contrasts, 108
ϕ x174, 156
bibliography, 234
Photographic-superposition, 196, 210, 216, 222
method, 196, 199, 216, 220
Photography, RNA:DNA heteroduplexes and, 157-158
Photosynthesis, 2
Photosynthetic membrane, freeze-etching of, 2
deep-etching of, 8, 11-14, 15, 16, 17
freeze-fractured relationship with, 14-29
freeze-fracture studies of, 4-5
deep-etching relationship with, 14-29
membrane preparation for, 5
thylakoid membrane:
fracture faces of, 5-8, 9, 10
organization, 2-4
Photosystem I, 22-23, 26
Photosystem II, 22-26
Pili, bibliography, 231
Poly (A), 158
Polynucleotides, *see* RNA:DNA heteroduplexes
Power spectrum, 199
Protamines, chromatin structure and, 37
Protein:
bibliography of respiratory, 231
DNA binding, 180-181
visualized on ribonucleic acid, 81-84
Psoralen, RNA:DNA heteroduplexes and, 155

"Quantasomes," 16

"Radial loop model," chromosome organization and, 56, 57
Radiation damage, 200, 205-206, 223
Reannealing, of heteroduplexes, *see* RNA:DNA heteroduplexes
Reconstruction apparatus, optical filtering and, 214
Resolution, *see* High voltage electron microscopy, resolution in
Resolving power:
of electron microscope, 91-92
thick specimens and depth of field, 123
Respiratory proteins, 231
Restriction endonuclease, DNA hybridization and, 152-153
Ribonuclease contamination, 71-72
Ribonucleic acids, *see* RNA
Ribosomal genes:
bibliography, 231
nuclease digestion, 41
packing ratio, 38
RNA polymerase, 38
transcriptionally active chromatin and, 38-42
Ribosomal RNAs, 179
R-loops:
DNA:RNA heteroduplexes, 149-150
hybridization and, 154, 155-156
interpretation, 158-162
structures, 142, 143, 144
RNA, 68-69
avidin:biotin, 180
biotin conjugate, 179-180
enrichment of, 178-179
ferritin for labeling, 181
formamide-spreading technique, 69-71
denaturation, 76-77
formaldehyde denaturation, 75
glyoxal treatment, 73, 75
heteroduplex mapping, 75-76
high-formanide method, 74
urea-formanide method, 73, 74-75
hybridization, 152-153

length measurements, 77-78
mercurated, 179
promoters for transcription, 177
protein visualization on, 81-84
reagents, 71-72
contaminants, 72
ribonuclease contamination, 71-72
ribosomal, 179
secondary structure, 78-81
splicing, 141, 162, 174
studies, 84-85
supplies, 72, 74
RNA:DNA heteroduplexes, 141, 181-182
adenovirus-2 application, 172-178
data analysis:
confidence intervals, 168-169
data presentation, 165-168
errors, 169-171
orientation, 163-165
quantitation, 171-172
R-loops, 158-162
RNA single-stranded DNA, 162-163
standard deviation, 168-169
density gradient centrifugation, 178
electron microscope procedures:
mounting samples on grids, 156
photography, 157-158
specimen evaluation, 156-157
tracing, 157-158
formation, 143
gel filtration, 178-179
gene enrichment:
of DNA, 179-180
of RNA, 178-179
hybridization:
buffer, 152
combination procedures, 155-156
DNA and, 152-153
formamide for, 151
gene enrichment and, 178-180
glassware for, 151
methods, 153-156
optimal conditions for, 148-149
rate, 146
R-loop formation and, 154, 155-156
RNA and, 152-153
RNA: single-stranded DNA
heteroduplex and, 154-155
labeling methods, 181
polynucleotide denaturation and
reannealing:
hybridization, 148-149
R-loops, 149-150
thermal stability of double helices, 144-148
RNA: single-stranded DNA structures, 143, 144, 150-151, 154-155
direct visualization, 180-181
interpretation, 162-163
stability of, 142
technique development, 141-144
Kleinschmidt technique, 142, 156
R-loop structures, 142, 143, 144
RNA polymerase, 38, 40
RNA: single-stranded DNA heteroduplex, *see* RNA:DNA heteroduplexes,
Rotational filtering, 196, 197, 216, 219, 220
Rotational symmetry, 204-205

Scaffold, as chromosome organization, 57
Scattering:
elastic, 223
electron, 94, 115, 124, 125
inelastic, 223
multiple, 194-195, 213, 216, 217, 219, 223
Shadowing, rotary, 71
Shadowing replication technique, 27
Sheet, image processing of, 203-204
three-dimensional reconstruction and, 216, 217
Silk fibroin, non-ribosomal transcription and, 42
Simian virus 40 minichromosomes, 34, 36
Specimens:
image processing, 202-205
preparation, 200, 222
unstained, 201, 205, 206, 222
Sperm, chromatin structure, 37
Spherical aberration, 195, 207, 212, 224
limit, 102
Spherical structures, three-dimensional reconstruction and, 216, 217
Spreading technique technique:
drop-spreading, 71
formamide, 69-71. *See also* RNA
Miller, 33, 34, 38
Spread preparations, 33
Stains, 72
Srandard deviation, RNA:DNA heteroduplexes and, 168-169
Strand separation temperature, 145
Superbeads, higher order structure and, 52

Supercoil, chromatin fragments as, 49
Support films, 200, 201, 222-223
SV 40, 156
Symmetry, 192, 193, 199, 202, 210, 213, 215, 216, 217-218, 219, 220, 224
  rotational, 204-205

Thickness, measurement of specimen, 99, 100
Thickness dependence, resolution in thick specimens and, 127-133
Thick specimens, resolution in, 94
  see also High-voltage electron microscope, resolution in
Thin sectioning, 204, 206-207, 218, 222
  resolution in, *see* High-voltage electron microscope, resolution in
Three-dimensional crystals, 204
  three-dimensional reconstruction and, 216, 217
Three-dimensional image reconstructions, 199, 221
Three-dimensional reconstruction, 192, 193, 196, 198, 202, 203, 204, 205, 208, 216, 217, 219, 222
Thylakoid membrane, *see* Photosynthetic membrane, freeze-etching of
Top-bottom effect:
  specimen damage and, 136
  voltage dependence and, 125, 127
Tracing, RNA:DNA heteroduplexes and, 157-158

Ultrasound data, transcriptionally active chromatin and, 45-46
Unstained specimen, 201, 205, 206, 222
Urea-formamide method, 73, 74-75

Virus:
  bibliography, 231-235
  three-dimensional reconstruction of, 205
Visibility, image, 105, 106
  Fresnel fringe and, 109

X-ray diffraction, 199, 201, 204, 209, 213, 217, 225